Impact of Digital Twins in Smart Cities Development

Ingrid Vasiliu-Feltes
University of Miami, USA

A volume in the Advances in Civil and Industrial Engineering (ACIE) Book Series

IGI Global
PUBLISHER of TIMELY KNOWLEDGE

Published in the United States of America by
IGI Global
Engineering Science Reference (an imprint of IGI Global)
701 E. Chocolate Avenue
Hershey PA, USA 17033
Tel: 717-533-8845
Fax: 717-533-8661
E-mail: cust@igi-global.com
Web site: http://www.igi-global.com

Library of Congress Cataloging-in-Publication Data

Names: Vasiliu-Feltes, Ingrid, 1971- editor.
Title: Impact of digital twins in smart cities development / Ingrid Vasiliu-Feltes, editor.
Description: Hershey PA : Engineering Science Reference, [2022] | Includes bibliographical references and index. | Summary: "This book defines Digital Twins as virtual replicas of physical entities that are created in software and that can be represented by devices, people, processes or systems and offer insights into domains such as ethics, data governance, cyber-security, inclusion, diversity and sustainability"-- Provided by publisher.
Identifiers: LCCN 2022007265 (print) | LCCN 2022007266 (ebook) | ISBN 9781668438336 (h/c) | ISBN 9781668438343 (s/c) | ISBN 9781668438350 (ebook)
Subjects: LCSH: Smart cities. | Cities and towns--Computer simulation. | Information modeling. | Digital twins (Computer simulation) | City planning.
Classification: LCC TD159.4 .I47 2022 (print) | LCC TD159.4 (ebook) | DDC 307.760285--dc23/eng/20220502
LC record available at https://lccn.loc.gov/2022007265
LC ebook record available at https://lccn.loc.gov/2022007266

This book is published in the IGI Global book series Advances in Civil and Industrial Engineering (ACIE) (ISSN: 2326-6139; eISSN: 2326-6155)

British Cataloguing in Publication Data
A Cataloguing in Publication record for this book is available from the British Library.

Advances in Civil and Industrial Engineering (ACIE) Book Series

ISSN:2326-6139
EISSN:2326-6155

MISSION

Private and public sector infrastructures begin to age, or require change in the face of developing technologies, the fields of civil and industrial engineering have become increasingly important as a method to mitigate and manage these changes. As governments and the public at large begin to grapple with climate change and growing populations, civil engineering has become more interdisciplinary and the need for publications that discuss the rapid changes and advancements in the field have become more in-demand. Additionally, private corporations and companies are facing similar changes and challenges, with the pressure for new and innovative methods being placed on those involved in industrial engineering.

The **Advances in Civil and Industrial Engineering (ACIE) Book Series** aims to present research and methodology that will provide solutions and discussions to meet such needs. The latest methodologies, applications, tools, and analysis will be published through the books included in **ACIE** in order to keep the available research in civil and industrial engineering as current and timely as possible.

COVERAGE

- Productivity
- Quality Engineering
- Construction Engineering
- Hydraulic Engineering
- Urban Engineering
- Coastal Engineering
- Ergonomics
- Optimization Techniques
- Materials Management
- Engineering Economics

The Advances in Civil and Industrial Engineering (ACIE) Book Series (ISSN 2326-6139) is published by IGI Global, 701 E. Chocolate Avenue, Hershey, PA 17033-1240, USA, www.igi-global.com. This series is composed of titles available for purchase individually; each title is edited to be contextually exclusive from any other title within the series. For pricing and ordering information please visit http://www.igi-global.com/book-series/advances-civil-industrial-engineering/73673. Postmaster: Send all address changes to above address.

Titles in this Series

For a list of additional titles in this series, please visit:
https://www.igi-global.com/book-series/advances-civil-industrial-engineering/73673

AI Techniques for Renewable Source Integration and Battery Charging Methods in Electric Vehicle Applications
S. Angalaeswari (Vellore Institute of Technology, India) T. Deepa (Vellore Institute of Technology, India) and L. Ashok Kumar (PSG College of Technology, India)
Engineering Science Reference • copyright 2023 • 345pp • H/C (ISBN: 9781668488164) • US $260.00 (our price)

Artificial Intelligence and Machine Learning Techniques for Civil Engineering
Vagelis Plevris (Qatar University, Qatar) Afaq Ahmad (University of Engineering and Technology Taxila, Pakistan) and Nikos D. Lagaros (National Technical University of Athens, Greece)
Engineering Science Reference • copyright 2023 • 315pp • H/C (ISBN: 9781668456439) • US $250.00 (our price)

Urban Life and the Ambient in Smart Cities, Learning Cities, and Future Cities
H. Patricia McKenna (AmbientEase, Canada)
Engineering Science Reference • copyright 2023 • 261pp • H/C (ISBN: 9781668440964) • US $215.00 (our price)

Remapping Urban Heat Island Atlases in Regenerative Cities
Hisham Abusaada (Housing and Building National Research Center, Egypt) Abeer Elshater (Ain Shams University, Egypt) and Marwa Khalifa (Ain Shams University, Egypt)
Engineering Science Reference • copyright 2022 • 267pp • H/C (ISBN: 9781668424629) • US $250.00 (our price)

Clean Technologies and Sustainable Development in Civil Engineering
Raluca- Andreea Felseghi ("Ştefan cel Mare" University of Suceava, Romania) Nicoleta Cobîrzan (Technical University of Cluj-Napoca, Romania) and Maria Simona Raboaca (National Research and Development Institute for Cryogenic and Isotopic Technologies, Râmnicu Vâlcea, Romania)
Engineering Science Reference • copyright 2022 • 290pp • H/C (ISBN: 9781799898108) • US $250.00 (our price)

For an entire list of titles in this series, please visit:
https://www.igi-global.com/book-series/advances-civil-industrial-engineering/73673

IGI Global
PUBLISHER of TIMELY KNOWLEDGE

701 East Chocolate Avenue, Hershey, PA 17033, USA
Tel: 717-533-8845 x100 • Fax: 717-533-8661
E-Mail: cust@igi-global.com • www.igi-global.com

Editorial Advisory Board

Table of Contents

Detailed Table of Contents

E-government portals are now playing an important role in facilitating citizens' lives by executing services at any time and from any location. Such a way of providing services results in great benefits for citizens and agencies, particularly within the COVID-19 pandemic context. Over the past few years, a number of best practices have been identified for designing and developing e-government portals but have been documented piecewise across the literature. A key approach to facilitating access to such best practices is their integration into a structured maturity model tailored to a domain of application. A digital twin is a digital representation of a physical object or system. However, data from the e-government, which is a part of what is called smart city, can be transferred to a digital twin to work with other data from other systems, such as IoT systems. Thus, these insights can be used to make better decisions.

The purpose of this chapter is to offer the readers an overview about how a data space is defined, how digital twins powered by data spaces can help us build inclusive sustainable smart cities, how emerging technologies can enhance data spaces, how data spaces impact digital twins and smart cities, and to share high level insights related to initiatives that have been adopted at the European Union level in terms of data spaces as business value creation and sustainability for a common path towards a sustainable data-driven ecosystem.

Section 2
Industry Perspectives

There are several industries that have shown significant advancements in using digital twin capabilities. Healthcare, life sciences, and fintech constitute a powerful trifecta that can profoundly impact society at a global scale. The design and deployment of digital twins is complex and intimately connected to other digital technologies. However, in order to be successful, we will also need to overcome some of the challenges faced with any emerging technologies and mitigate against the risks of maleficent use. Specifically, ethics, cybersecurity concerns, and cross-industry standards will be on the agenda of most enterprises wishing to deploy digital

open and close-ended questions. A total of 163 forms were distributed in different locations within the smart community sites. A descriptive analysis on the usage was carried out to access the initiatives, and recommendations are suggested at the end of the chapter.

Foreword

A "Digital Twin" is an in silico (or virtual) algorithmic, dynamic, and adaptive evolving system that is a representation of a physical entity, process, or environment. It is modeled on a variety of real-world empirical data sources including historical and, more recently, real-time high dimensionality and high frequency streaming sensor data. Digital Twins are built with a variety of physic-based and data driven mathematical modeling techniques including machine learning, optimization, simulation, and more.

It is important to note that a Digital Twin is not a simulation in a strict sense of the word. A Digital Twin is a digital representation of things from the real world and in real time, whereas a simulation is a simulacrum of the operation of a real-world process or system over time. A simulation is often developed from idealized data and is not informed and cannot adapt to new real-world data. This is a critical distinction, as digital twins can be used not only to understand the behavior or performance of a system but, given that they are representative of the real-world and are informed by real world and real time data, they are capable of informing and predicting changes and unprogrammed performance or events including the ability to address issues of emergent behavior. Systems with many components often exhibit emergent behaviors; i.e., behaviors that arise from the relationships between the components rather than from the components themselves. Understanding complex interactions between components and identifying and addressing emergent behavior is a profoundly data driven process, often necessitating new modalities of analysis and data collection. Digital Twins uniquely address the issues of data collection, processing and analysis and thus can be used to predict and anticipate novel and unseen events, especially anomalies.

As industries move beyond "digitization," they need to begin the process of "digitalization." Digitalization is the adoption of digital technologies both organizationally and operationally. One is about data and the other is about process. Processes and workflows, including operational decision making, will need to rely on "algorithmic automation" by applying machine learning, artificial intelligence, high dimensionality, and high frequency data. Digitalization and automation in decision making will be the foundation for the development of new products and services. The core technology critical to this transformation will be the development, adoption, and implementation of Digital Twin technologies across industries.

Recently, a new digital twin consortium has been established that aims to deploy digital twin technology in new markets as well as in the development of smart cities. Designing smart cities, smart communities, and smart ecosystems powered by optimal digital twin deployments is a vision that currently only futurists can entertain and requires some time to reach large-scale adoption. However, it is incumbent upon us as a society to educate and train future generations on how to leverage digital twin technologies in order to optimize our daily lives as well as increase our efficiency, productivity, and safety.

Impact of Digital Twins in Smart Cities Development aims to provide insights regarding the global landscape for current digital twin research and deployments and highlights some of the challenges and opportunities faced during large-scale adoptions. It covers a broad range of key topics such as ethics, data governance, cybersecurity, legal, compliance, as well various industry and regional perspectives, and sustainability and would be an ideal resource for urban planners, engineers, policymakers, industry leaders, scientists, economists, academicians, practitioners, researchers, instructors, and students.

Mark Wolff
NeoGeminae Inc., USA & University of Miami, USA

Preface

My first encounter with digital twin technology was while searching for an alternative to improve safety in clinical trials and medical procedures. I was intrigued by their potential and began an in-depth research of the underlying technology infrastructure, to better understand the full spectrum of tools that were included in this broader terminology. I quickly realized that the applications for digital twin tools were industry-agnostic and offered a unique solution for the design, deployment, and monitoring of a seemingly endless array of products, processes, services, etc.

However, my decision to write this book was instant once I read about the possibility to deploy digital twins for smart cities, which had already been an area of special interest to me from an academic, technology, business, and investment, as well as a sociology point of view.

Since the beginning of this editorial journey, I have witnessed major scientific advances, with an exponential increase in phenomenal use cases, such as the development of a digital twin of the human heart and lungs, the design of smart buildings, planning of smart cities, and even the creation of a digital twin dedicated to the study of planet Earth. This was possible via technologies built by NVIDIA and dedicated to tackling climate change.

Upon sharing my enthusiasm with colleagues in various industries, I realized that the concept of digital twins was abstract for many, expressed concerns about digital trust, while a few mirrored my excitement. It became clear to me that perhaps the book should address each of these facets and offer a balanced view of the potential, as well as challenges and opportunities we are facing when deploying digital twin technologies.

The final decision about the title and focus on digital twin deployments for smart cities was during the research for another article. The more I read, I realized that smart cities were the ultimate mega-use case that fully leveraged not only the technical capabilities, but also their environmental, social and governance (ESG) potential.

The final structure of the book aims to offer readers a general section with a high-level overview of the legal, regulatory, and data governance perspectives, followed by section two which highlights specific industry examples (space, fintech, health tech, construction, etc). The third section showcases a few use cases from various regions of the world.

Chapter 1 highlights how the nascent Digital Twin is a computer program which analyses and con-trolls the physical system or asset as well as simulates it in a real-world condition through data transmission. It is akin to the pairing technology used by NASA to unknot the operational, maintenance and repair issues of systems which weren't physically nearby. Therefore, the technology helps to predict how the physical system or asset reacts under changing circumstances. Consequently, in the present chapter, the authors have attempted to discuss, analyze, and critically evaluate the ethicolegal parameters underlying the application of the DT technology

Chapter 2 offers insights into why E-government portals are now playing an important role in facilitating citizens' lives by executing ser-vices at any time and from any location. Such a way of providing services results in great benefits for citizens and agencies, particularly within the COVID-19 pandemic context. Over the past few years, a number of best practices have been identified for designing and developing e-government portals, but have been documented piecewise across the literature. A key approach to facilitating access to such best practices is their integration into a structured maturity model tailored to a domain of application. are considered an integrated source of such best practices that should be used to enhance the quality of portals. A digital twin is a digital representation of a physical object or system. However, data from the e-government - which is a part of what is called smart-city - can be transferred to a digital twin to work with other data from other systems, such as, IoT systems. Thus, these insights can be used to make better decisions.

The purpose of Chapter 3 is to offer the readers an overview about how a data space is defined, how digital twins powered by data spaces can help us build inclusive sustainable smart cities, how emerging technologies can enhance data spaces, how data spaces impact Digital Twins and Smart Cities and share high level insights related to initiatives that have been adopted at the European Un-ion level in terms of data spaces as business value creation and sustainability for a common path towards a sustainable data-driven ecosystem.

Chapter 4 showcases several industries that have shown significant advancements in using digital twin capabilities. Healthcare, life sciences and fintech constitute a powerful trifecta that can profoundly impact society at a global scale. The design and deployment of digital twins is complex and intimately connected to other digital technologies However, in order to be successful, we will also need to overcome some of the challenges faced with any emerging technologies and mitigate against the risks of maleficent use. Specifically, ethics, cybersecurity concerns, and cross-industry standards will be on the agenda of most enterprises wishing to deploy digital twins.

By a combined deployment of digital twins with all other emerging technologies we could potentially build a highly performant global digital twin ecosystem Futurists would even venture to envision the existence of state of the art Smart Cities where digital healthcare twins, digital life sciences twins, digital finance twins would be part of our daily life.

In Chapter 5 the author highlights that smart cities include the intelligent networking and smart operation of existing ones in accordance with the latest energy standards, SDGs, climate strategies and common good. The building sector causes almost 40 percent of global CO_2 emissions and is therefore the key to central climate neutrality by 2050. Innovative building technologies in existing buildings enable intelligent ecosystems within the 5th industrial revolution. Experience from the practical application provides signposts for the use of the latest digital technologies for the cities of tomorrow, with more efficiency, comfort and safety for everyone who lives and works in them.

Chapter 6 is focused on applications of digital twins for the space industry. This includes everything from the soles of athletic shoes to developing groundbreaking technologies like remote sensing, broadband connectivity, and high-performance computing used by modern enterprises all over the world. In truth, space supplies the critical infrastructure upon which every other infrastructure is dependent. Once considered the final frontier and only accessible to a select few — space is now home to the most dynamic and innovative environment on the planet and beyond. Cultivating the talent to fuel and drive bold innovations like digital twins will be one of the greatest challenges we face. Workforce shortages and skills deficits across geographies, disciplines and demographics threaten not just future advancement but also expanding economic opportunity.

Chapter 7 talks about the new living and urban worlds of tomorrow and the sustainable design of space and society require even more efficiency, comfort and security. Technological progress and safe design of smart cities can only be achieved with the help of a will to innovate and technical know-how of the partnering network constituting the building project team. Digital twins, AI, Machine Learning, Deep Learning, Intelligent Sensor Systems, Intelligent Networking between buildings, Artificial Intelligence of Things, Clouds represent highly efficient technologies to both optimize fire safety and fire protection, increase human and building operation safety as part of the implementation of a holistic and comprehensive building protection solution in smart cities. A model says more than a thousand words: State-of-the-art fire protection systems, a customized automatized building solution, can be designed, planned, tested, individually adapted and adjusted with the help of digital twins and AI, e.g. to optimize early fire detection, targeted alarms and evacuation, and to reduce costly false alarms.

In Chapter 8 the authors offer insights regarding how current practices based on digital twin technology can help the development of smart cities in China. This chapter will explore the roles of the government in promoting the application of digital twin technology in the development agenda of industrial infrastructure and smart cities through policies (i.e. institutions). How do these institutions promote the development of digital twin technology in various industries? How these developments will in turn affect stakeholders in various sectors. This chapter will investigate the impact of China's digital twin technology re-search and institutional changes on the sustainable development of smart cities. This chapter will use bibliometric analysis to perform statistical and correlation analysis on the digital twin / blockchain / AI / IoTs literature included in the Web of Science (WOS) database. Cases from China will be used for comparison. This chapter will provide suggestions for the development of digital twin technology-based ecosystems in emerging economies.

Chapters 9 and 10 offer smart cities regional perspectives from Lundu and Putraja. Stated in Southwest Sarawak, a Malaysian state on Borneo. Lundu is declared as one of the smart community initiatives that were launched by the Malaysian Communications and Multimedia Commission (MCMC) in February 2016. Whilst this paper intended to understand and map with development index of a Smart Community in Lundu, so that the involved stakeholders can be informed about the necessary criteria which enable citizens to be deemed as smart. A quantitative research method has been employed whereby a self-administrated questionnaire was developed to gather information from the respondents which consisted of different sections with both open and close-ended questions. A total of 150 forms were distributed in different locations within the smart community site. Four key components in International Growth Centre (IGC) development index which are Humanware, Technoware, Infoware, and Valueware were used as a reference point in developing the Smart Community Index (SCI) and recommendations on infrastructure and monitoring improvement were suggested at the end of Chapter 9.

Putrajaya is a Malaysian federal administrative center and is called also City in the Garden. Due to this factor, this paper intended to access the insights of the Putrajaya smart community initiatives since a quantitative research approach has been employed where a self-administrated questionnaire was developed to gather information from the respondents which consisted of different sections with both open and close-ended questions. A total of 163 forms were distributed in different locations within the smart community sites. A descriptive analysis on the usage was carried out to access the initiatives and recommendations were suggested at the end of Chapter 10.

My hope is that this book stimulates the reader's imagination and the desire to learn more about all other industrial applications where digital twins have a transformative impact, and can change the paradigm, such as transportation, autonomous vehicles, energy, agriculture, atmospheric sciences, defense systems, etc.

Perhaps my next book will address the next frontier of digital twin technology and examine the exponential impact it can have during our transition from web 3.0 to web 4.0. By then, maybe the book will describe use cases that combined digital twin technology with quantum computing, satellite internet and human-computer interfaces, or other emerging technologies.

Ingrid Vasiliu-Feltes
University of Miami, USA

Acknowledgment

I would like to express my gratitude to a wonderful group of professionals that inspired and encouraged me to embark on this editorial journey, as well as to those that shared their expertise and insights during all stages of this book. Their reflections and opinions generated increased awareness among our audience, pointed out several ongoing challenges we are facing globally, and encouraged us to be proactive in our quest to change the paradigm for future generations.

I am extremely grateful to all co-authors for their important contributions and to our publisher for the support provided during this process.

Our editorial board played a crucial role in ensuring the successful completion of this book and I would like to extend my deepest gratitude to all our members.

It has been a privilege to work with such talented individuals, and I am confident that their contributions will be of great value to anyone seeking to deepen their understanding of how digital twin technologies can impact smart cities developments

Section 1
Global Perspectives

Chapter 1
Digital Twins:
Legal and Regulatory Perspectives - Emerging Trends and Challenges

Mahendra Kumar Bhandari
Infinity Law Tech Educational Services, India

Diya Sarkar
Kalinga Institute of Industrial Technology, India

Prafulla Chandra Mishra
Kalinga Institute of Industrial Technology, India

ABSTRACT

The nascent digital twin (hereinafter termed as DT) is a computer program that analyses and controls the physical system or asset as well as simulates it in a real-world condition through data transmission. It is akin to the pairing technology used by NASA to unknot the operational, maintenance, and repair issues of systems that weren't physically nearby. Therefore, the technology helps to predict how the physical system or asset reacts under changing circumstances. Consequently, in the chapter, the authors discuss, analyze, and critically evaluate the ethico-legal parameters underlying the application of the DT technology.

INTRODUCTION

The 21st century, an era flourishing with the rise of the most trending and disruptive technologies, evidently has brought massive relevance to the technology of digital twins (DT). DT is a recent taxonomy but conceptually the technological foundation

DOI: 10.4018/978-1-6684-3833-6.ch001

dates to the 1960's. Initially, it revolutionized the aerospace industry and latterly, DT's significance has grown multiple folds with the advent of Cybernetics, Big data, Artificial Intelligence and Robotics. It is presumed that the growing application of Nanotechnology (especially the Internet of Bio-Nano Things) (Almazrouei, Shubair, and Saffre, 2018). has potential like Midas's touch in the health sector, military applications, and environmental fields. The terms 'Virtual Twin' (Stark and Damerau, 2019). are synonymous while 'Virtual Prototype', 'Hybrid Twin Technology', and 'Digital Asset Management are relative terms to DT (Stark and Damerau, 2019).

Supposedly, DT can be called a virtual clone (Palla, 2022; Gartner, 2020). of a physical object (e.g., a windmill) that virtually simulates. Such a simulation can take place even before the real object is designed or deployed. This technology has been deployed upon real objects which weren't physically proximate at compound systemic stages by various groups from time to time for assessing, simulating, and controlling it. In short, a DT is a digital duplicate of a physical object or asset. The present Industry 4.0 phase looks at this technology as a key to unlocking unrequited space between the physical and virtual worlds (Althavale and Althavale, 2021).

Undoubtedly, DT is a remarkable edge of technological evolution that capacitates technological perfection. Thus, the scope of DT is wide-ranging and diverse. It discharges accurate predictions of the real object in real-time concomitantly. It not only alongside showcases the exact performance of the real asset, but also it is a means to comprehend the performance of the asset model, by predicting its behavioral pattern and optimizing its operations and services.

The DT technology has been applied to simple infrastructures and has come out with brilliant benefits and precision at the manufacturing level. However, DT is predicted to have massive potential to manage machine infrastructures under not only complex systems but also multiple systems at multiple levels at the same time. Perhaps DT technology knows no bounds with the advent of AI, Cybernetics, Nanotechnology, and other disruptive technologies. It is spreading wings across multiple objects which astonishingly rise beyond machines and their designs. There are fundamental attempts currently in progress that involves "What Ifs" and DT to raise the sustainability of the Planet, enhance the quality of life of humans by breakthrough into bioscience as well as implement smart city/nation for effective governance with the extensive use of DT.

Every new technology comes with its pros and cons and DT is not an exception to it. A technology standing alone is generally harmless, the concern begins while implementation and further blending with other technologies. Hence, in the case of DT obvious challenges, both ethical and legal come to the shore. The indicative potential of DT creates an influx of competition by companies, industries, government concerns, and agencies. It is inevitable to face fundamental trials and confines when complex dynamic model systems (like geographical twins, weather, climate, life, behavior, health, etc.) backed by uncertainty are built. On one hand, the gross machine learning infrastructures grasp trillion information and pose challenges like data privacy, data ownership; data interoperability; confidentiality, and liability concerns. On the other hand, the growing complexities pose a conflict of interest and intensify security challenges. The objective of this chapter is to throw light upon the regulatory challenges regarding the mounting use of DT technology and attract the attention of policymakers, lawyers, engineers, academicians, researchers, and scholars so that the technology is judiciously put to use.

A BRIEF HISTORY OF DIGITAL TWIN

The simulation developed as a key tool for understanding the scientific phenomenon. It all began with a simulation that allowed the effective handling of complex systems. For complex systems like spaceships and rockets, it was extremely complex to understand the interactions and infrastructure by analytical means. An example of a simulation use case is NASTRAN (NASA Structural Analysis), a Finite Element Analysis (FEA) software developed for NASA for the aerospace industry.

Perhaps, NASA's Apollo Mission during the 1960s marked the conceptual inception of the program DT technology which indeed involved simulation in twin forms, however, the success of the rescue program was the result of the efforts of the intelligent NASA engineers. The plan involved the building of a pair of identical spacecraft i.e., the earth model and the space model. The earth model was the twin which could precisely reflect the model in space. During the mission, the twin on earth was used to simulate the spacecraft and it could accurately reflect and predict the actual status of the space model in operation to a far extent. This approach effectively assisted the astronauts in making accurate decisions during emergencies. Up to this stage, the twins were still physical.

In 1991, the concept of DT was first used in David Gelernter's book, 'The Mirror Worlds'. Thereafter, the concept of virtual digital representation was proposed in the manufacturing sector by Professor M. Grieves in 2003. He explained during a Product Life Cycle course at the University of Michigan that it was an equivalent of a physical product, which was a digital copy of one or a set of specific devices that can represent a real device in abstract form (Grieves, 2016). This program could be used for testing under real and simulated conditions by way of information and data analysis of the device *per se*. This model was then known as the 'Mirrored Spaced Model'(MSM). Subsequently, in 2006 it came to be known as the 'Information Mirror Model'(IMM). The IMM had elements like virtual space, real space, and nexus between them both. As it consisted of the fundamental elements of the DT, it is known as the 'embryonic form' of the DT. Prof. Grieves and his NASA colleague, Sir John Vickers authored a paper based on the concept of Digital Twin for a Defense Manufacturing Conference (DMC), 2010.

In 2011, Prof. Grieves introduced the term DT in the book titled "Virtually perfect: driving innovative and lean products through product lifecycle management". In it, Sir John Vickers has described the conceptual model of DT involving three entities (1) real space entity products; (2) virtual space virtual products; and (3) data and information interface between real space and virtual space. In the same year, the DT was proposed and developed by the US Air Force Research Laboratory to solve the maintenance and life prediction of an aircraft under a future convoluted service atmosphere. Subsequently, DT entered the planning and implementation stage from the conceptual stage.

In 2012 NASA formally introduced DTs in the public domain. Recent advances in information technology are enabling further evolution and popularity of DTs which have been discussed in the chapter below.

DEFINING DIGITAL TWIN

In generic terms, the DT simply is a technology that connects physical and virtual spaces. There is still no exhaustive definition of the term 'DT' and hence there are possibilities to explore varied definitions across different domains. The common element of all definitions centralizes to the point of digital data and digital mirroring.

During the evolutionary stage, the technology of the twin model lacked the digital element, it was still physical. It was a model that reflected the operation of the real object in real space itself. At that point when NASA put the technology to use, it is observed that the twin was an exact physical model that reflected the operation of the real object in real-time by way of simulation. Precisely the physical twin technology model then had more or less a mirror appearance in terms of geometry, size, content, structure, functions, or performances along with its macro and micro properties. The twin reflected the real-time operation by means of simulation. However, with the

Figure 1. How the physical asset/network/process is modeled as a digital replica to form a DT

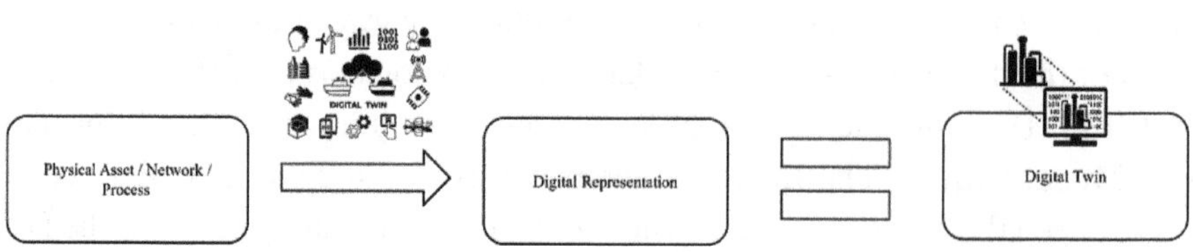

sharp progress of computer science and latest technological evolution, the practical development of the concept of DT is growing several folds.

In the Whitepaper on DTs, the concept model of DT was defined to consist of three segments: a) physical asset in real space; b) virtual asset in virtual/digital space; c) the nexus between the virtual and real assets together tied with data and information (Grieves, 2014).

The technology's increasing demand has led to proposing a segmented definition which extends the previous definition with additional elements of data and services. There are multiple definitions of DT and every definition more or less throws light upon a specific trait (like based on lifecycle, data-driven cloud platform, digital counterpart of a real model, cyber physical system, integrates interdisciplinary virtual product models, specific task-based function, represents a unique asset, etc.) (Stark and Damerau, 2019) of DT. Tuegel's definition of DT highlights the lifecycle perspective. He opined DT to be a cradle-to-grave model of an aircraft structure's ability to meet mission requirements, including sub-models of the electronics, the flight controls, the propulsion system, and other subsystems (Tuegel, 2012).

DT is a dynamic virtual representation of a concurrent physical product. DT integrates complete electronic information and data generated during the lifetime of the product it is paired to. Such virtual mirroring of the product is possible not only for a mature product but also during its phase of ideation. The DT bridge the gap between virtual and real space to predict, understand and optimize the real object for its unhindered function, as well as accomplish the purpose for which the asset is created. It is a product digital counterpart of a physical product (Rios et al., 2015). Such virtual model comprises of data-based methods and primarily simulation techniques. It can be called an inevitable backbone of industry in near future.[(Digital Twin, n.d.).]

DT refers to the processes and methods for describing and modeling the characteristics, behavior, formation process, and performance of physical objects using digital technology, and can also be referred to as DT technology. The DT model refers to a virtual model that completely corresponds to and is consistent with the physical entities in the real world and can simulate its behavior and performances in a real-time environment in real time (Digital twins for small satellites, n.d.).

DT is a computer programs that use data to create simulations that predicts how a product, or a process will perform. The program of DT is involved in virtual representation of the asset in operation and in the process, it generates continuous data about the asset, it evaluates the state of the asset through the data generated and predicts the future behavior which results into optimization of the operation with a refined control (*What is digital twin technology and how does it work?,* n.d.). A DT lays the foundation of the intelligent manufacturing system (He and Bai, 2020).

Figure 2. The primary functions of digital twin to mirror a real object

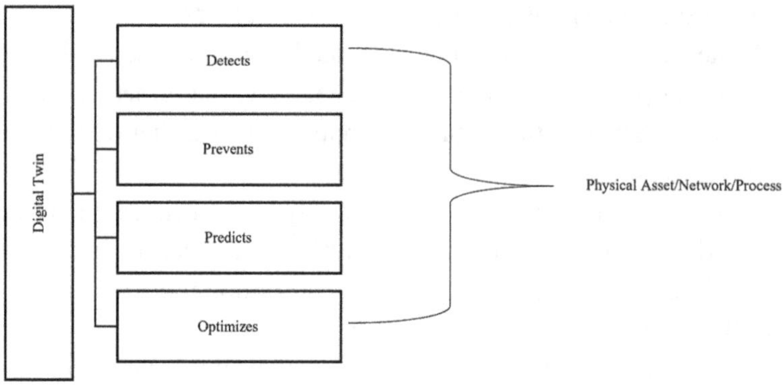

The most important enlightening a significance of DT is that it realizes the feedback from the physical system to the digital model of cyberspace.

Thus, it is a process of reverse thinking in the industrial field. Also, one must understand that shifting physical twins into DT in this cyber era was inevitable. The digital and physical world can be truly harmonized throughout the entire lifecycle. Various types of simulation, analysis, data accumulation, and mining based on digital models, and even the application of artificial intelligence, can ensure its applicability to real-world physical systems (Banyai and Felice, 2020). This is the significance of DT to intelligent manufacturing. The intelligence of an intelligent system must be first perceived, modeled, and then analyzed. Today, DT is looked upon as a potent technology for accurate modeling of a real production system. In short, DT has evolved over time and today a DT can mirror an element; a system of elements; or a system of systems.

FEATURES DIGITAL TWIN

The twin model has several characteristic features including Interactive design tool; virtuality; stores processes; maps data; simulates and predicts; 3-Dimensional; dynamic; data driven; multi-physical; hierarchical and multi-scale; realistic; computability; and multidisciplinary.

Therefore, DT is an information space that reflects a physical product on a digital mapping model. Each physical product uniquely corresponds to the DT model. It not only elucidates the geometric properties (like size, shape, texture, tolerance, etc.) of the physical product but also explains various physical properties (like thermodynamics, material properties, stress analytics, etc.) of the products as well. The product DT is multi-disciplinary and reflects multi-scalability. It represents the

Figure 3. The cumulative functioning features of a DT

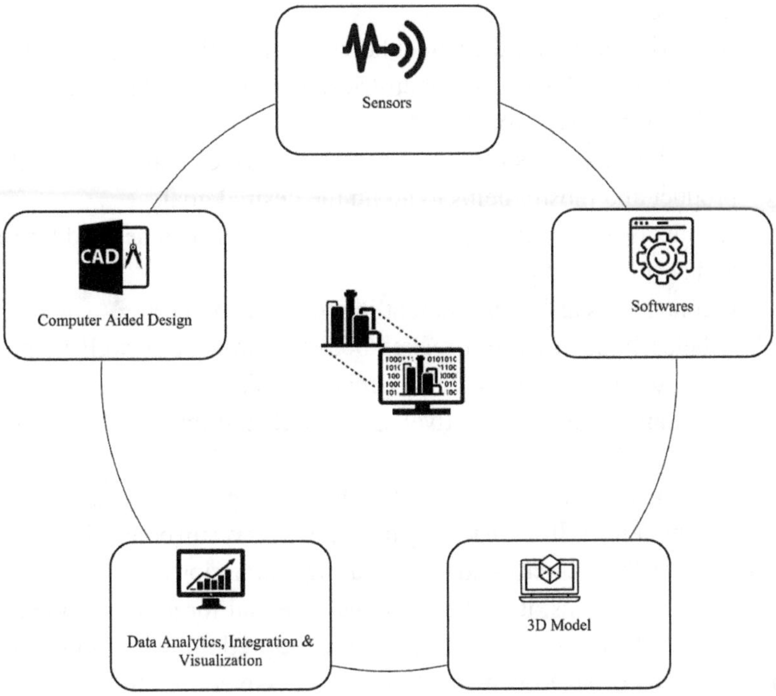

real product both at macro as well as micro levels. It is hierarchical as it classifies different components that the final product is comprised of. The DT model is dynamic as it constantly changes and improves by way of unceasing interaction with the real product. It is very realistic with high degree of actuality in terms of content, nature and appearance. By using simulations, calculations, and analysis it computes the real status and working of the physical object with the application of probabilistic statistics.

SIGNIFICANCE OF DIGITAL TWIN TECHNOLOGY

The pandemic and the post-pandemic era have obligated the human race to adopt new ways of life and work. DT technology has acquired greater relevance and impact in the present setting of digital existence. Earlier as well the technology had potential usability but, in the present, it apparently is more meaningful. The above-discussed features of DT are in section 1.3. makes the technology most sought and reliable during the present times. It has effectively transformed the methods of how the physical asset's infrastructures were built, maintained, and managed without any prejudice.

1. **Substantiating Digital Twin:** The DT technology and its significance can be mapped as follows:
    a. It is a first-in-its-kind channel between the digital and physical world.
    b. It enhances the longevity of industries and enterprises through its corrective and preventive approach.
    c. It predicts and fixes processes in advance which enhances the life of the product and mostly helps to attain the desired goal.
    d. It collects data, which is vital for the future that can be used for research and development purposes.
    e. It contributes as key ingredient in modern engineering.
    f. It helps build trust and makes both the product as well as processes transparent through Informed decision-making.
    g. It guarantees higher effectiveness through multiple checks and balances.

DT is a powerful means to meet the ends of the present industry as well as organizational challenges. It significantly institutes a powerful bond with the physical object to detect intrinsic cracks and predict design alterations.

On a case-by-case basis, it offers a unique potential for greater efficiency and cost-effectiveness. Today, DTs are considered so the predominant in business sector that it is observed to have been mentioned successively for three years among Gartner's Top 10 Strategic Technology Trends (Gartner, 2020). Apart from it being economical, DT manifests unique services such as on-site diagnostics, prescriptive maintenance, or operation optimization. The technology has grown to a logarithmic capacity building by enhancing the efficiency, scalability, and usability of simulation model, as well as data models.

DIGITAL TWIN USE CASES

As discussed above, before DTs, physical twins were used in aerospace engineering. But since the introduction of DTs technicians can predict and foresee an upcoming problem which ensures the safety and a qualitative approach. The DT technology has multiple applications including industry, engineering, manufacturing, design, and construction. The DT of a product is helpful in determining the variables like quality and performance of the real product by virtually experimenting for the purposes of improvement of design and optimizing the performance of the product.

Thus, DT technology has vital potential to enhance research and development of the assets with the help of enormous data it can generate. Due to its data repository skill, even before beginning production, DT helps monitor and reflect the asset before it exists. This helps improve efficiency and improves performance during

Figure 4. The plusses of DT in various sectors

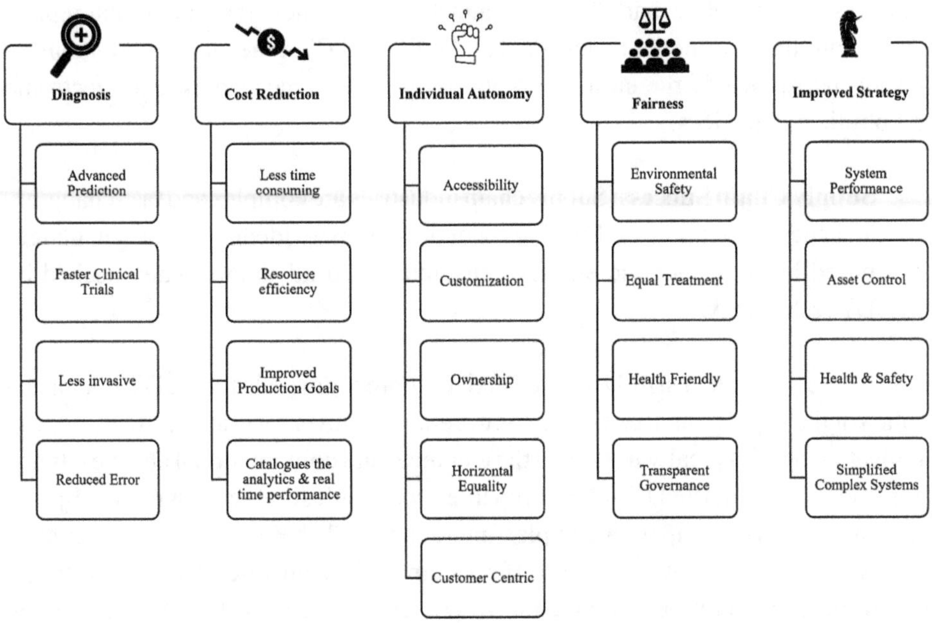

and post after manufacturing process. With DTs, the final processing or the end-of-life decisions for the products help better utilize the resources and make informed decisions. It helps prevent setbacks due to sudden functional failures.

Hence, with enormous benefits come enormous use cases. It is evident that it can be used across a broad gamut of industries and services (Kasyanau, 2022; Gartner, 2020). Thus, before installing new machines for manufacturing, the companies can use DT to simulate the impact and benefits of the new machine. It can fine-tune the production as well as it can receive real time information which can be used to predict beforehand. However, DTs are being widely used in variety of sectors (like Manufacture, Automotive, Healthcare, Retail, Disaster management, Smart cities etc.) for different applications and purposes. Some of the use cases are elaborated as under:

1. **Manufacturing systems and services:** The rise of the fourth industrial revolution and the enhanced adoption of big data is steering toward data and data driven manufacturing strategies. Thus, DT is the most trending data driven manufacturing concept enabling manufacturers and various business sectors to simulate products to build high-quality, cost effective, and quicker products. The DT are being actively used in the **manufacturing process** to come out with improved products. It also helps in elevating the defects during the process which prevents loss and improves the brand.

For example, in the automobile sector, it improves not only the performance level but also gears up better marketing and launch of the vehicle through digital replica even before the real model is designed and built. A DT in the automobile industry is a virtual replica of the entire automobile, software, mechanics, and electronic and physical behaviors.

2. **Supply Chain Success:** Supply chain industries are complex and fast-changing, and the application of DT does wonders. It is evident that the challenges posed by the pandemic era have ensured added advantage when backed by DT technology.

DT have been successfully applied and it is unrecognizable today. DT is a virtual replica of the supply chain consisting of warehouses, inventory, logistics, etc.an,d its application has drastically improved the technical and computational abilities. In the supply chain sector, the DT helps companies cope with simulation with all objects/ assets as well as cope up with complex interconnected chains within a shorter span of time (Pace, 2019). According to the LLamasoft retail research reports, in the present times the companies which have proactively adopted the DT technology, are swimming ahead in troubled waters (Jakovljevic, 2019).

3. **Healthcare Service Management:** In the health sector, smart products with sensors have the capacity to assess various health conditions of an individual. Thus, the person's digitally paired device can gather real-time information which in turn helps predict latent health conditions (e.g., blood pressure, oxygen levels, sugar levels, heart rate, etc.) and improve individual health. In the health sector, DTs are used to develop digital health data. For instance, generating a digital replica of a hospital enhances resource optimization (e.g., Detecting Covid 19 cases, bed shortages, staff availability and scheduling, managing wards); risk management, etc. Another interesting development is the growing need for human DT. Such twins are required for modeling organs and single cells or genetic constructs of an individual. Meanwhile, doctors are modeling outpatient specific DTs of lungs to make decisions with precision about ventilator use when treating a COVID-19 patient.

In addition, it covers lifestyle habits and physiological traits for generating personalized treatment and medication. Such personalized data serves as the basis of clinical trials and research data. Also, with advanced digital modeling doctors discover the pathology even before the illnesses are evident. They are also using DT to prepare for difficult surgeries (Gabor et al., 2020; Björnsson et al., 2020). DT for chemical substances and drugs help the chemists modify or redesign particle size and chemical compositions. Similarly, DT of a medical device helps test the success or failure rates of the product, reduces costs, and improves performance and safety. Hence eventually, DT technology is expected to capture the human body, its functions and processes s at the micro (Internet of Bio-Nano Things) (Celik, Salama, and Eltawil, 2022) as well as macro (Internet of Bodies) (Lee, 2020). level. The application of DT technology in the health sector perhaps will bring a breakthrough in fighting incurable diseases (and like cancer, AIDS) to mapping brain activity (Björnsson et al., 2020). In the article, "Fantastic voyage: Live long enough to live forever", Kurzweil and Grossman suggest that humans might overcome death with the use of DT in health sector.

4. **Construction and Smart Production Management:** Construction companies with the help of DTs find it easier to understand the real-time performance of building infrastructure. The accumulation of data comes out to be a ready reference for planning and designing future construction plans.
5. **Expansion of the use cases:** Other than the commonly used sectors, the DT offers better results for stakeholders like improved satisfaction and experience for customers. In short, it eases verification, avoids failure, and predicts customer demands. The DT today is gaining more importance and gradually taking over larger projects starting from disaster management (DT can help combat the challenges of global climate change), smart city DT (e.g., Shanghai) to National DT (e.g., UK). Interestingly, the National DT Program (NDTp) which is run by the Centre for Digital Built Britain envisages to launch a high-quality national ecosystem of connected DTs to build a better national environment; to ensure secure resilient data sharing and management of information; to provide coordination and alignment among key players. With the help of digital framework task group (DFTG). The NDT offers to benefit society; the economy; the business which reduces uncertainty and improves business; benefits the environment with the help of increased recycling and greater efficiency (*National Digital Twin Programme*, n.d.). The simulator analytics of a National twin/ Smart City DT involves managing city health, test accessibility options, handle emergency responses; assess the new construction etc., Singapore has modelled out a complete DT of the city-nation for tracking climate, traffic, pollution, and the city layouts.

Destination Earth (DestinE) is aiming to develop a highly accurate digital model of the Earth to monitor and predict the interaction between natural phenomena and human activities in light of climate change. DestinE, a part of the Green Deal and Digital Strategy of European Commission, is looking forward to achieving a unique objective of the twin transition 'green and digital'. Hence, the European

Commission and other implementing entities viz., the European Space Agency (ESA), European Centre for Medium-Range Weather Forecasts (ECMWF), and European Organization for the Exploitation of Meteorological Satellites (EUMETSAT) signed the contribution Agreement to start the preliminary phase of implementation by mid of 2024 (*National Digital Twin Programme,* n.d.).

The potential use cases indicate that the DT applications have become an essential part of everything from smart city planning to improve human body systems and creating a digital replica of the earth with sustainable goals. Undoubtedly, it is and will rapidly grow within the next few years.

ETHICAL CONFRONTS AND LEGAL CONUNDRUMS OF DIGITAL TWINS

The DT has multiple ethical and legal implications because of its complex nature and upcoming applications. But as understood from the above discussion, DT has no looking back. It will grow with the speed of digitization and with the growth of complex models of DT, the ethical and legal questions shall keep growing. Several questions of ethics arise in all use sectors of DT. However, the question is, how seriously the ethical challenges are taken and what are the after steps? In practice, it is generally seen that whenever there is a disaster or even a question of profit, any ethical principle may be overridden. At times even laws may be overruled. Hence,

Figure 5. Ethical and legal concerns that should be addressed and balanced

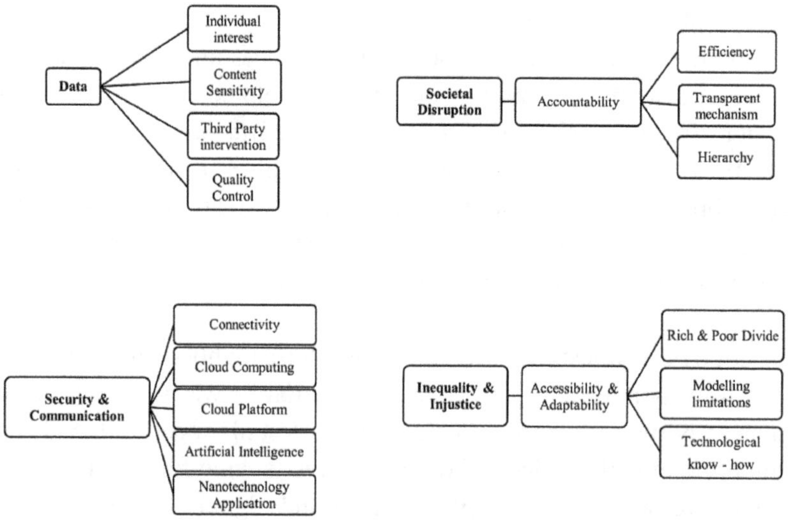

the application of DT technology has brought revolutionary advances and it can be found that DT technology applications can pose ethical dilemmas like influence, competition, conflict, discrimination, inequality, etc.

To identify the ethical challenges, the present section outlines the DT application zones as follows:

1. **Planning and Optimization Zone:** It is among the earliest areas where DTs were used. These tools of choice in general reasonably make sure that the goal-setting for-profit and utility maximization is done right. Under defined task structures, DTs work unhindered and efficient. However, when all important processes are deeply assessed, such processes involve challenges like:
    a. *Computational complexities:* Certain problems are computationally cumbersome and thus, finding proper solutions in real-time becomes unrealistic. To curb such challenges, simplifications must be done offline.
    b. *Delayed Planning:* At times, requirements like market availability, market demand, or supply-related issues may require quick decision-making, or the implementation of decisions to suit the nature of the requirement may cause challenges like instability and chaos.
    c. *System Breakdown:* There may arise breakdown situations like the breakdown of machines, electricity blackouts, production plant failure, or lack of surplus supply. Such turbulent times challenge the predictability and precision of DT. Under graver circumstances like cyber security attacks, the setup may face data compromise issues.
    d. *External Breakdown:* Externally there may accrue costs unexpectedly for situations like economic crises, litigations, insolvency, etc. Interestingly, the pandemic era caused severe breakdowns because of lockdowns, such recurrences can also add to such external breakdown factors. Such challenges may lead to damages as well as resource exploitation (Dasgupta and Ehrlich, 2013).
2. **Infrastructural and Geographical Zone:** It is relatively simple to apply DT under these zones because of three reasons, (1) it changes very little over time; (2) it comprises of measurable units of materials and structures, and (3) large quantities of measurable data give more appropriate average results. The challenges include the following:
    a. *Usage Patterns:* In cases of infrastructures and geographies, often no heed is given to usage patterns. It is when the infrastructural allocations are made, it specifies the qualified users. Incidentally, people use, occupy, or interact in infrastructural spaces in ways not anticipated. The gap between expected and unexpected forms of infrastructure creates ethical problems like injustice, inequality, conflict of interest, and competition.

It is, therefore, pertinent to create shared allocations to avoid ethical violations.

b. *Accessibility Rights:* The data collection based on usage patterns eases the classification of people's status and interest. Further, by deploying the technology to run smart cities, smart spaces, and even smart nations, ethical concerns like privacy, security and equality arise. On a sectoral basis, the data collected may be used to allow or disallow access to goods and services, even before giving the individual the right to choose or determine.

3. **Digital Twin Cognitive Computing Zone:** Cognitive computing is where an attempt is made to put together applications which preexist as concepts and then these tools are used to solve industrial problems and bring about Zero-Defect Manufacturing (ZDM). This discipline in order to give accurate results require enormous amount of sensitive data.

   a. *Data driven:* There is greater utility of cognitive DT in manufacturing and interestingly the recent advances are reaching milestones in human mind and body mechanisms which are data driven. However, such applications not only have greater corrective measures in the field of genetics, brain functioning, organ health etc. but also comes with certain risks. The technology which are represented by numbers is not as complex as the same technology with non-numeric data. Thus, a high threat is posed upon a society as it may turn inhumane.

   b. *Objectifying human beings*: The cognitive DT poses real threat upon the human autonomy and freedom by overtaking human ability to reflexivity and self-organizability (Durán and Fernández-Dols, 2021). The technology if not tamed with precision may treat humans like robots or objects by substituting, re-setting or discarding. Such inappropriate applications will negate humans the right to life and dignity. Such intervention into the biological existence of humans would not be far from the stage of a world "Matrix" where the individuals would be bounded by virtual technologies. The threats of such time include the replacement of individuals completely by DT, a complete state of posthumanism and the absence of a fair society.

4. **Digital Twin and the internet of Bio-Nano things (IoBNT) zone:** The technology can influence body functions ranging from immune system support genetic engineering, brain controlling, bio-hybrid implants, and drug delivery system (Freitas, 2006; Patra et al., 2013). In the world facing sustainability threats on account of population explosion how feasible is a technology to extend life expectancy. Thus, the maximization of DT application in the health sector will create a novel form of capitalism that would make the

facility available to the economically affluent and at the same time control an individual's right to determine health-related concerns. A capitalist-centric approach can even control the regulatory regime and can lead to gross violations of human rights while rendering inequality and injustice. If taken a careless approach, such DT measures to health will takeover an individual's choice to live or die. The progress of nanoparticle-based technology can be used in treatments and diagnosis. It is referred as the Internet of Bodies (IoB). The present DT application in the light of IoB has the possibility to surveil human body function and interfere with body functions (e.g., gene editing). The rise of Nanotechnology, the underlying technology in the IoB raises grave apprehensions as such particles can be effortlessly absorbed to the human body through air, water, vaccinations, medicines, food, etc. The ethical dilemma that may arise is the idea of "individual consent" or the "right to be informed" in advance of such deposition. Furthermore, militarization (Akyildiz, Brunetti and Blazquez, 2008) is a grave concern again with nanotechnology. Such militarization can be done through every means starting from food, water, medication, etc. If not belled in time it can lead to mass destruction and crimes against humans.

5. **Digital twin and the planetary health zone:** The human health and health of civilization implicate the planetary health (Whitmee et al., 2015). It is the health of human civilization and the state of the natural systems on which it depends (Horton and Lo, 2015). This will combine the Internet of Things (IoT) with light-based data transmission, Quantum Computing, and Artificial Intelligence.

With such powerful recourse, there it can be assumed that the use of nanotechnology and 'smart dust' to collect data wirelessly in unimaginable amounts and unimaginable sources from the world. This could then act as a means to identify everyone through an electronic ID which would release readable data concerning anything and anybody possibly on earth. Such civilization-saving measures on the negative front can lead to war and depopulating.

These solutions can serve the purpose of making humans controlled and manageable perhaps like things, in the race of uniformizing human conduct, it will destroy the very essence of human dignity, which is the foundation of democracies around the world and a key component of the UN Charter of Human Rights. A data-driven approach will cost human beings their freedom, dignity, culture, creativity, and emotions (*Digital Threats to Humans and Society*, 2022). Conclusively, if the DT technology is not reasonably applied, it may end up automating the society to extreme levels which can inevitably undermine human intelligence, and can further cause a negation of innovation, diversity, and much more.

EMERGING TRENDS AND LEGAL IMPLICATIONS OF DIGITAL TWINS

The DT with its wide scope of applications have considerable legal implications. It poses risks that differ based on its nature and its contractual needs. They raise concerns in terms of liability, ownership, privacy, cyber security, malicious use, responsibility for data accuracy, assurance, governance, trust, liability, ongoing maintenance, and more.

1. *Data Ownership:* Data ownership remains a puzzle when it comes to determining the true owner viz., data sharer or DT model owner. The legal clarity for the need to determine the rightful owner of the data held within the DT model raises concerns about data ownership. The agencies, parties or individuals contributing into the data pool which attract copyright laws should be recognized. However, in complicated situations like conflict-of-interest situations of joint ownership may be recommended on a factual basis (*Digital Twin and its Legal Implications*, 2021). But the rights of joint owners may not be clear unless there are specific contractual provisions.

The need for the DT to incorporate materials in line with the intellectual property becomes crucial when the question of data retention and over what period of time accrues. Either such a twin model must be continually updated to incorporate the changing data to stay in line with copyright laws or simply through contractual obligation allow the DT to function within the permitted use. To ease the data ownership and control concerns, further simplification by granting a license for use of data during the lifetime of the asset or DT and *vice versa.*

2. *Data sharing and Privacy:* As discussed above, DT has enormous benefits for any organization or industry. However, such benefits can add value only in case of interactions between reliable sources from a security-minded approach. Fundamentally, the DT needs to be secure and trustworthy. In DT the question of privacy concerns needs to be addressed on the levels of data processing and data protection. Potentially, there can be risks of IP theft, system access, non-compliance, and loss of informational integrity due to cyber-attacks. The concern of privacy is very critical because it is location and fact-based. The privacy laws depend upon the data type and the data collection procedures as well as the means (Behrendt, 2018). In the present times, by algorithmic approach, data can be transferred unethically from an object to a subject, from machines to humans, etc. The *cognitive computing discipline* is trying to derive features that can detect and process accountability (Durán and Fernández-Dols, 2021).

Eventually, the DTs are expected to capture body structures, functions, and processes. Nanotechnology may offer solutions by exposing the body to nanoparticles or nanobots. In return, it may allow one to read out activity patterns. Relevant technologies have become known under names such as the "Internet of Bodies" or the "Internet of (Bio-)Nano-Things" which pose grave questions of personal data protection and the question of consent.

3. *Liability:* The question of liability becomes very complicated when there are hierarchies. At differential levels, there are separate models for collecting and processing data. Thus, if there comes an error from the permutation and combination of data from multiple sources rather than an ecosystem of DT interconnected together, where lies the question of liability? Apart from the concern of error, what if due to a single deviation the functioning of all other interconnected models is affected? The data heap and multilayering make it difficult to locate where and with whom the liability lies. As a matter of precaution, when there are multiple parties and data sources, security and accuracy measures need to be right in place. At a point when different parties depend on the accuracy of the data, trust issues may come up. But it is not feasible unless the technology is fully grown. At this juncture, on the case-by-case basis one needs to address, and the concerns may have to face economic setback due to payment of damages or to reinstatement. The Gemini principles emphasize heavily on trust, openness, quality, clarity of purpose, and more. Conclusively, the complexities involved in the liability context can be best handled through clear- cut functions, purpose, and contractual obligations (NEC suite of contracts). It will foster trust between the parties in question and encourage data sharing under secure platforms and maintain the quality of the same.

CONCLUSION

The concerns of security are crucial with any technology. And gigantic technology like DT has its unique set of challenges. It is still an emerging technology, there is less clarity about its contributions to mankind (Singh et al., 2018). It is perceived to have great potentials but in the process of its implementation it, may have to face ambiguous complications (like data related problems, expensive investments, lack of rules and regulatory measures, etc.) at differential levels.

What has been understood so far is that DT is a combination of several technologies including 3-dimensional simulations, IoT, AI, big data, machine learning, and cloud computing. In the future it can be merged with other technologies as well (Luiz,

Meyendorf, and Mrad, 2017). The technologies it can be combined with are also in an evolutionary stage. The laws and regulations about them are yet to be formalized. Hence it is challenging to accurately access the approximate growth of DT.

Cost is also a concern. The complete procedure of creating a DT involves developing ultra-high-fidelity computers and its simulation. It is time-consuming as well as labor- intensive which also makes it an expensive investment (Gabor et al., 2020; Björnsson et al., 2020). Gartner and Marc Halpern, at the PDT Europe conference in Gothenburg, Sweden expressed worries regarding the costs and time-related aspects of DT. Even though Halpen was positive about the basic content and structure of DT, he still added that bringing DT will take time and resources beyond imagination. Industries should take note of this concern before implementing DT.

There are unending DT infrastructures discussed and there is a need to define a stable structure for DT in the industrial level. The absence of standards for device communication and data collection can lead to the quality of data being compromised and resultantly degrade the performance of DT (*Common Pitfalls to Digital Twin Implementation,* n.d.). Data is one of the most prominent concerns of DT. It revolves around privacy, ownership, transparency, and confidentiality. Holding data and its sharing depends on the company policies. Lack of proper policy measures on data sharing internally or externally by an organization can end up in data silos departmentally which can be detrimental. But, in the light of cyber security being a challenge, having data silos makes DT less vulnerable to cybercrime (Gagné, 2020). Again, data interoperability is challenging. It is all about how the data among different DTs is generated and shared at different hierarchies. Each level generates a unique set of data which can create complex data set relationships (Gagné, 2020). Lack of proper data governance and management is a major concern as well.

However, IPC-2552 is a US-based interoperable framework to create smart value chains. It is said to be the first international standard for DT product, manufacturing, and lifecycle frameworks (Peterson, 2021; Gagné, 2020). IPC is expected to help with process optimization which in turn reduces losses. Supposedly, it can be said to be a master plan for effective operationalization of DT. Though the international standard is a much-awaited step taken towards a systematic development of DT technology, but the regulatory approaches still remain vague. Thus, to cope with the regulatory concerns as discussed above, *firstly,* there is need for "Global standard norms" for safe and secured application of DT, which will help take note of accountability of failures, actions as well as negligence. *Secondly,* to further germinate the DT potentially potential, it is required to create a "regulatory sandbox". Such experimental tech-based framework will develop a safe and secure application of DT. *Thirdly,* an effective monitoring mechanism must be in place to zero down all the tectonic shifts. Last but not the least, the foreseeable threats to freedom, equality, and exploitation appeal for a strong consensus on "adopting ethical standards" for a continued sustainable and humane ecosystem.

Finally, like on all great issues, much remains to be said.

ACKNOWLEDGMENT

We express our deep gratitude to Dr. Ingrid Vasiliu-Feltes, (University of Miami, FL) for her constant guidance and support. We are thankful to Infinity Online Global Academy (IOGA) for the inputs, the University of Munich, Germany and Kalinga Institute of Industrial Technology, India for the library source materials. Lastly many thanks to all who have contributed in some way or the other.

This research received no specific grant from any funding agency in the public, commercial, or not-for-profit sectors.

REFERENCES

Akyildiz, I. F., Brunetti, F., & Blázquez, C. (2008). Nanonetworks: A new communication paradigm. *Computer Networks*, *52*(12), 2260–2279. doi:10.1016/j.comnet.2008.04.001

Almazrouei, E., Shubair, R. M., & Saffre, F. (2018). The internet of nano things: Concepts and applications. *Clinical Orthopaedics and Related Research*, 22–27. https://arxiv.org/pdf/1809.08914.pdf

Athavale, V. A., & Athavale, A. (2021) Digital twin- a key technology driver in industry 4.0. *Engineering Technology, 4*(1). 555628. doi:10.19080/ETOAJ.2021.04.555628

Banyai, T., & Felice, F. D. (Eds.). (2020). *Industry 4.0-Impact on Intelligent Logistics and Manufacturing*. IntechOpen. doi:10.5772/intechopen.76554

Behrendt, V. D. (2018). A Digital Twin-based Privacy Enhancement Mechanism for the Automotive Industry. *International Conference on Intelligent Systems (IS)*, 272-279, 10.1109/IS.2018.8710526

Björnsson, B., Borrebaeck, C., Elander, N., Gasslander, T., Gawel, D. R., Gustafsson, M., Jörnsten, R., Lee, E. J., Li, X., Lilja, S., Martínez-Enguita, D., Matussek, A., Sandström, P., Schäfer, S., Stenmarker, M., Sun, X. F., Sysoev, O., Zhang, H., & Benson, M.Swedish Digital Twin Consortium. (2020). Digital twins to personalize medicine. *Genome Medicine*, *12*(1), 1–4. doi:10.118613073-019-0701-3 PMID:31892363

Celik, A., Salama, K. N., & Eltawil, A. M. (2022). Internet of Bodies: A Systematic Survey on Propagation Characterization and Channel Modeling. *IEEE Internet of Things Journal*, *9*(1), 321–345. doi:10.1109/JIOT.2021.3098028

Common Pitfalls to Digital Twin Implementation. (n.d.). https://www.compunneldigital.com/infographic/common-pitfalls-to-digital-twinimplementation/#gref

Dasgupta, P., & Ehrlich, P. R. (2013). Pervasive Externalities at the Population, Consumption and Environment Nexus. *Science*.

Digital Threats to Humans and Society. (2022, June 29). FuturICT. Retrieved from http://futurict.blogspot.com

Digital Twin and its Legal Implications. (2021, August 5). Abou Naja. Retrieved from https://abounaja.com/blogs/digital-twin-legal-implications

Digital Twin. (n.d.). Siemens Digital Industries Software. Retrieved from https://www.plm.automation.siemens.com/global/en/our-story/g lossary/digital-twin/24465

Digital twins for small satellites. (n.d.). Stratosphere. Retrieved from https://stratosphere-technologies.com/digital-twins-for-smal l-satellites/

Durán, J. I., & Fernández-Dols, J. M. (2021). Do emotions result in their predicted facial expressions? A meta-analysis of studies on the co-occurrence of expression and emotion. *Emotion (Washington, D.C.)*, *21*(7), 1550–1569. doi:10.1037/emo0001015 PMID:34780241

Freitas, R. A. Jr. (2006). Pharmacytes: An ideal vehicle for targeted drug delivery. *Journal of Nanoscience and Nanotechnology*, *6*(9-10), 2769–2775. doi:10.1166/jnn.2006.413 PMID:17048481

Gabor, T., Belzner, L., Kiermeier, M., Beck, M. T., & Neitz, A. (2016). A simulation-based architecture for smart cyber-physical systems. *Proceedings of the 2016 IEEE International Conference on Autonomic Computing (ICAC)*, 374–379. 10.1109/ICAC.2016.29

Gagné, M. R. (2020). *Digital Twins, Another Reason to Worry about the iot and Data Security*. https:// irishtechnews.ie/digital-twins-iot-and-data-securit y/

Gartner. (2021). *Gartner Top Strategic Technology Trends*. Available at: https://www.gartner.com/en/information-technology/insights/t op-technology-trends

Grieves, M. W. (2014). *A whitepaper, digital twin: manufacturing excellence through virtual factory*. https://www.3ds.com/fileadmin/PRODUCTS-SERVICES/DELMIA/PDF/Whitepaper/DELMIA-APRISO-Digital-Twin-Whitepaper.pdf

Grieves, M.W. (2016). *Origins of the digital twin concept*. doi:10.13140/RG.2.2.26367.61609

He, B., & Bai, K.J. (2020). Digital twin-based sustainable intelligent manufacturing: A review. *Springer, Adv. Manuf., 9*, 1–21. doi:10.1007/s40436-020-00302-5

Horton, R., & Lo, S. (2015). Planetary Health: A new Science for Exceptional Action. *Lancet, 386*(10007), 1921–1922. doi:10.1016/S0140-6736(15)61038-8 PMID:26188746

Jakovljevic, P. (2019). *Llamasoft: A Heritage of Supply Chain Decision Science Applied to Interconnected Enterprise Level Decisioning*. Technology Evaluation Centers (TEC). Available at: https://www3.technologyevaluation.com/research/article/llamasoft-a-heritage-of-supply-chain-decision-science-applied-to-interconnected-enterprise-wide-decisioning.html

Kasyanau, A. (2022). What are digital twins and what can we use them for? *Forbes*. Retrieved 26 Jul 2022 https://www.forbes.com/sites/forbestechcouncil/2022/06/27/what-are-digital-twins-and-what-can-we-use-them-for/?sh=294f15c7fa6e

Lee, M. (2020). *The Internet of Bodies: Opportunities, Risks and Governance*. RAND Center for Global Risk and Security. https://www.rand.org/multimedia/video/2020/10/29/what-is-the-internet-of-bodies.html

Liu, Z., Meyendorf, N., & Mrad, N. (2017). The role of data fusion in predictive maintenance using digital twin. *Proceedings of the AIP Conference Proceedings*, 16–21.

National Digital Twin Programme. (n.d.). Centre for Digital Built Britain, University of Cambridge. Retrieved from https://www.cdbb.cam.ac.uk/what-we-did/national-digital-twin-programme

Pace, S. G. (2019). *The evolution of digital twins in supply chain*. National Library of Medicine. doi:10.118613073-019-0701-3

Palla, K. (2022). The rise of digital twin technology. *Forbes*. Retrieved 26 Jul 2022 https://www.forbes.com/sites/forbestechcouncil/2022/08/03/the-rise-of-digital-twin-technology/?sh=2082ccc02f97

Patra, D., Sengupta, S., Duan, W., Zhang, H., Pavlick, R., & Sen, A. (2013). In- telligent, self-powered, drug delivery systems. *Nanoscale*, *5*(4), 1273–1283. doi:10.1039/C2NR32600K PMID:23166050

Peterson, Z. (2021). *IPC-2551 Digital Twin Standards Released*. https://resources.altium.com/p/ipc-2551-digital-twin-standard-released

Rios, J., Hernandez, J. C., Oliva, M., & Mas, F. (2015). Product avatar as digital counterpart of a physical individual product: Literature review and implications in an aircraft. *Transdisciplinary Lifecycle Analysis of Systems.*, *2*. Advance online publication. doi:10.3233/978-1-61499-544-9-657

Singh, S., Shehab, E., Higgins, N., Fowler, K., Tomiyama, T., & Fowler, C. (2018). *Challenges of Digital Twin in High Value Manufacturing*. SAE Technical Paper; SAE International.

Stark, R., & Damerau, T. (2019) Digital twin. In CIRP Encyclopedia of Production Engineering. Springer. doi:10.1007/978-3-642-35950-7_16870-1

Tuegel, E. (2012, April). The airframe digital twin: some challenges to realization. In *53rd AIAA/ASME/ASCE/AHS/ASC structures, structural dynamics and materials conference 20th AIAA/ASME/AHS adaptive structures conference 14th AIAA* (p. 1812). 10.2514/6.2012-1812

What is digital twin technology and how does it work? (n.d.). TWI Global. Retrieved from https://www.twi-global.com/technical-knowledge/faqs/what-is-digital-twin

Whitmee, S., Haines, A., Beyrer, C., Boltz, F., Capon, A. G., de Souza Dias, B. F., Ezeh, A., Frumkin, H., Gong, P., Head, P., Horton, R., Mace, G. M., Marten, R., Myers, S. S., Nishtar, S., Osofsky, S. A., Pattanayak, S. K., Pongsiri, M. J., Romanelli, C., ... Yach, D. (2015). Safeguarding human health in the Anthropocene epoch: Report of The Rockefeller Foundation–Lancet Commission on planetary health. *Lancet*, *386*(10007), 1973–2028. doi:10.1016/S0140-6736(15)60901-1 PMID:26188744

ADDITIONAL READING

Geraci, R. M. (2012). *Apocalyptic AI: Visions of Heaven in Robotics, Artificial Intelligence, and Virtual Reality*. Oxford University Press.

Grieves, M. W. (2005). Product lifecycle management: The new paradigm for enterprises. *International Journal of Product Development*, 2(1/2), 71–84. doi:10.1504/IJPD.2005.006669

Hurwitz, J., Kaufman, M., & Bowles, A. (2015). *Cognitive computing and big data analytics*. John Wiley & Sons.

Tao, F., Zhang, M., & Nee, A. Y. C. (Eds.). Digital twin driven smart manufacturing. Academic Press.

KEY TERMS AND DEFINITIONS

Automation: A technological process which diminishes human input.

Digital Twin: A computer generated virtual clone of a physical object, process, or network. It map and mirrors the real object in virtual space.

Internet of Bio-Nano Things: A process of inducing bio-chemicals in a living organism (primarily humans) that interacts through cyberspace.

Internet of Bodies: The computer program or device that collects biological data or alter biological function of living things (primarily humans) through cyberspace.

Physical Environment: The space where the physical asset exists.

Product Life-Cycle Management: A strategic process of eliminating waste and inefficiency across complete journey of a product's life. It is an integrated information driven approach to all aspects of product's life including final disposal. It manages information about the product.

Virtual Space: The computer simulated environment that has no physical existence.

Chapter 2
An E-Government Portal Maturity Model:
A Detailed View and Web-Based Tool

Laila Cheikhi
Mohammed V University in Rabat, Morocco

Abdoullah Fath-Allah
Mohammed V University in Rabat, Morocco

Rafa E. Al-Qutaish
https://orcid.org/0000-0001-7186-2992
ADP, UAE

Ali Idri
Mohammed V University in Rabat, Morocco

Alain Abran
https://orcid.org/0000-0003-2670-9061
École de Technologie Supérieure, Canada

ABSTRACT

E-government portals are now playing an important role in facilitating citizens' lives by executing services at any time and from any location. Such a way of providing services results in great benefits for citizens and agencies, particularly within the COVID-19 pandemic context. Over the past few years, a number of best practices have been identified for designing and developing e-government portals but have been documented piecewise across the literature. A key approach to facilitating access to such best practices is their integration into a structured maturity model tailored to a domain of application. A digital twin is a digital representation of a physical object or system. However, data from the e-government, which is a part of what is called smart city, can be transferred to a digital twin to work with other data from other systems, such as IoT systems. Thus, these insights can be used to make better decisions.

DOI: 10.4018/978-1-6684-3833-6.ch002

INTRODUCTION

In traditional government portals through physician offices, much paperwork and employees are deployed by agencies to provide services to citizens. This requires on-site citizen presence. The emergence and deployment of e-government portals have resulted in great benefits to governmental agencies and citizens in terms of time savings and flexibility. Such e-portal services are provided electronically to citizens who no longer need to travel to the agencies: citizens can request and execute the services online through its portal (a web application or a mobile application).

Nowadays, the appearance of the COVID-19 pandemic in 2020 has led to increased use of e-government portals where citizens can request service anytime and anywhere. The agencies can then service the request online without any time and location constraints. This contributes to limiting COVID-19 virus contagion and the number of affected citizens. Furthermore, in the context of limited government funding, agencies should leverage the worldwide experience and best practices when building e-government portals. This chapter presents our model of e-government portals' best practices model (eGPBPM) and a corresponding measurement-based e-government portals' maturity model (MBeGPMM).

MAIN FOCUS OF THE CHAPTER

The design of eGPBPM is based on an extensive literature review to collect the best practices used by researchers and industry leaders in this domain (Fath-Allah et al., 2014b). The model is composed of five categories, together with 43 subcategories and 172 specific practices (Figure 1).

- **Back-end category:** best practices that can run in the background and are usually not seen by users (e.g., system, data processing, and business logic). It includes eight best practice subcategories and 33 specific practices.
- **Front-end web design category:** best practices when the user sees and interacts with the portal, especially those practices related to the interface or the design of the portal. It includes five best practice subcategories and 23 specific practices.
- **Front-end web content category:** best practices when the user sees and interacts with the portal, especially those practices related to the information and content of the portal. It includes 12 best practice subcategories and 48 specific practices.
- **External category:** best practices in general associated with the technical aspects of the portal, its marketing, and the involvement of the citizen in

Figure 1. eGPBPM Structure and Components

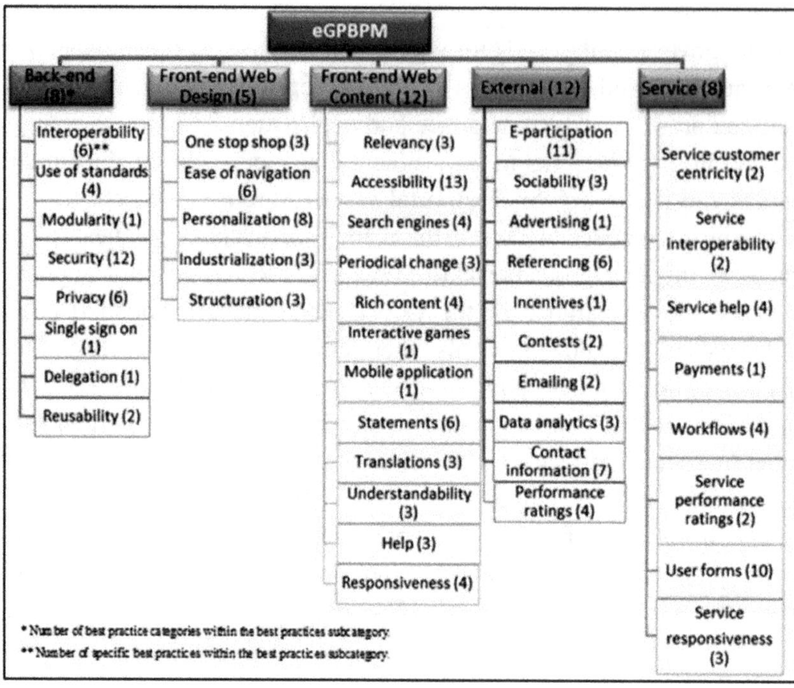

the e-government process. It includes 10 best practice subcategories and 48 specific practices.

- **Service Category:** best practices for services offered in the e-government portal It includes eight best practice subcategories and 28 specific practices.

This model provides a unified terminology and full coverage of e-government portals' best practices.

The MBeGPMM (Fath-Allah et al., 2014c), (Fath-Allah, Cheikhi, Al-Qutaish, & Idri, 2016) has been designed based on the analysis of many maturity models proposed in the literature (Fath-Allah et al., 2014a), (Fath-Allah et al., 2015) to address a number of their limitations, including missing best practices from the literature, inconsistent terminology across models, and undocumented calculation formulas limiting their reuse. MBeGPMM is aligned with eGPBPM, which also allows agencies to identify their portals' strengths and weaknesses, provide guidelines to address these weaknesses, and improve the maturity stage of portals. This maturity model includes four stages: presence, interaction, transaction, and integration. Each stage includes a set of best practice subcategories, each of which includes corresponding specific practices as well as measures. MBeGPMM and eGPBPM were evaluated by experts through a survey study, which demonstrated that the two models were valid and reliable (Fath-Allah, Cheikhi, Al-Qutaish, & Idri, 2016), (Fath-Allah, Cheikhi, Al-Qutaish, & Idri, 2016). Moreover, the use of the maturity model in practice and a real environment was tested through a case study (Fath-Allah, Cheikhi, Al-Qutaish, & Idri, 2016).

This chapter section is structured as follows. Section 2 provides a structural view of the MBeGPMM, followed by the procedural view of the model in Section 3. Section 4 provides the details of the model.

MBeGPMM: Structural View

This section presents the structure and components of MBeGPMM. It is a hierarchical framework with five levels (see Figure 2).

1. The maturity stages.
2. The best practices categories.
3. The best practices subcategories.
4. The specific practices.
5. The survey questions.

The first level of the MBeGPMM is related to the maturity stage. Based on the analysis of existing e-government maturity models from the literature with similarities and differences in terms of the number of stages, names, and contents (Fath-Allah et al., 2014a). . Still, they share some common stages related to the web availability of the portal, citizens' interactions with governments, citizens' transactions with

Figure 2. MEeGPMM Hierarchy (levels) and Components

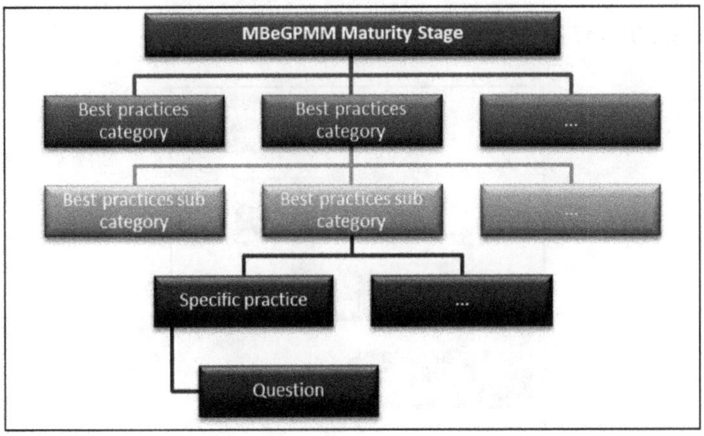

governments, and information sharing between agencies. MBeGPMM unified the naming as well as the content of the maturity models' stages based on their focus, which resulted in four stages:

Stage 1 (Presence): At this stage, the portal presents only static information or pages to citizens.
Stage 2 (Interaction): At this stage, the portal allows citizens to interact with the government.
Stage 3 (Transaction): At this stage, the portal allows citizens to complete transactions or execute online services on the web.
Stage 4 (Integration): At this stage, the portal allows information sharing across various agencies and the inclusion of citizens in the e-government process.

Each stage constituted the basis for the subsequent level (Figure 3). For instance, an e-government portal qualified as stage 2 should have deployed the best practices of stage 1 and all the best practices of stage 2.

The second and third levels of the MBeGPMM described in Figure 4 are related to the best practice categories and subcategories aligned with eGPBPM (see Figure 1). The mappings between the maturity stages and the eGPBPM five best practice categories (e.g., back-end, front-end web design, front-end web content, external and service) (Fath-Allah et al., 2014a), as well as with the 43 eGPBPM best practices subcategories have revealed that (Fath-Allah et al., 2014a):

Stage 1 (Presence) includes best practices that belong to front-end web design and content categories, with a set of 12 subcategories of best practices that should be considered.

Figure 3. MBeGPMM Maturity Stages

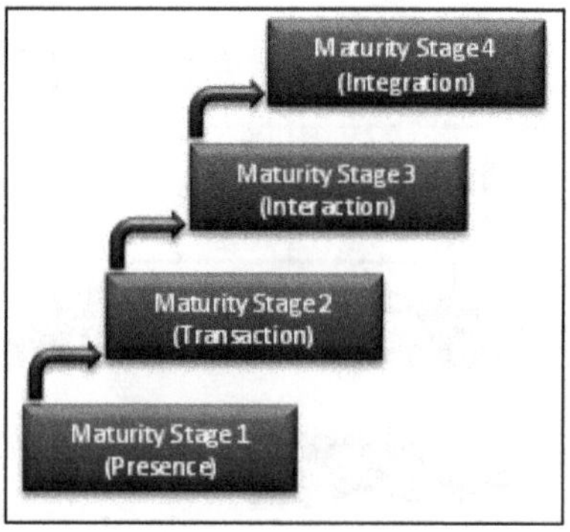

Figure 4. MBeGPMM Second and Third Levels of Best Practices Subcategories per Maturity Stages

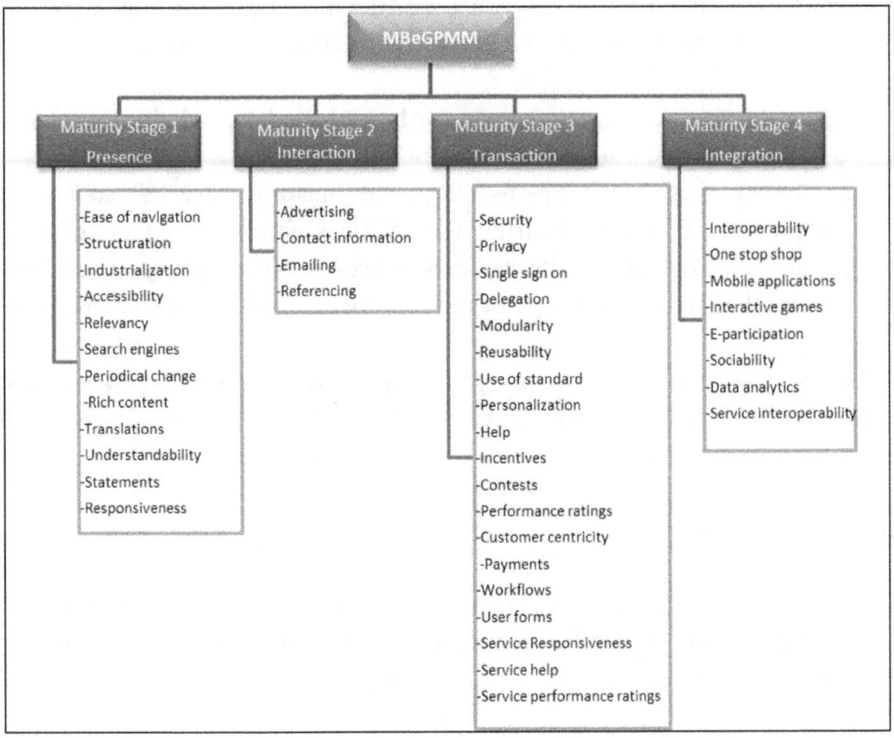

Stage 2 (Interaction) includes best practices that belong to the external category, with a set of four subcategories of best practices that should be considered.

Stage 3 (Transaction) includes best practices that belong to back-end, front-end web design and content, and external and service categories, with a set of 19 subcategories of best practices that should be considered.

Stage 4 (Integration) includes best practices that belong to back-end, front-end web design and content, and external and service categories, with a set of eight subcategories of best practices that should be considered.

The fourth level of the MBeGPMM is related to specific best practices. A set of 172 specific practices were collected from the literature for each best practice subcategory. Of these 172 specific practices, 55 belong to the present stage 1, 16 to the interaction stage 2, 71 to the transaction stage 3, and 30 to the integration stage 4.

The fifth level includes assessment questions that verify the presence or absence of those specific practices for an e-government portal. Details about the specific practices and assessment questions are provided in Section 3.

MBeGPMM: A Procedural View

This section presents the five steps of the MBeGPMM process, including their calculation formulae (Fath-Allah, Cheikhi, Al-Qutaish, & Idri, 2016):

1. Define the maturity stage of the portal to be evaluated. It is suitable for agencies to start with the first stage (e.g., the present stage).
2. Answer the assessment questions related to the maturity stage defined in step 1.
3. Compute the Compliance with the Best Practices of a sub-category (CwBPsC).
4. Compute the Compliance with the Best Practices of a Maturity Stage (CwBPMS).
5. Identify the missing best practices to be implemented for the studied stage.

Once implemented, agencies can repeat the process for the studied stage to reach a high level of compliance with its best practices. Then, agencies can repeat the process for the next maturity stage and complete their survey.

Calculation Method

Each best practice subcategory contains a set of assessment questions that verify the presence/absence of specific practices within this subcategory. The scoring of assessment questions can be one of the following formats:

- Binary scoring (1|0) was used to verify the presence/absence of best practice.
- Percentage scoring such as ratios or the scores obtained in automatic Web tools.
- Equally weighted scoring for assessment questions with multiple responses.

In calculation, it is assumed that the assessment questions and best practice subcategories are equally weighted.

- CwBPsC for all the best practices subcategories belonging to the selected maturity stage can be calculated by equally weighing the survey questions' score for each best practice subcategory, using equation (1).

$$CwBPsC = \left(\sum_{i=1}^{n} SQ\left(i\right) \right) / \; n \tag{1}$$

Where:

CwBPsC = compliance with best practices of a subcategory

n = Number of assessment questions in the corresponding best practice subcategory.
SQ = Assessment question score

- CwBPMS can be calculated by equally weighing the corresponding best practices subcategories' CwBPsC of the selected maturity stage, using equation (2).

$$CwBPMS = \left(\sum_{i=1}^{m} CwBPsC\left(i\right) \right) / \ m \tag{2}$$

Where:
MSCwBP = maturity stage compliance with best practices.
m = number of subcategories in the corresponding maturity stage
SCwBP = compliance with best practices in a subcategory.

MBeGPMM: A Detailed Content

This section provides the detailed content of the MBeGPMM using the 43 best practice subcategories presented in Figure 1. The best practices subcategory is a set of specific best practices in a subcategory linked to a category that should be implemented together to satisfy a set of specific goals.

Each best practices subcategory is expressed in terms of the following features:

- The best practices category includes a set of best practice subcategories that are considered important and linked to the category's purpose. The five main categories (Figure 1) are back-end, front-end web design, front-end web content, external, and service.
- The maturity stage group is a set of best practices that are considered important for the portal to reach this maturity stage. The four stages (Figure 4) are 1 for presence, 2 for interaction, 3 for transaction, and 4 for integration.
- The purpose is to provide general information about the best practices subcategory and its usefulness.
- The introductory notes provide a summary of the main specific practices covered by the best practices subcategory.
- The specific goal (SG) provides summarized information about the tasks that the portal should tackle to satisfy the best practices subcategory purpose.
- Specific practice (SP) is a required activity that must be present in the portal to achieve the corresponding specific goal. It is useful in the evaluation to identify the extent to which the best practices subcategory is met.
- The assessment question (SQ) allows verification of the presence/absence of a specific practice.

For illustration purposes, the "ease of navigation" and the "structuration" best practices subcategories are detailed in the following subsections. The other best practice subcategories are provided in the appendix. All 43 best practice subcategories and their corresponding assessment questions are included in the web-based tool described in Section 5.

Ease of Navigation

Best Practices Category and Maturity Stage

A front-end web design best practices category at maturity stage 1 (presence).

Purpose

The purpose of the ease of navigation best practice subcategory is to increase customer satisfaction and e-government adoption by allowing citizens to easily navigate and perform their tasks in the portal.

Introductory Notes

The ease of navigation is related to the user's ability to navigate the e-government portal easily and perform the required activities in a user-friendly manner. For instance, making the portal intuitive, user-friendly, and visually pleasing, or even having a clear menu structure are all considered important aspects of ease of navigation.

Specific Practices (SP) by Specific Goals (SG)

- SG1: Have a clear menu structure that allows users to navigate easily.
- SP 1.1: Show the current menu level for the users, so that they know where they are in the portal and what other services are available to them (Oakland, 1991).
- SP 1.2: Maintain the main menu on all pages of the portal (Rodríguez et al., 2008).
- SG2: Offering easy access to information and services without needing to go to other pages.
- SP 2.1: Provide a list of frequently searched items (Peters et al., 2004).
- SP 2.2: Provides a list of the most used services (ex. on the home page (Seifert & McLoughlin, 2007), (Georgescu & Georgescu, 2008), (UN E-Government Survey 2012, 2012), (West, 2007).
- SP 2.3: Provide a service catalog indicating the list of services available to citizens in the portal (Capgemini, 2010).

- SG3: Increasing accessibility and ease of use to make it easier for citizens to access e-government.
- SP 3.1: Offer a common look or similar sites in appearance to make it easier for citizens to access e-government portals (West et al., 2004).

Assessment Questions

- SQ 1.1: Does the menu show the current menu level and other pages or services available? (yes/no binary scoring)
- SQ 1.2: Is the main menu maintained on the rest of the pages? (yes/no binary scoring)
- SQ 2.1: Is there a list of frequently searched items in the portal? (yes/no binary scoring)
- SQ 2.2: Is there a list of the most used services on the portal or home page? (yes/no binary scoring)
- SQ 2.3: Does the portal offer a common look or similarity in appearance with other portals from the same or other jurisdictions? (yes/no binary scoring)
- SQ 3.1: Is there a service catalog indicating the list of services available to citizens in the portal? (yes/no binary scoring)

Structuration

Best Practices Category and Maturity Stage

A front-end web design best practices category at maturity stage 1 presence).

Purpose

The purpose of the structuration best practices subcategory is to help users and search engines find web pages easily.

Introductory Notes

Structuration is related to organizing websites in a structured way. This means that before starting to write web pages, time is spent designing the structure of the website. An example is the use of sitemaps, which provide a general overview of the structure of the portal.

Specific Practices (SP) by Specific Goals (SG)

- SG1: Give a general overview of the entire structure of the website and help the user stay oriented and know where the user is at each level of the portal and what other options are available to the user.
- SP 1.1: Before starting to write pages of the portal, the designer(s) should write the structure of the website (Oakland, 1991). It should be a common practice to figure out the global architecture of the e-government portal before starting to write the web pages, that is, 'think before you leap, then think again' (Oakland, 1991).
- SP 1.2: Use of HTML sites. They describe the entire structure of the website and include all the content or pages available on the website. In addition, HTML site maps help citizens find web pages easily.
- SG2: Help the search engines find Web pages easily.
- SP 2.1: Use of XML sitemaps These are designed for search engines, that is, they are machine-readable. As the HTML sitemaps, the XML sitemaps also describe the whole structure of the portal and its pages. Moreover, they help the search engine find web pages easily.

Assessment Questions

- SQ 1.1: Before starting writing pages of the portal, does the designer(s) spend time writing the structure of the website? (yes/no binary scoring)
- SQ 1.2: Is there a sitemap available in the portal? (yes/no binary scoring)
- SQ 2.1: Is there an XML sitemap for search engines available in the portal? (yes/no binary scoring)

SOLUTIONS AND RECOMMENDATIONS

Web-based Tool

A new web-based tool was developed to automate the data-collection process of this maturity model. This tool was built using MYSQL, PHP, and HTML. This web-based tool can be accessed at https://www.ipharmacie.ma/dash/sign-in.php. It allows e-government portals' Webmasters to login using their usernames and passwords. Figure 5 displays the login page of this tool.

Once the user is connected to the web-based tool, the vertical menu provides a number of options, as shown in Figure 6.

Figure 5. Login page of the Web-based tool

Figure 6. Web-based tool menu

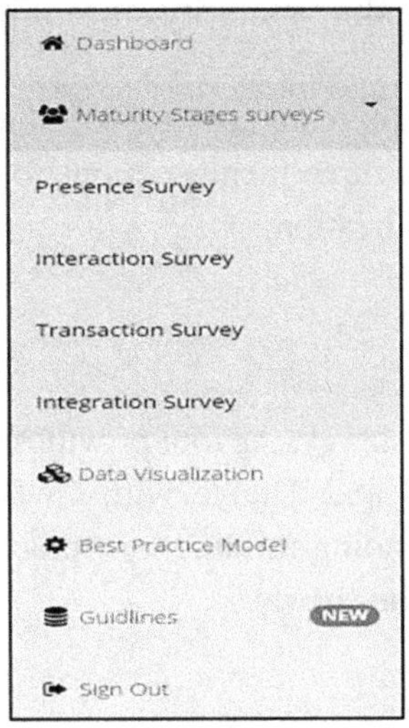

- the dashboard option provides general information for the users,
- The maturity stage surveys present the assessment questions related to each maturity stage (presence, interaction, transaction, and integration).
- the data visualization and best practices model tabs provide graphs and charts,
- the Guidelines represent the missing best practices identified from the survey questions, and
- the Sign Out allows the users to log out.

Process

The MBeGPMM allows agencies to identify the maturity stage of their e-government portals and provides guidelines and best practices for agencies to move to an upper stage of maturity. The first step in this process is to define the maturity stage to be verified (the first time the stage is 1). Next, the assessment questions corresponding to the maturity stage being verified can be executed by the portal's Webmaster. The tool then calculates the CwBPMS at the maturity stage. It also provides guidelines and best practices to move to a higher stage of maturity. Once the best practices are implemented, the Webmaster can select a higher stage and complete its corresponding survey.

Input and Output

The input to the web-based tool is the set of four surveys: presence, interaction, transaction, and integration surveys. Each survey verifies the presence/absence of best practices in its corresponding maturity stage. Figure 7 shows an example of the present survey.

The following subsections explain the web-based tool output. This includes the output of the surveys and the three menu sections: dashboard, data visualization, and best practice model visualization.

Survey Output

After the completion of each survey, the tool provides a summary of the execution. This includes the CwBPMS and a table with CwBPsC of each best practices subcategory and its corresponding missing best practices, if any. Figure 8 shows an example of the output of the survey.

Figure 7. Screen Shot of the Presence Survey

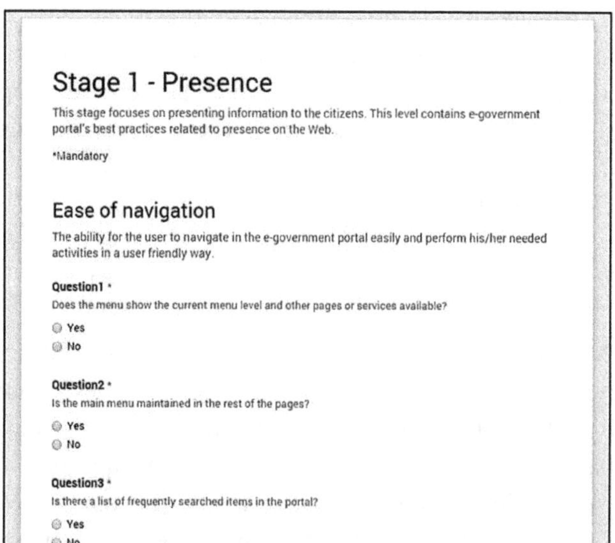

Stage 1 - Presence

This stage focuses on presenting information to the citizens. This level contains e-government portal's best practices related to presence on the Web.

*Mandatory

Ease of navigation

The ability for the user to navigate in the e-government portal easily and perform his/her needed activities in a user friendly way.

Question1 *

Does the menu show the current menu level and other pages or services available?

Yes

No

Question2 *

Is the main menu maintained in the rest of the pages?

Yes

No

Question3 *

Is there a list of frequently searched items in the portal?

Yes

No

Figure 8. Presence Survey Output

Overall quality level of maturity stage 1 (Presence): 53%
Details:

Subcategory	Grade	Guidlines and missing best practices
Ease of navigation	67%	- Maintain the main menu is all the pages of the portal. - Provide a list of most used services (ex. in home page).
Structuration	33%	- Before starting writing pages of the portal, the designer(s) should spent time in writing the structure of the Website. - Use of XML sitemaps.
Industrialization	67%	- Follow Web Content Accessibility Guidelines WCAG 1.0 or 2.0 (ISO/IEC 40500).
Relevancy	33%	- Grouping information by target groups. - Have information at the right level of detail.
Accessibility	50%	- Following the WCAG 1.0 or 2.0 guidelines (ISO/IEC 40500). - Having listen features to hear text of a page. - Making the portal accessible via mobile devices. - Making the portal accessible with all browsers and systems. - Monitoring HTML broken links. - Not overdoing pictures. - Have either a TTY (text telephone) or TDD (telephonic device for the deaf) phone number. - Use or implement a responsive template that changes the size depending on the mobile device resolution.

Figure 9. Example of the dashboard section

Your current maturity stage is: **Presence**, with a **53%** completion.
The overall quality level of your portal is: **15.4%**.

You have some missing best practices to be completed to move to the **Interaction stage** .

Maturity Stages Quality Level

#	Maturity Stage	Quality Level
1	Presence	53
2	Interaction	7.1
3	Transaction	0
4	Integration	1.5

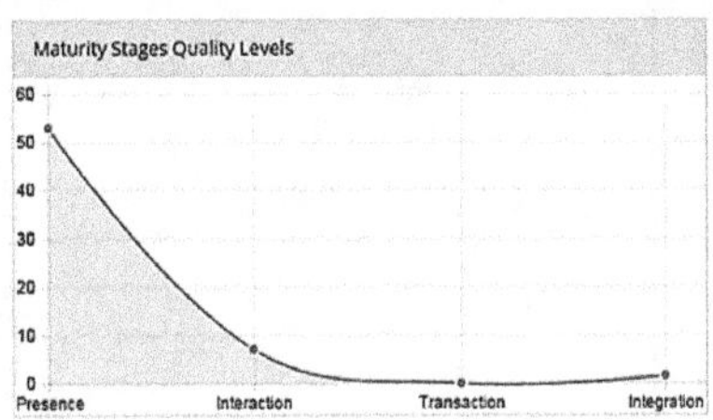

Dashboard

This section provides a general overview of the tool. This includes the current maturity stage with percentages of completion (CwBPMS), the overall portal compliance with best practices (this is a weighted average of the four maturity stages' CwBPMS), whether there are missing best practices to be implemented to move to the next stage, CwBPMS for each maturity stage. Figure 9 shows a view of the dashboard section.

Data Visualization

This section also displays information regarding the CwBPMS for each maturity stage. This information is displayed in the Pie and Line charts. Figure 10 displays these two charts.

Figure 10. Pie and Line charts for the Data Visualization section

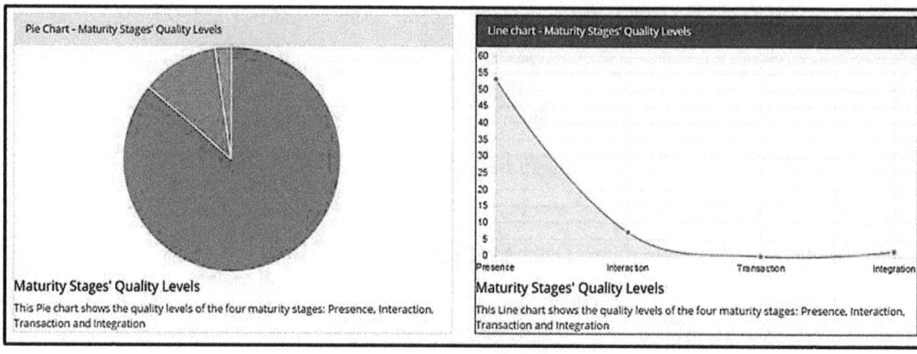

Figure 11. SCwBP of the presence maturity stage

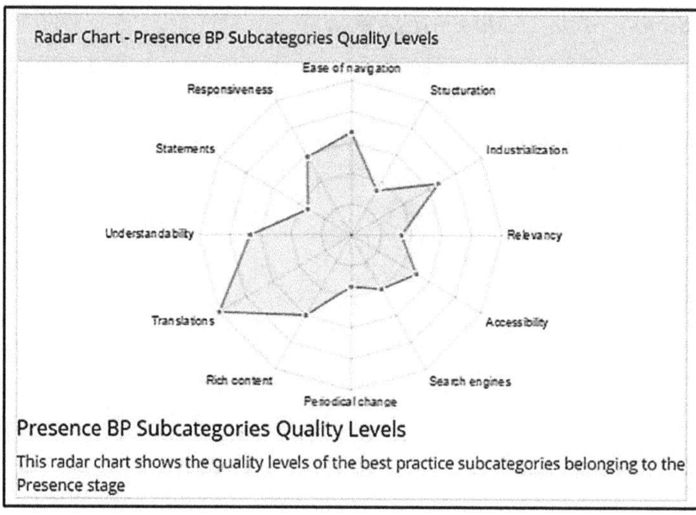

In addition, this section displays radar charts for the CwBPsC belonging to each maturity stage (presence, interaction, transaction, and integration). Figure 11 displays this information for the maturity stage.

Best Practice Model Visualization

This section provides a summary of the compliance with best practices from the eGPBPM perspective. This means that the web tool shows information regarding compliance with best practices for each best practices category (backend, web content, web design, external, and service). Figure 12 shows a Doughnut chart with

Figure 12. Best practices Categories' Compliance with Best Practices

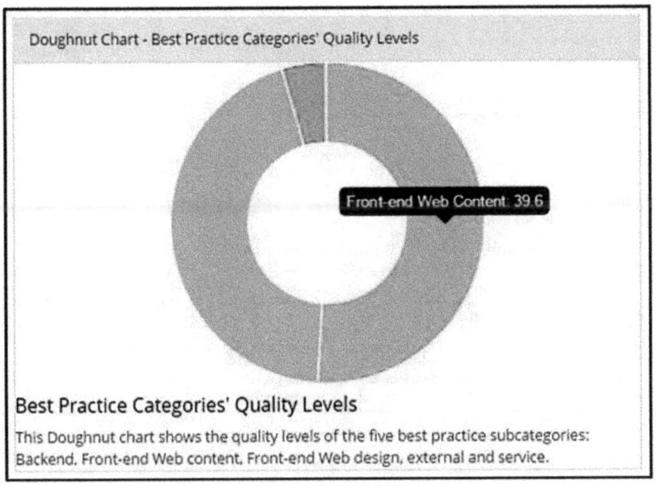

Figure 13. Front-end Web Content Best practices Subcategories' Compliance with Best Practices

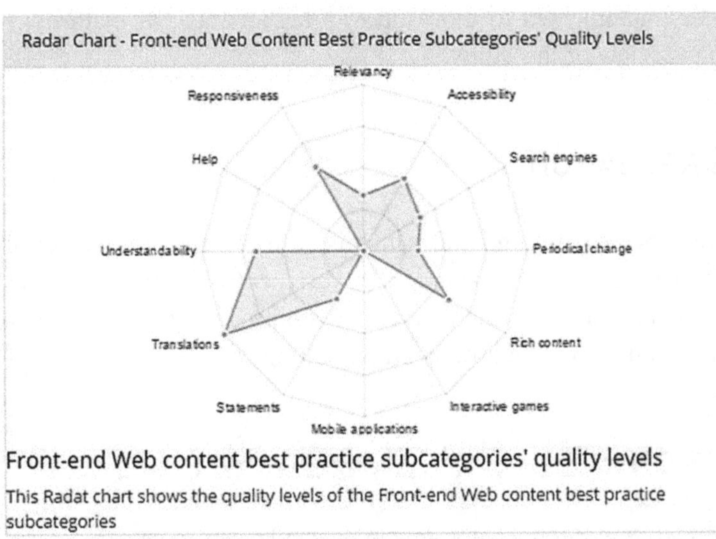

this information calculated by equally weighting the best practice subcategories belonging to each category.

Moreover, this section provides radar charts displaying the compliance with best practices for each best practice subcategory for each best practice category. Figure 13 shows an example of the front-end web content category.

Figure 14. Overview of the guidelines section

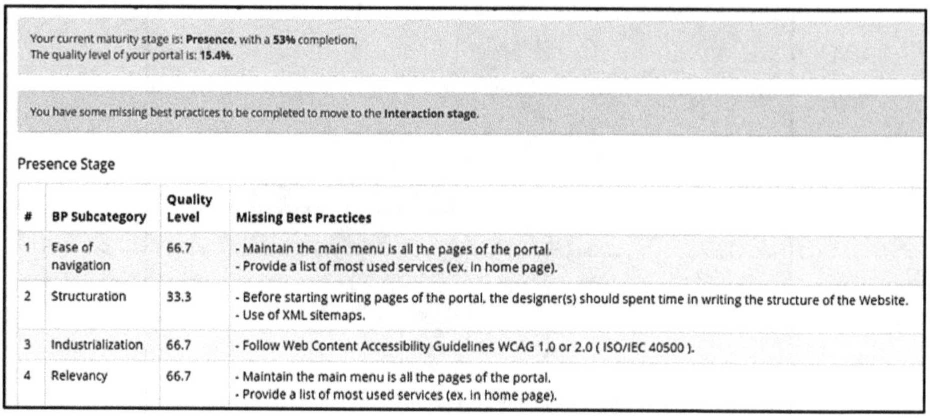
Your current maturity stage is: **Presence**, with a **53%** completion.
The quality level of your portal is: **15.4%**.

You have some missing best practices to be completed to move to the **Interaction stage**.

Presence Stage

#	BP Subcategory	Quality Level	Missing Best Practices
1	Ease of navigation	66.7	- Maintain the main menu is all the pages of the portal. - Provide a list of most used services (ex. in home page).
2	Structuration	33.3	- Before starting writing pages of the portal. the designer(s) should spent time in writing the structure of the Website. - Use of XML sitemaps.
3	Industrialization	66.7	- Follow Web Content Accessibility Guidelines WCAG 1.0 or 2.0 (ISO/IEC 40500).
4	Relevancy	66.7	- Maintain the main menu is all the pages of the portal. - Provide a list of most used services (ex. in home page).

Guidelines

This section provides information on the current stage of maturity for e-government portals. Moreover, the section also provides information on the missing best practices (if any) to be implemented to move up to a higher stage of maturity. Figure 14 presents an overview of this study.

FUTURE RESEARCH

Furthermore, when collecting data, which is based on the presence or absence of specific practices, the involvement of e-government officers is needed. In future work, the aim is to automate the data collection process to allow agencies to rate their portals' maturity without human intervention.

CONCLUSION

A detailed view of the proposed maturity model is presented next together with a web-based tool to automate the data collection process and facilitate the use and output of the maturity model. The availability of this MBeGPMM material provides agencies with insights and tools for designing and building e-government portals of good quality.

The MBeGPMM is built based on the eGPBPM, which is a hierarchical and staged model. The first level includes four stages: presence for stage 1, interaction for stage 2, the transaction for stage 3, and integration for stage 4. Each stage includes a set of best practice subcategories in the second level, and each subcategory includes its corresponding specific practices in the third level. The fourth level provides assessment questions that allow verification of the presence or absence of specific practices. The use of the MBeGPMM allows agencies to identify the maturity stage of the e-government portal and the missing best practices at any stage of maturity using the output of the survey. The formulae used to calculate the maturity stage compliance with the best practices were also provided.

The content of the MBeGPMM is detailed for each best practice subcategory using the best practices category and maturity stage, purpose, introductory notes, specific goals, specific practices, and survey questions. This model is based on a list of structured best practices that allow designers to build portals that satisfy worldwide e-government portals' best practices and the agencies to evaluate their portals in order to improve them. Thus, this model moves beyond the benefits of the actual maturity models and provides the process and calculation methodology.

The process of the MBeGPMM can be done manually, and a web-based tool to automate and facilitate the process and calculations was developed. Also, an overview of this tool is provided, which describes the input (e.g., the four surveys of presence, interaction, transaction, and integration), and the output of the tool (e.g., surveys' output, dashboard, maturity stage visualization, best practice model visualization, and guidelines).

The resulting maturity model is important not only for researchers but also for organizations, agencies, and governments to enhance their e-government portals' quality. Thus, the authors do not claim that this was the final model. Thus, the proposed MBeGPMM can be considered a starting point for a maturity model for e-government portals, which requires feedback from the software engineering community of researchers as well as practitioners from academia and industry to improve it.

Agencies and governments are encouraged to use this model to evaluate the maturity of their portals to identify their weaknesses and strengths. It also provides them with guidelines to improve the quality of services to citizens. Furthermore, by performing many case studies on different e-government portals, agencies can conduct comparative studies among the portals. The insights gained will allow the MBeGPMM to reach a certain maturity and be adopted by the industry; this is the motivation behind making public the details of the maturity model.

REFERENCES

Accenture. (2003). eGovernment leadership: Engaging the customer. Accenture.

Al-Khouri, A. M. (2011). *An innovative approach for e-government transformation.* arXiv preprint arXiv:1105.6358.

Alasem, A. (2009). An Overview of e-Government Metadata Standards and Initiatives based on Dublin Core. *Electronic Journal of E-Government, 7*(1), 1–10.

Alsmadi, I., Al-Kabi, M., & Wahbeh, A. (2012). A Conceptual Organization for Websites Metrics and E-Government Websites: A Case Study. *International Journal of Information Communication Technologies and Human Development, 4*(1), 93–117. doi:10.4018/jicthd.2012010107

Andreasson, K., Millard, J., & Snaprud, M. (2012, October). Evolving e-government benchmarking to better cover technology development and emerging societal needs. In *Proceedings of the 6th International Conference on Theory and Practice of Electronic Governance* (pp. 430-439). 10.1145/2463728.2463811

Assar, S., Boughzala, I., & Isckia, T. (2011). eGovernment trends in the Web 2.0 era and the open innovation perspective: an exploratory field study. In *Electronic Government: 10th IFIP WG 8.5 International Conference, EGOV 2011, Delft, The Netherlands, August 28–September 2, 2011 Proceedings, 10*, 210–222.

Benchmark Measurement of European eGovernment Services, 8th Benchmark Measurment. European Commission. (2009). Rep. Prep. Capgemini Rand Eur. IDC Sogeti DTI.

Berntzen, L., & Olsen, M. G. (2009, February). Benchmarking e-government-a comparative review of three international benchmarking studies. In *2009 Third International Conference on Digital Society* (pp. 77-82). IEEE. 10.1109/ICDS.2009.55

Bhattacharya, D., Gulla, U., & Gupta, M. P. (2012). E-service quality model for Indian government portals: Citizens' perspective. *Journal of Enterprise Information Management, 25*(3), 246–271. doi:10.1108/17410391211224408

Caldwell, B., Cooper, M., Reid, L. G., Vanderheiden, G., Chisholm, W., Slatin, J., & White, J. (2008). Web content accessibility guidelines (WCAG) 2.0. *WWW Consortium (W3C), 290*, 1-34.

Capgemini, I. (2010). *Digitizing public services in Europe: Putting ambition into action: 9th benchmark measurement.* European Commission.

Choudrie, J., & Weerakkody, G. G. V. (2004). Evaluating Global e-Government Sites: A View using Web Diagnostic Tools. *Electronic Journal of E-Government*, *2*(2), 105–114.

DeLone, W. H., & McLean, E. R. (1992). Information systems success: The quest for the dependent variable. *Information Systems Research*, *3*(1), 60–95. doi:10.1287/isre.3.1.60

DeLone, W. H., & McLean, E. R. (2003). The DeLone and McLean model of information systems success: A ten-year update. *Journal of Management Information Systems*, *19*(4), 9–30. doi:10.1080/07421222.2003.11045748

e-Government of Korea Best Practices. (2014). Ministry of Public Administration and Security and National Information Society Agency. Retrieved from https://www.mois.go.kr/html/site/eng/eng_mov_2.html

eGovernment: International Best Practices. (2008). Forfás.

Fath-Allah, A., Cheikhi, L., Al-Qutaish, R. E., & Idri, A. (2014a). E-government maturity models: A comparative study. *International Journal of Software Engineering and Its Applications*, *5*(3), 71–91.

Fath-Allah, A., Cheikhi, L., Al-Qutaish, R. E., & Idri, A. (2014b). E-government portals best practices: a comprehensive survey. *Electronic Government, an International Journal, 11*(1/2), 101-132.

Fath-Allah, A., Cheikhi, L., Al-Qutaish, R. E., & Idri, A. (2014c, October). Towards a best practice based e-government portals maturity model. In *Proceedings of the International Conference on Computer Science and Information Systems (ICSIS'14)* (pp. 100-106). Academic Press.

Fath-Allah, A., Cheikhi, L., Al-Qutaish, R. E., & Idri, A. (2015). E-Government Portals Maturity Models: A Best Practices' Coverage Perspective. *Journal of Software*, *10*(7), 805–824. doi:10.17706//jsw.10.7.805-824

Fath-Allah, A., Cheikhi, L., Al-Qutaish, R. E., & Idri, A. (2016). A Measurement-Based E-Government Portals' Maturity Model. In *International Perspectives on Socio-Economic Development in the Era of Globalization* (pp. 129–149). IGI Global. doi:10.4018/978-1-4666-9908-3.ch008

Fath-Allah, A., Cheikhi, L., Al-Qutaish, R. E., & Idri, A. (2016, August). A Validation of a Measurement Based E-Government Portals' Maturity Model. In *2016 42th Euromicro Conference on Software Engineering and Advanced Applications (SEAA)* (pp. 100-107). IEEE.

Georgescu, M. R., & Georgescu, I. (2008). Do we need a powerful e-government? *Communications of the IBIMA*, *5*, 242–249.

Government on the Web II. (2002). National Audit Office. Retrieved from https://www.nao.org.uk/wp-content/uploads/2002/04/0102764.pdf

Henriksson, A., Yi, Y., Frost, B., & Middleton, M. (2007). Evaluation instrument for e-government websites. *Electronic Government, an International Journal, 4*(2), 204-226.

Hiller, J. S., & Bélanger, F. (2001). Privacy strategies for electronic government. *E-government, 200*, 162–198.

Iribarren, M., Concha, G., Valdes, G., Solar, M., Villarroel, M. T., Gutiérrez, P., & Vásquez, Á. (2008). Capability maturity framework for eGovernment: A multi-dimensional model and assessing tool. In *Electronic Government: 7th International Conference, EGOV 2008, Turin, Italy, August 31-September 5, 2008 Proceedings, 7*, 136–147.

Kaisara, G., & Pather, S. (2011). The e-Government evaluation challenge: A South African Batho Pele-aligned service quality approach. *Government Information Quarterly, 28*(2), 211–221. doi:10.1016/j.giq.2010.07.008

Kim, D. Y., & Grant, G. (2010). E-government maturity model using the capability maturity model integration. *Journal of Systems and Information Technology, 12*(3), 230–244. doi:10.1108/13287261011070858

Klischewski, R., & Askar, E. (2010, October). Success factors of developing G2G services: the case of Egypt. In *Proceedings of the 4th International Conference on Theory and Practice of Electronic Governance* (pp. 152-160). 10.1145/1930321.1930354

Kumar, V., Mukerji, B., & Butt, I. (2007). Factors for Successful E-Government Adoption: A Conceptual Framework. *Electronic. Journal of E-Government, 5*(1), 63–76.

Layne, K., & Lee, J. (2001). Developing fully functional E-government: A four stage model. *Government Information Quarterly, 18*(2), 122–136. doi:10.1016/S0740-624X(01)00066-1

Lee, G., & Kwak, Y. H. (2012). An open government maturity model for social media-based public engagement. *Government Information Quarterly, 29*(4), 492–503. doi:10.1016/j.giq.2012.06.001

Main Features Reviewed by the United Nations e-Government Survey. (2008). United Nations.

Makolm, J. (2002). Best practices in e-government. In Electronic Government. Springer. doi:10.1007/978-3-540-46138-8_60

Moon, M. J. (2002). The evolution of e-government among municipalities: Rhetoric or reality? *Public Administration Review*, *62*(4), 424–433. doi:10.1111/0033-3352.00196

Oakland, J. S. (1991). The European centre for total quality management. *Total Quality Management*, *2*(3), 305–308. doi:10.1080/09544129100000035

Page Speed Insights. (2015). Google. Retrieved from http://developers.google.com/speed/ pagespeed/insights/

Papadomichelaki, X., & Mentzas, G. (2012). e-GovQual: A multiple-item scale for assessing e-government service quality. *Government Information Quarterly*, *29*(1), 98–109. doi:10.1016/j.giq.2011.08.011

Peters, R. M., Janssen, M., & Van Engers, T. M. (2004, March). Measuring e-government impact: existing practices and shortcomings. In *Proceedings of the 6th international conference on Electronic commerce* (pp. 480-489). 10.1145/1052220.1052281

Posch, K. C., Posch, R., Tauber, A., Zefferer, T., & Zwattendorfer, B. (2011). *Secure and privacy-preserving eGovernment—best practice Austria*. Springer Berlin Heidelberg. doi:10.1007/978-3-642-19391-0_19

Rodríguez, R. A., Welicki, L., Giulianelli, D. A., & Vera, P. M. (2008, December). Measurement framework for evaluating e-governance on municipalities websites. In *Proceedings of the 2nd international conference on Theory and practice of electronic governance* (pp. 381-387). 10.1145/1509096.1509176

Rohleder, S. J. (2004). *The Government Executive Series. eGovernment Leadership: High Performance*. Maximum Value.

Seifert, J. W., & McLoughlin, G. J. (2007, July). *State e-government strategies: Identifying best practices and applications*. Library of Congress.

Shahkooh, K. A., Saghafi, F., & Abdollahi, A. (2008, April). A proposed model for e-Government maturity. In *2008 3rd International conference on information and communication technologies: From theory to applications* (pp. 1-5). IEEE. 10.1109/ICTTA.2008.4529948

Siau, K., & Long, Y. (2005). Synthesizing e-government stage models–a meta-synthesis based on meta-ethnography approach. *Industrial Management & Data Systems*, *105*(4), 443–458. doi:10.1108/02635570510592352

Simple answers about life in Switzerland. (2014). Retrieved from https://www.ch.ch/fr/

Sørum, H. (2011). An empirical investigation of user involvement, website quality and perceived user satisfaction in eGovernment environments. *Electronic Government and the Information Systems Perspective: Second International Conference, EGOVIS 2011, Toulouse, France, August 29–September 2, 2011 Proceedings*, 2, 122–134.

Standard, I. (2002). *Ergonomics of human-system interaction–usability methods supporting human-centred design. ISO standard tr 16982: 2002*. International Organization for Standardization.

Tinholt, D., & van der Linden, N. (2013). *Public services online 'Digital by default or by detour?'–Assessing user centric eGovernment performance in Europe–eGovernment benchmark 2012*. Academic Press.

UN, DESA. (2012, June). *Compendium of Innovative E-government Practices*. United Nations.

UN E-Government Survey 2012: E-Government for the People. (2012). United Nations.

WCAG 2 Overview. (2012). W3C. Retrieved from https://www.w3.org/WAI/intro/wcag

Webmasters. (n.d.). Google. Retrieved from https://www.google.com/webmasters/

West, D. (2007). Global perspectives on e-government. *Governance and information technology: From electronic government to information government*, 17-32.

West, D. M. (2004). E-government and the transformation of service delivery and citizen attitudes. *Public Administration Review*, *64*(1), 15–27. doi:10.1111/j.1540-6210.2004.00343.x

West, D. M., Deitch, A., Vigón, M. A. P., & Sobisch, F. (2004). *Development of e-government strategies in Chile Canada and Brazil*. Academic Press.

What is Plain Language? (2014). Plain Language Action and Information Network (PLAIN). Retrieved from https://www.plainlanguage.gov/about/definitions/

Windley, P. J. (2002). *eGovernment maturity.* USA Wind. Technolometria.

X-Road. (2014). E-Estonia. Retrieved from https://e-estonia.com/components/x-road

Yahoo Yslow for Chrome. (2015). Yahoo. Retrieved from https://developer.yahoo.com/yslow/

APPENDIX 1

1. Industrialization

- **Best practices category and maturity stage**: A front-end web design best practices category at maturity stage 1 (presence).
- **Purpose:** The purpose of industrialization is to follow the industry guidelines for website presentation.
- **Introductory Notes:** Industrialization is related to industry guidelines in web design. An example is the recommendations in ISO/IEC 40500 Web Content Accessibility Guidelines (WCAG) 2.0. Another example is the use of cascading style sheets (CSS) to separate the content and style of webpages.

2. Relevancy

- **Best practices category and maturity stage**: A front-end web content best practices category at maturity stage 1 (presence).
- **Purpose:** The purpose of the relevancy best practices subcategory is to provide relevant information to citizens. In other words, citizens should not know the organization responsible for each type of information. However, citizens can easily and quickly find relevant and useful information.
- **Introductory Notes:** Relevancy is related to organizing the portal's web content according to citizens' needs. This means that information should not be organized from an organizational perspective; however, the information should be relevant to citizens or citizens. An example is grouping information by theme and target groups.

3. Accessibility

- **Best practices category and maturity stage**: A front-end web content best practices category at maturity stage 1(presence).
- **Purpose:** The purpose of the accessibility best practices subcategory is to make the portal accessible for people with disabilities (such as hearing or vision deficiencies) or without.
- **Introductory Notes:** Accessibility is related to the degree to which the e-government portal is accessible to people with disabilities. An example is compliance with the ISO/IEC 40500 - Web Content Accessibility Guidelines (WCAG) 2.0. Other examples include employing accessibility web experts, having 'listen' features, and Audio Completely Automated Public Turing test to tell Computers and Humans Apart (Audio CAPTCHA).

4. Search Engines

- **Best practices category and maturity stage**: A front-end web content best practices category at maturity stage 1(presence).
- **Purpose:** The purpose of the search engine best practices subcategory is to allow citizens to easily find information and services.
- **Introductory Notes:** Search engines allow citizens to search for the portal's content and services. It is important to provide powerful search engines with advanced search options, such as videos, news, or laws, to narrow and make many filters on the search.

5. Periodical Change

- **Best practices category and maturity stage:** A front-end web content best practices category at maturity stage 1(presence).
- **Purpose:** The purpose of the periodical change best practices subcategory is to provide citizens with timely information.
- **Introduction Notes:** Periodical changes are related to changing and updating the e-government portal's content periodically. This can be achieved by having expiry dates and a team of editors responsible for changing the content of the portal periodically, for instance, within the last three months.

6. Rich Content

- **Best practices category and maturity stage:** A front-end web content best practices category at maturity stage 1(presence).
- **Purpose:** The purpose of the rich content best practices subcategory is to offer citizens a broad range of information.
- **Introductory Notes:** Rich content is related to offering a broad range of information such as codes, laws, news, regulations, publications, press releases, databases, websites, interactive maps, etc. Another example is the use of videos in reception and organizing them through life events. These videos are easy to understand and explain the procedures and steps to accomplish citizens' needs and tasks depending on the event or theme in an easy and efficient way of learning.

7. Translations

- **Best practices category and maturity stage**: A front-end web content best practices category at maturity stage 1 (presence).

- **Purpose:** The purpose of the translation best practices subcategory is to make the portal accessible to all populations.
- **Introductory Notes:** Translations are related to offering users the possibility to translate the portal's content into different languages. An example translates the portal to local dialects, the second language, and English. This helps make the portal accessible to all populations.

8. Understandability

- **Best practices category and maturity stage**; front-end web content best practices category at maturity stage 1(presence).
- **Purpose:** The purpose of the understandability best practices subcategory is to offer clear and easily understandable information to citizens.
- **Notes:** Understandability is related to offering information or content in the portal that can easily be understood by citizens in the first reading.

9. Statements

- **Best practices category and maturity stage**: A front-end web content best practices category at maturity stage 1(presence).
- **Purpose:** The purpose of the statement best practices subcategory is to be transparent on data usage in order to increase citizens' trust, protect the portals' data from any improper use, and protect the agency from any liability or responsibility.
- **Notes:** The Statements Best Practices subcategory is related to offering privacy and security statements (with clear information on how data is processed and the steps the user needs to take if his data are misused), disclaimers (with no responsibility for the accuracy of information), and copyrights in the e-government portal.

10. Responsiveness

- **Best practices category and maturity stage:** A front-end web content best practices category at maturity stage 1 (presence).
- **Purpose:** The purpose of the responsiveness best practices subcategory is to have a high-performance portal, that is, with a short response execution time.
- **Introductory Notes:** Responsiveness is related to a short response time when navigating in the portal. This is considered important because pages that take a large amount of time to load may give an impression to the citizens that the portal is improperly functioning.

11. Contact Information

- **Best practices category and maturity stage**: An external best practices category at maturity stage 2 (interaction).
- **Purpose:** The purpose of the contact information best practices subcategory is to allow citizens to interact with the agency.
- **Introductory Notes:** Contact information is related to offering means for citizens to interact with agency. An example is having a 'contact us' link with an email form and publishing contact information (such as email, phone, address, and key persons to contact).

12. Advertising

- **Best practices category and maturity stage**: An external best practices category at maturity stage 2 (interaction).
- **Purpose:** The purpose of the advertising best practices subcategory is to advertise the portal and its services in different media channels.
- **Introductory Notes:** Advertising promotes the services provided by e-government portals. This increases citizens' awareness of the portal and its services. For instance, advertising the portal and its services in TV, radio, brochures, and the Internet.

13. Emailing

- **Best practices category and maturity stage**: An external best practices category at maturity stage 2 (interaction).
- **Purpose:** The purpose of the emailing best practices subcategory is to advertise the portal and its services using email.
- **Introductory Notes:** Emailing is related to advertising the portal and its services via email. An example is push technology, which allows citizens to register their emails and receive newsletters (such as email subscriptions and newsletters).

14. Referencing

- **Best practices category and maturity stage:** An external best practices category at maturity stage 2 (interaction).
- **Purpose:** The purpose of the referencing best practices subcategory is to optimize the portal and its pages for search engines.

- **Introductory Notes:** Referencing is related to enhancing search engine optimization (SEO) to make the portal accessible and well optimized and ranked in the search results of search engines such as Google and Yahoo.

15. Security

- **Best practices category and maturity stage**: A back-end best practices category at maturity stage 3 (transaction).
- **Purpose:** The purpose of the security best practices subcategory is to allow citizens to make secure transactions and authentications and make the portal resistant against hacking attacks.
- **Introductory Notes:** Security is related to allowing citizens to make transactions in a secure or safe way and keep their records confidential. An example is the use of digital signatures for encryption and authentication, such as securing emails and submitted forms, and identifying users.

16. Privacy

- **Best practices category and maturity stage**: A back-end best practices category at maturity stage 3 (transaction).
- **Purpose:** The purpose of the privacy best practices subcategory is to keep citizens' data and preferences private from any disclosure.
- **Introductory Notes:** Privacy is related to keeping citizens' data and preferences private from any disclosure. For example, avoiding the use of cookies and hypertext markup language (HTML) get forms, and compliance with the platform for privacy preferences (P3P) are considered important aspects of privacy.

17. Single Sign On

- **Best practices category and maturity stage**: A back-end best practices category at maturity stage 3 (transaction).
- **Purpose:** The purpose of the single sign in the best practices subcategory is to provide citizens with access to all services in the portal with a single authentication.
- **Introductory Notes:** Single sign on (SSO) or signing in only one time allows citizens to login into many applications only once without having the need to login another time. This should give citizens access to all services and systems with a single login.

18. Delegation

- **Best practices category and maturity stage**: A back-end best practices category at maturity stage 3 (transaction).
- **Purpose:** The purpose of the delegation best practices subcategory is to allow citizens to act on behalf of other citizens or companies.
- **Introductory Notes:** Acting on behalf of a person or company. This means that a company or person can delegate tasks to a delegate by creating a delegation for him in the portal. The delegate can then take the role of this person or company that delegates him to execute services on behalf of them.

19. Modularity

- **Best practices category and maturity stage**: A back-end best practices category at maturity stage 3 (transaction).
- **Purpose:** The purpose of the modularity best practices subcategory is to allow reusing services or components, and thus changing small functionalities does not break the whole functionality.
- **Introductory Notes:** Modularity is related to building the portal in a modular way or module by module. An example is the use of a service-oriented architecture (SOA). The SOA focuses on breaking down the application into small modules called services. Each service is responsible for small functionality.

20. Reusability

- **Best practices category and maturity stage**: A back-end best practices category at maturity stage 3 (transaction).
- **Purpose:** The purpose of the reusability best practices subcategory is to save time and profit from the reuse of open source modules, services, and components instead of building them from scratch.
- **Introductory Notes:** This relates to reusing Web 2.0, components and services that are already available, such as Google calendar and WordPress. This means that before starting the development of any new module, it is important to check if the module is already available as an open source.

21. Use of Standards

- **Best practices category and maturity stage**: A back-end best practices category at maturity stage 3 (transaction).

- **Purpose:** The purpose of the use of standard best practices subcategory is to enable cross-platform interoperability and facilitate interoperability efforts (eGovernment: International Best Practices, 2008), (Posch et al., 2011), and enable flexibility and ease of change in requirements and technology (eGovernment: International Best Practices, 2008).
- **Introductory Notes:** This is related to the use of standards, especially the open source ones in the implementation of the e-government portal, such as programming, design, and interoperability standards. An example is the use of a simple object access protocol (SOAP), extensible markup language (XML), JavaScript object notation (JSON), and Dublin core metadata.

22. Personalization

- **Best practices category and maturity stage:** A front-end web design best practices category at maturity stage 3 (transaction).
- **Purpose:** The purpose of the personalization best practices subcategory is to allow citizens to personalize the portal's content and services according to their needs.
- **Introductory Notes:** This is related to the fact of offering the possibility of the citizen personalizing and customizing the portal's functionalities according to his/her needs. An example is the possibility to access a 'MyPage' section, or the possibility to tailor the content of the portal according to citizens' requirements.

23. Help

- **Best practices category and maturity stage:** A front-end web content best practices category at maturity stage 3 (transaction).
- **Purpose:** The purpose of the help best practices subcategory is to help the user navigate in the portal.
- **Introductory Notes:** This is related to offering help to guide citizens navigating in the portal and performing their required tasks. An example is the use of Frequently Asked Questions (FAQs) and guided tours to explain the navigation process.

24. Incentives

- **Best practices category and maturity stage:** An external best practices category at maturity stage 3 (transaction).

- **Purpose:** The purpose of the incentives best practices subcategory is to encourage citizens to use e-government portals.
- **Introductory Notes:** This is related to offering privileges for online users over traditional users. An example is offering free of charge services for online users to encourage them to opt for the Web channel over traditional channels or offices. This can promote and increase online service usage.

25. Contests

- **Best practices category and maturity stage**: An external best practices category at maturity stage 3 (transaction).
- **Purpose:** The purpose of the contests' best practices subcategory is to encourage citizens to create services.
- **Introductory Notes:** This is related to organizing competitions for the public to create services. In this way, users become producers of services that promote e-participation and help create innovative services. This can be encouraged by making available public data.

26. Performance Ratings

- **Best practices category and maturity stage**: An external best practices category at maturity stage 3 (transaction).
- **Purpose:** The purpose of this best practice subcategory is to allow citizens to rate the portal and to get feedback from them on the overall performance of the portal.
- **Introductory Notes:** This is related to offering citizens the possibility to rate the portal and its content. An example is the use of feedback icons and surveys for rating the portal and complaint box. Another example is the measurement of public participation.

27. Service Customer Centricity

- **Best practices category and maturity stage**: A service best practices category at maturity stage 3 (transaction).
- **Purpose:** The purpose of this best practice subcategory is to assess the needs of citizens to design services accordingly.
- **Introductory Notes:** This is related to ensuring that the services are designed for the citizen and not for the organization, thus helping to build citizen-centric services. This can be achieved via observatory or ethnographic work and eye-tracking techniques, such as observing users trying e-government services (during or after implementation).

28. Service Help

- **Best practices category and maturity stage**: A service best practices category at maturity stage 3 (transaction).
- **Purpose:** The purpose of this best practice subcategory is to help citizens execute online services.
- **Introductory Notes:** This is related to offering help for citizens on how to execute e-government services. For example, providing help for users when filling online forms, providing FAQ sections for services, and providing interactive assistance such as videos explaining how to execute e-government services.

29. Payments

- **Best practices category and maturity stage:** A service best practices subcategory at maturity stage 3 (transaction).
- **Purpose:** The purpose of this best practice subcategory is to offer a way for citizens to complete online monetary transactions.
- **Introductory Notes:** This is related to offering the citizens the ability to pay in the portal via credit/debit cards or electronic banking. For example, offering a wide range of payments, ranging from taxes to fines and fees, and the possibility of paying for e-government services that need payments.

30. Workflows

- **Best practices category and maturity stage**: A service best practices category at maturity stage 3 (transaction).
- **Purpose:** The purpose of this best practice subcategory is to provide citizens with information concerning service delivery timelines and the workflow or process steps of their applications.
- **Introductory Notes:** This is related to providing citizens with information regarding service delivery timelines and process steps of their applications (i.e., workflow). An example is the possibility that citizens can track the status of their online applications. This can be achieved using emails or dashboards, where citizens can login and see the status of all the e-government services that they have executed.

31. Service Performance Ratings

- **Best practices category and maturity stage:** A service best practices category at maturity stage 3 (transaction).

- **Purpose:** The purpose of this best practice subcategory is to obtain feedback from users and measure their level of satisfaction with e-government services, with the final aim of improving those services.
- **Introductory Notes:** This is related to offering citizens the possibility to rate the portals' services. An example is the use of emoticons to score the services upon using them. This makes it possible for users to rate the performance of e-government services.

32. User Forms

- **Best practices category and maturity stage:** A service best practices category at maturity stage 3 (transaction).
- **Purpose:** The purpose of this best practice subcategory is to provide suitable and easy-to-use user forms.
- **Introductory Notes:** This is related to ensuring that the forms submitted and filled by citizens during the execution of an e-government service are suitable and easy to use. Examples include the use of mandatory fields, reduction of menus, possibility of previewing the form before submission, and showing the stage of completion of the form.

33. Service Responsiveness

- **Best practices category and maturity stage**: A service best practices category at maturity stage 3 (transaction).
- **Purpose:** The purpose of this best practice subcategory is to ensure that the services are responsive and designed in an efficient way to help users complete it in a short amount of time.
- **Introductory Notes:** This is related to ensuring that the service response time is short and acceptable. For example, this can be achieved by measuring the response time of services and analyzing the code and algorithms for performance enhancements.

34. Interoperability

- **Best practices category and maturity stage**: A back-end best practices category at maturity stage 4 (integration).
- **Purpose:** The purpose of this best practice subcategory is to allow agencies to share information with each other.
- **Introductory Notes:** This is related to joining governments to work together and exchange information. This includes two types of interoperability: vertical

interoperability across departments from the same jurisdictions and horizontal interoperability across departments from various jurisdictions of governments.

35. One Stop Shop

- **Best practices category and maturity stage:** A front-end web design best practices category at maturity stage 4 (integration).
- **Purpose:** The purpose of this best practice subcategory is to make the portal a single point of entry for all e-government services.
- **Introductory Notes:** This is related to a single point of entry for all e-government services. This means that citizens should be able to go to a single point of access to use all e-government services. This is considered the preferred delivery model for e-government portals.

36. Mobile Applications

- **Best practices category and maturity stage**: A front-end web content best practices category at maturity stage 4 (integration).
- **Purpose:** The purpose of this best practice subcategory is to offer mobile applications to citizens.
- **Introductory Notes:** This is related to offering applications such as iPhone Operating System (iOS) or Android apps that can be downloaded and installed on mobile devices such as smartphones and tablets. These can be downloaded from Apple and Google stores.

37. Interactive Games

- **Best practices category and maturity stage**: A front-end web content best practices category at maturity stage 4 (integration).
- **Purpose:** The purpose of this best practice subcategory is to offer a way of entertainment for users.
- **Introductory Notes:** This is related to offering games for users, where they can play alone or with/against each other. For example, decision games allow citizens to increase their knowledge and interactive games that allow citizens to learn in a fun and educational way.

38. E-participation

- **Best practices category and maturity stage:** An external best practices category at maturity stage 4 (integration).

- **Purpose:** The purpose of this best practice subcategory is to involve citizens in the e-government process.
- **Introductory Notes:** This is related to the involvement of citizens in the e-government process. This can be achieved using techniques such as allowing citizens to communicate with government officials or with each other (using tools such as chat rooms, comment forms to ask questions and receive answers, blogs, and discussion forums), seeking feedback from citizens, integrating the feedback in the portal, and involving the citizen in the policy-making process.

39. Sociability

- **Best practices category and maturity stage:** An external best practices category at maturity stage 4 (integration).
- **Purpose:** The purpose of this best practice subcategory is to engage citizens in social media websites, with the aim of extending e-participation, strengthening the relationship with citizens, and improving service usage and adoption.
- **Introductory Notes:** This is related to the use of social media websites such as Facebook, Twitter, WordPress, YouTube, and Flickr. This is considered important because it helps to extend participation and strengthens the relationship with citizens.

40. Data Analytics

- **Best practices category and maturity stage:** An external best practices category at maturity stage 4 (integration).
- **Purpose:** The purpose of this best practice subcategory is to analyze data for insights.
- **Notes:** This is related to using the data for insights. Examples include measuring the portals' performance (number of visitors, page views, etc.) using tools such as Google analytics, and use of sentiment analysis, which allows the analysis of online conversations on social websites such as Facebook and Twitter.

41. Service Interoperability

- **Best practices category and maturity stage:** A service best practices category at maturity stage 4 (integration).

- **Purpose:** The purpose of this best practices subcategory is to integrate services with other services or systems from the same or different jurisdictions.
- **Introductory Notes:** This is related to joining e-government services to work together and exchange information. An example is the horizontal and vertical integration of e-government services with other departments from the same or different jurisdictions.

APPENDIX B

Referencing

- **Best practices category and maturity stage:** An external best practices category at maturity stage 2 (interaction).
- **Purpose:** The purpose of the referencing best practices subcategory is to optimize the portal and its pages for search engines.
- **Introductory Notes:** Referencing is related to enhancing search engine optimization (SEO) to make the portal accessible and well optimized and ranked in the search results of search engines such as Google and Yahoo.

Table 1. Industrialization

Specific Practices (SP) by Specific Goal (SG)	Assessment Questions (Yes/No Binary Scoring)
· **SG1: Follow industry guidelines.** · SP 1.1: Follow ISO/IEC 16982 (Standard, 2002) on ergonomics of human system interaction and usability methods. · SP 1.2: Follow the ISO/IEC 40500 - Web Content Accessibility Guidelines (WCAG) 2.0 (WCAG 2 Overview, 2012).	· SQ 1.1: Are the guidelines of ISO/ IEC 16982 on ergonomics of human system interaction and usability methods followed? · SQ 1.2: Are the ISO/IEC 40500 guidelines followed?
· **SG2: Separate the content and style of Webpages for easy maintenance.** · SP 2.1: Use of CSS (Choudrie & Weerakkody, 2004). The CSS are files that contain the style or layout of the Web pages.	· SQ 2.1: Is the portal using CSS?

Table 2. Relevancy

Specific Practices (SP) by Specific Goal (SG)	Assessment Questions (Yes/No Binary Scoring)
· **SG1: Group information from a citizen perspective (not from an organizational perspective) in a way that helps citizens to find it.** · SP 1.1: Grouping information by themes or life events (Capgemini, 2010), (Benchmark Measurement of European eGovernment Services, 2009), such as: health, education, marriage. or agriculture, etc. depending on the portal features. · SP 1.2: Grouping information by target groups (West, 2007), (Capgemini, 2010), (Benchmark Measurement of European eGovernment Services, 2009), such as: youth, elderly, students and workers (UN E-Government Survey 2012, 2012).	· SQ 1.1: Is the information grouped by theme or life event (ex. mothering, marriage, birth, etc.)? · SQ 1.2: Is the information grouped by target groups (ex. students, workers, job seekers, etc.)?
· **SG2: Make information relevant to the citizen, detailed and efficient.** · SP 2.1: Is the information at the right level of detail (Sørum, 2011), that is with documents with detailed information about laws.	· SQ 2.1: Is the information at the right level of detail?

Table 3. Accessibility

Specific Practices (SP) by Specific Goal (SG)	Assessment Questions (Yes/No Binary Scoring)
· **SG1: Making the Website accessible.** · SP 1.1: Following the ISO/IEC 40500 – WCAG 2.§0 (WCAG 2 Overview, 2012), (Caldwell et al., 2008) guidelines for making Web content more accessible. · SP 1.2: Employing accessibility Web experts, such as employing a group of experts suitable for disabled people (UN, 2012). · SP 1.3: Making the portal accessible via mobile devices (UN E-Government Survey 2012, 2012). · SP 1.4: Making the portal accessible with all browsers and systems (Papadomichelaki & Mentzas, 2012). · SP 1.5: Not using inline multimedia elements (Choudrie & Weerakkody, 2004), (Oakland, 1991), as some people may not have the appropriate plugins (Choudrie & Weerakkody, 2004), (Oakland, 1991). · SP 1.6: Monitoring HTML broken links (Oakland, 1991), (West, 2007). Broken links may give an impression to the user that Website is improperly functioning. · SP 1.7: Not overdoing pictures (Oakland, 1991), as they slow down the browsers. · SP 1.8: Use or implement a responsive template that changes the size depending on the mobile device resolution.	· SQ 1.1: What is the portal's grade in compliance with WCAG 2.0 guidelines using the e-accessibility checker? (percentage scoring) · SQ 1.2: Does the web team contain accessibility Web experts? · SQ 1.3: Is the portal accessible via mobile devices? (Android/ IOS /Windows devices 'Phones&Tablets', equally weighted scoring) · SQ 1.4: Is the portal accessible with all browsers? (Chrome/Internet Explorer/Firefox, equally weighted scoring) · SQ 1.5: Is the portal not using inline multimedia elements that need special plugins (such as: flash)? · SQ 1.6: Are HTML broken links monitored? · SQ 1.7: Are pictures not overdone? · SQ 1.8: Is the portal's template responsive i.e. changes the size depending on the mobile device resolution (phone, tablet)?
· **SG2: Making the Website accessible for people with vision deficiencies.** · SP 2.1: Having AUDIO CAPTCHAs; a test used to differentiate humans from computers which should also be available in an audio format (UN E-Government Survey 2012, 2012). · SP 2.2: Having 'listen features' to hear text of a page (UN E-Government Survey 2012, 2012). This allows hearing text of a Web page for people with visual disabilities. · SP 2.3: Possibility to change font size and colors of the pages for the people with vision problems (UN E-Government Survey 2012, 2012), (Benchmark Measurement of European eGovernment Services, 2009).	· SQ 2.1: Is there AUDIO CAPTCHAs available in the portal? · SQ 2.2: Are there listen features to hear text of a page available in the portal? · SQ 2.3: Is there a possibility to change font size and colors of the pages for the people with vision problems?
· **SG3: Making the Website accessible for people with hearing deficiencies.** · SP 3.1: Having Sign language videos (UN E-Government Survey 2012, 2012). · SP 3.2: Having either a TeleTYpe (TTY) or Text telephone (TT) or Telephonic Device for the Deaf (TDD) phone number (West, 2007)	· SQ 3.1: Are there sign language videos available in the portal? · SQ 3.2: Is there a TTY or TT or TDD phone number available?

Table 4. Search engines

Specific Practices (SP) by Specific Goal (SG)	Assessment Questions (Yes/No Binary Scoring)
· **SG1: Allow citizen to search for the portal content and services.** · SP 1.1: Offering search engines (Oakland, 1991), (Berntzen & Olsen, 2009), (Capgemini, 2010). · SP 1.2: Offering many search options in the search engine (Oakland, 1991), (Capgemini, 2010), (UN E-Government Survey 2012, 2012). For instance, filtering by image, videos, news, or themes, etc. · SP 1.3: Using metadata in conjunction with search engines, since metadata can be used to enhance content and service availability by making them easier to find (Peters et al., 2004).	· SQ 1.1: Is there a search engine available in the portal? · SQ 1.2: Are there various search options on the search, such as, filtering by image, videos, or news? · SQ 1.3: Is metadata used in conjunction with search engines?
· **SG2: Show the search results according to their relevancy.** · SP 2.1: The Website search engine rank the results according to relevance of the users' query (Kaisara & Pather, 2011).	· SQ 2.1: Does the Website search engine rank the results according to relevance of the users' query?

Table 5. Periodical change

Specific Practices (SP) by Specific Goal (SG)	Assessment Questions (Yes/No Binary Scoring)
· **SG1: Monitoring the expiry and last update dates for each information or article within the portal** · SP 1.1: Having expiry dates or review dates (Oakland, 1991). This is important so that the Web site team can update certain Web pages (Oakland, 1991). · SP 1.2: Having date of the last update (Choudrie & Weerakkody, 2004).	· SQ 1.1: Are there expiry dates or review dates of the portal's pages? · SQ 1.2: Is there a date of the last update available in the portal's pages?
· **SG2: Updating the portal's information periodically** · SP 2.1: Having a team of editors responsible of the portal's timeliness (UN, 2012) and updating the contents in a timely manner.	· SQ 2.1: Is there a team of editors responsible of the portal's timeliness?

Table 6. Rich content

Specific Practices (SP) by Specific Goal (SG)	Assessment Questions (Yes/No Binary Scoring)
· **SG1: Offering a broad range of information in the portal.** · SP 1.1: Offering a broad range of information such as codes, laws, news, regulations, publications, press releases, databases, audio/video clips, Webcasts, and interactive maps (UN E-Government Survey 2012, 2012), (UN, 2012), (Georgescu & Georgescu, 2008).	· SQ 1.1: Are there broad ranges of information available in the portal? (codes, laws, news, etc., equally weighted scoring)
· **SG2: Offering videos in the portal.** · SP 2.1: Having videos in reception (Oakland, 1991) since their presence is considered important (Berntzen & Olsen, 2009). · SP 2.2: Having a series of videos organized by themes or life events (Oakland, 1991). · SP 2.3: Having videos that explain the procedures to accomplish citizen's needs depending on the life event (Simple answers about life in Switzerland, 2014).	· SQ 2.1: Are there videos in reception? · SQ 2.2: Are there a series of videos organized by themes or life events? · SQ 2.3: Are there videos that explain the procedures to accomplish citizen's needs depending on the life event?

Table 7. Translations

Specific Practices (SP) by Specific Goal (SG)	Assessment Questions (Yes/No Binary Scoring)
· **SG1: Making the portal accessible for foreigners.** · SP 1.1: Translating the portal to English is an e-government feature (Berntzen & Olsen, 2009).	· SQ 1.1: Is the portal's content translated to English?
· **SG2: Making the portal accessible for natives.** · SP 2.1: Translating the portal to local dialects (UN E-Government Survey 2012, 2012) or all the local dialects of the country (Capgemini, 2010). · SP 2.2: Translating the portal to all official languages of the country (Capgemini, 2010).	· SQ 2.1: What is the percentage of available local dialects translations over all the local dialects of the country? (percentage scoring) · SQ 2.2: What is the percentage of available official languages translations over all official languages of the country (percentage scoring)

Table 8. Understandability

Specific Practices (SP) by Specific Goal (SG)	Assessment Questions (Yes/No Binary Scoring)
· **SG1: Offering clear and easily understandable information.** · SP 1.1: Write in plain language (What is Plain Language?, 2014). This means that the user can understand the contents directly from the first time. For example, the US Plain Language initiative (What is Plain Language?, 2014). · SP 1.2: Offering content that can easily be understood (West et al., 2004), (DeLone & McLean, 1992), (DeLone & McLean, 2003), (Sørum, 2011). · SP 1.3: Easy to remember Uniform Resource Locator (URL) (Papadomichelaki & Mentzas, 2012).	· SQ 1.1: Is the portal's content written in plain language? · SQ 1.2: Is the portal's content easily understood? · SQ 1.3: Is the portal's URL easy to remember?

Table 9. Statements

Specific Practices (SP) by Specific Goal (SG)	Assessment Questions (Yes/No Binary Scoring)
· **SG1: Increasing citizens' trust and be transparent on data usage.** · SP 1.1: Having security statements (Kumar et al., 2007), (Makolm, 2002), (Benchmark Measurement of European eGovernment Services, 2009), (Georgescu & Georgescu, 2008). · SP 1.2: Having privacy statements (Kumar et al., 2007), (Makolm, 2002), (Benchmark Measurement of European eGovernment Services, 2009), (Georgescu & Georgescu, 2008), (Capgemini, 2010). · SP 1.3: Privacy statements should provide steps the user needs to do if his data is misused (Benchmark Measurement of European eGovernment Services, 2009). · SP 1.4: Providing information how data is retrieved, stored, and shared among other governmental departments (Kumar et al., 2007), (Makolm, 2002).	· SQ 1.1: Is there a security statement in the portal? · SQ 1.2: Is there a privacy statement in the portal? · SQ 1.3: Is the privacy statement providing steps the user needs to do if his data is misused? · SQ 1.4: Are the statements providing information how data is retrieved, stored, and shared among other governmental departments or third parties? ·
· **SG2: Protecting the portals' data against unauthorized reuse.** · SP 2.1:Having copyrights, which are considered tips for a successful e-government portal (Oakland, 1991).	· SQ 2.1: Are there copyrights available in the portal?
· **SG3: Protecting the agency from taking liability or responsibility.** · SP 3.1:Having disclaimers with no responsibility for the accuracy on information; they are tips for a successful e-government portal (Oakland, 1991).	· SQ 3.1: Are there disclaimers (with no responsibility for the accuracy on information) available in the portal?

Table 10. Responsiveness

Specific Practices (SP) by Specific Goal (SG)	Assessment Questions (Yes/No Binary Scoring)
· **SG1: Enhancing the performance of Webpages.** · SP 1.1: The Yahoo tool Yslow can be used to enhance performance of the pages (Alsmadi et al., 2012), (Yahoo Yslow for Chrome, 2015). · SP 1.2: The google tool Page Speed Insights can be used to enhance the speed of the pages (Page Speed Insights, 2015). · SP 1.3: Possibility to view only the text of the pages without the use of style sheet (Choudrie & Weerakkody, 2004). · SP 1.4: Not overdoing pictures as they slow down browsers (Oakland, 1991).	· SQ 1.1: What is the portal's grade in the Yahoo tool Yslow? · SQ 1.2: What is the portal's grade in Page speed tool (user experience should not be taken into account) "http://developers.google.com/speed/pagespeed/insights/"? · SQ 1.3: Is it possible to view only the text of the pages without the use of style sheet? · SQ 1.4: Are pictures not overdone as they slow down browsers?

Table 11. Contact Information

Specific Practices (SP) by Specific Goal (SG)	Assessment Questions (Yes/No Binary Scoring)
· **SG1: Allow citizens to interact with the agency or its officials** · SP 1.1: Contact information (email, phone, address, key persons to contact) (West, 2004), (West, 2007), (Rohleder, 2004), (Windley, 2002). · SP 1.2: The contact details of relevant government officials are clearly posted (Kaisara & Pather, 2011). · SP 1.3: 'Contact us' link provides an easy access to contact the agency. · SP 1.4: Presence of information regarding user's possibility to complain (Tinholt & van der Linden, 2013).	· SQ 1.1: Is there contact information available in the portal? (email, phone, etc., equally weighted scoring). · SQ 1.2: Are the contact details of relevant government officials clearly posted, like emails? · SQ 1.3: Is there a 'contact us' link in the portal? · SQ 1.4: Is there information/way regarding user's possibility to complain?
· **SG2: Allow citizens to understand the organizational structure of the agency** · SP 2.1: Presence of the organizational structure and chart, the names and titles of head of departments/functions (Tinholt & van der Linden, 2013).	· SQ 2.1: Is there an organizational structure or chart with the names and titles of head of departments/functions?
· **SG3: Guarantee a timely interactivity with citizens** · SP 3.1: Presence of turnaround time mechanisms to respond to submitted requests (Main Features Reviewed by the United Nations e-Government Survey, 2008). · SP 3.2: Timely responses to mails (Government on the Web II, 2002).	· SQ 3.1: Is there turnaround time mechanisms to respond to submitted requests? · SQ 3.2: Are there timely responses to mails?

Table 12. Advertising

Specific Practices (SP) by Specific Goal (SG)	Assessment Questions (Yes/No Binary Scoring)
· **SG1: Advertising the portal and its services.** · SP 1.1: Advertising the Websites in printed brochures, television, radio and internet; this helps citizens to be aware of the services and information available to them (West et al., 2004).	· SQ 1.1: Is the portal advertised in: printed brochures/ television/radio/internet? (equally weighted scoring)

Table 13. Emailing

Specific Practices (SP) by Specific Goal (SG)	Assessment Questions (Yes/No Binary Scoring)
· **SG1: Advertising the portal via emails.** · SP 1.1: Use of email systems (Government on the Web II, 2002), (West, 2004); the emails can be used to advertise the portal and its services using email marketing software. · SP 1.2: Use of push technology (email subscriptions and newsletters) (Government on the Web II, 2002), (West, 2004); this allows citizens to register their email addresses to receive newsletters.	· SQ 1.1: Is there a use of email systems? · SQ 1.2: Is there a use of push technology (i.e. email subscriptions and newsletters)?

Table 14. Referencing

Specific Practices (SP) by Specific Goal (SG)	Assessment Questions (Yes/No Binary Scoring)
· **SG1: Making the portal optimized for search engines.** · SP 1.1: Set up links with majority of search engines (Papadomichelaki & Mentzas, 2012). This might be specific for every search engine. · SP 1.2: Applying search engine optimization. · SP 1.3: Using write keywords in title of the pages to make sure the citizens can access the portals' pages easily. · SP 1.4: Using URL rewriting to help improving specific URLs for certain search queries. · SP 1.5: Submitting sitemaps to search engines using tools such as: Google Webmaster Tools (Webmasters,). · SP 1.6: The Website 'marketing.grader.com' assesses the current SEO practices (Alsmadi et al., 2012).	· SQ 1.1: Is there links set up with majority of search engines? · SQ 1.2: Are SEO applied? · SQ 1.3: Are the right keywords used in titles for all pages or for most important pages? · SQ 1.4: Is URL rewriting set for most important pages? · SQ 1.5: Are sitemaps submitted to all search engines? · SQ 1.6: Are tools to measure and enhance SEO being used such as 'marketing.grader.com'?

Table 15. Security

Specific Practices (SP) by Specific Goal (SG)	Assessment Questions (Yes/No Binary Scoring)
· **SG1: Increase adoption and trust in e-government by making secure transactions and authentications.** · SP 1.1: Use of digital signatures for security or encryption (Posch et al., 2011) such as to secure emails and submission of forms. · SP 1.2: Use of digital signature for authentication (Posch et al., 2011) such as to identify users.	· SQ 1.1: Are digital signatures used for security or encryption? · SQ 1.2: Are digital signature used for authentication?
· **SG2: Making digital signatures equivalent to the handwritten signatures.** · SP 2.1: Digital signatures must be printed and equivalent to handwritten signatures. This means that they can be used not only for authentication and identification but also as an alternative to the handwritten ones (Posch et al., 2011). · SP 2.2: Possibility to verify the signer is a public authority in printed papers relying on text-based signatures (Posch et al., 2011).	· SQ 2.1: Can digital signatures be printed and be equivalent to handwritten signatures? · SQ 2.2: Is there a possibility to verify the signer is a public authority in printed papers?
· **SG3: Making sure the portal is always available.** · SP 3.1: Having recovery mechanism (e-Government of Korea Best Practices, 2014). · SP 3.2: Having real-time backup mechanism (e-Government of Korea Best Practices, 2014).	· SQ 3.1: Is there a recovery mechanism in place? · SQ 3.2: Is there a real-time backup mechanism in place?
· **SG4: Resistance against hacking attacks.** · SP 4.1: Limit Authorizations of database users. · SP 4.2: Install patches and updates of open source modules and Web 2.0 components regularly as they contain security upgrades. · SP 4.3: Granting minimum file system permissions (read/write/execute) for files. · SP 4.4: Limit files system permissions to configuration files that contain passwords. · SP 4.5: Resistance to SQL injections can be checked using Firefox add-on 'SQL Inject Me'. Other Web vulnerability scanners in the Web can be used to check other vulnerabilities. · SP 4.6: Implement security precautions depending on the environment used. For instance, for PHP environment, one have to maintain the PHP.INI file to disable PHP functions that are not used and might be considered as a threat.	· SQ 4.1: Are database users granted minimum authorizations (for instance, avoid delete table)? · SQ 4.2: Are patches and updates of open source modules and Web 2.0 components installed or upgraded regularly? · SQ 4.3: Are files granted minimum system permissions (read/write/execute)? · SQ 4.4: Do configuration files containing passwords have minimum system permissions (read/write/execute)? · SQ 4.5: Is SQL injection being inspected using tools such as the Firefox add-on 'SQL Inject Me'? · SQ 4.6: Are security precautions implemented depending on the environment used?

Table 16. Privacy

Specific Practices (SP) by Specific Goal (SG)	Assessment Questions (Yes/No Binary Scoring)
· **SG1: Keeping citizens' data and preferences private.** · SP 1.1: Follow P3P guidelines (Choudrie & Weerakkody, 2004) which is a protocol developed by W3C that allows the portals to explain their intended use of the users' data. · SP 1.2: Avoid cookies (Choudrie & Weerakkody, 2004), which are set by the portal designers to store users' information. · SP 1.3: Making sure the portal is operational by disabling cookies (Henriksson et al., 2007). · SP 1.4: Avoid the use of HTML get forms (Choudrie & Weerakkody, 2004), as they are a potential security risk · SP 1.5: Use a role-based authorization for officials (Makolm, 2002); that is, employees within the government will have an access to only the data they are allowed to process. · SP 1.6: Not sharing personal information with others (third party) (Papadomichelaki & Mentzas, 2012), that is, citizens' data won't be shared with other companies or stakeholders.	· SQ 1.1: Are P3P privacy guidelines being followed? · SQ 1.2: Are cookies avoided in the portal? · SQ 1.3: Is the portal operational when cookies are disabled? · SQ 1.4: Are HTML get forms avoided? · SQ 1.5: Is there a role-based authorization for officials? · SQ 1.6: Is personal information of citizens not shared with third parties?

Table 17. Single Sign On

Specific Practices (SP) by Specific Goal (SG)	Assessment Questions (Yes/No Binary Scoring)
· **SG1: Providing access to all services and systems with a single login.** · SP 1.1: Implementing single sign on to all services provided in the portal to the users (UN E-Government Survey 2012, 2012).	· SQ 1.1: Is there a single sign on in place?

Table 18. Delegation

Specific Practices (SP) by Specific Goal (SG)	Assessment Questions (Yes/No Binary Scoring)
· **SG1: Ensuring the continuity of the citizens' tasks when they are busy or simply unavailable.** · SP 1.1: Implementing delegation mechanisms (Posch et al., 2011). This means that citizens can delegate other people to perform their activities or execute services on behalf of them.	· SQ 1.1: Are there delegation mechanisms in place?

Table 19. Modularity

Specific Practices (SP) by Specific Goal (SG)	Assessment Questions (Yes/No Binary Scoring)
· **SG1: Achieve reusability and facilitate maintainability and interoperability efforts.** · SP 1.1: Breaking down an application into modules called services, each module is responsible of a small functionality. An example is the use of SOA.	· SQ 1.1: Is there a use of Service Oriented Architecture or something similar (application is broken down into modules called services)?

Table 20. Reusability

Specific Practices (SP) by Specific Goal (SG)	Assessment Questions (Yes/No Binary Scoring)
· **SG1: Save time and by reusing Web 2.0 and open source modules.** · SP 1.1: Reusing Web 2.0 modules (or open source) that are already developed and available for use (Assar et al., 2011). For instance, if a calendar or a blog already exist and do fulfill the requirements of the portal, then no need to build them from scratch.	· SQ 1.1: Is there a check of the existence of open source modules before starting their implementation?
· **SG2: Save time by reusing services and components (even from agencies from the same or different jurisdictions).** · SP 2.1: Reusing services and components if a similar service has already been developed by another agency from the same or different jurisdiction.	· SQ 2.1: Before developing new components or services, does the agency check if other agencies have the component or service already available?

Table 21. Use of Standards

Specific Practices (SP) by Specific Goal (SG)	Assessment Questions (Yes/No Binary Scoring)
· **SG1: Facilitate interoperability efforts, ease of change in requirements and prevent the fact of 're-architecting' as problems arise.** · SP 1.1: Use of Web services. This was used in the Tax eFiling system in Norway (eGovernment: International Best Practices, 2008). · SP 1.2: Use of SOAP. This was used in the Altinn project (eGovernment: International Best Practices, 2008). · SP 1.3: Use of XML or JSON. As an example, XML was used in the Altinn project (eGovernment: International Best Practices, 2008). · SP 1.4: Use of Dublin core metadata. Dublin core metadata standard is the most widely used among the other meta data standards (Alasem, 2009).	· SQ 1.1: Are Web services being used? (yes/no binary scoring) · SQ 1.2: Is SOAP being used? (yes/no binary scoring) · SQ 1.3: Is XML or JSON being used? (yes/no binary scoring) · SQ 1.4: Is Dublin core metadata or any equivalent metadata standard being used?

Table 22. Personalization

Specific Practices (SP) by Specific Goal (SG)	Assessment Questions (Yes/No Binary Scoring)
· **SG1: Offering personal access to the citizens.** · SP 1.1: Offer personal accounts (Main Features Reviewed by the United Nations e-Government Survey, 2008), (Capgemini, 2010); that is, the citizens can create a personal account in the portal where they can login. · SP 1.2: Offer online access to citizens' data with the possibility to modify the data. This means that the citizens can access their private data and modify it (Tinholt & van der Linden, 2013). · SP 1.3: Having personalized sections specific for citizens like 'My pages' or 'My messages' which has resulted in an increased number of visitors (Benchmark Measurement of European eGovernment Services, 2009).	· SQ 1.1: Is there a possibility for citizens to create personal accounts? · SQ 1.2: Is there an online access to citizens' data? · SQ 1.3: Are there personalized sections for citizens like 'My pages'?
· **SG2: Offering the citizens the possibility to tailor and customize the services manually and automatically.** · SP 2.1: Customizing services of interest by citizens (UN E-Government Survey 2012, 2012); this can be done manually. · SP 2.2: Services should be customized automatically according to user profiles and requirement (age, gender, education level, occupation, etc.) (Assar et al., 2011). For example, there might be some services only for elderly, and the youth should not see these services as it is not relevant for them.	· SQ 2.1: Can the services of interest be customized by citizens depending on their interest? · SQ 2.2: Are the services been customized automatically according to users' profiles and requirements (such as: age, education level, and occupation)?
· **SG3: Offering the citizens the possibility to tailor and customize the portal's content and themes manually and automatically.** · SP 3.1: Possibility to tailor the portals' content and customize the favorite themes manually depending on the citizen interest. · SP 3.2: Possibility to tailor the portals' content or themes depending on citizens' profiles (such as: gender and age). This can be done automatically. An example is to customize the themes depending on the gender. This means that, themes such 'Getting pregnant' should be available only for women.	· SQ 3.1: Can the citizens tailor the portal's content and themes according to their interest? · SQ 3.2: Is there a possibility to tailor the portal's content and themes automatically depending on citizens' profiles (such as: gender and age)?
· **SG4: Allowing citizens to store their information in the portal.** · SP 4.1: Offering storage space for the citizens that can be purchased to store users' documents ranging from certificates to diplomas (Benchmark Measurement of European eGovernment Services, 2009).	· SQ 4.1: Is there a storage space for the citizens?

Table 23. Help

Specific Practices (SP) by Specific Goal (SG)	Assessment Questions (Yes/No Binary Scoring)
· **SG1: Helping the user at any time.** · SP 1.1: Provide live support like click to chat (Capgemini, 2010)	· SQ 1.1: Is there a live support like click to chat?
· **SG2: Answering the citizens' common questions.** · SP 2.1: Provide help or/and FAQs features (Benchmark Measurement of European eGovernment Services, 2009), (UN E-Government Survey 2012, 2012), (Capgemini, 2010). FAQs produce answers to the user's question quickly.	· SQ 2.1: Are there FAQs available in the portal?
· **SG3: Providing citizens with help explaining the navigation process.** · SP 3.1: Use of interactive assistance like videos, guided tours or downloadable manuals explaining the navigation process (West, 2007), (Capgemini, 2010), (Bhattacharya et al., 2012).	· SQ 3.1: Is there interactive assistance like videos, downloadable manuals or guided tours explaining the navigation process?

Table 24. Incentives

Specific Practices (SP) by Specific Goal (SG)	Assessment Questions (Yes/No Binary Scoring)
· **SG1: Encouraging the citizens to use the portals instead of visiting the governmental offices.** · SP 1.1: Offering privileges for online users to encourage the citizens to opt for the Web portals over the traditional ones or offices. For instance, offer an extended period for filling taxes or profit from filling their taxes for free and getting refund in half of the normal duration (UN E-Government Survey 2012, 2012).	· SQ 1.1: Are online users privileged over traditional users?

Table 25. Contests

Specific Practices (SP) by Specific Goal (SG)	Assessment Questions (Yes/No Binary Scoring)
· **SG1: Encouraging citizens to create services.** · SP 1.1: Organizing competitions for the public to create services, which encourages participation and exploits the problem-solving skills of the citizens. Therefore, citizens are not only simple consumers of services, but become producers. This helps achieving great service usage (UN E-Government Survey 2012, 2012). · SP 1.2: Making public data available to encourage participation and collaboration (Assar et al., 2011), because citizens can use this data to create innovative services.	· SQ 1.1: Are competitions organized for the public to create services? · SQ 1.2: Is public data made available?

Table 26. Performance Ratings

Specific Practices (SP) by Specific Goal (SG)	Assessment Questions (Yes/No Binary Scoring)
· **SG1: Getting feedback from citizens on the global portal's performance.** · SP 1.1: Presence of feedback icons provides a way for citizens to share their experience with governments and provide feedback to the governments (Benchmark Measurement of European eGovernment Services, 2009). · SP 1.2: Use of surveys for rating the portal (e.g. overall satisfaction) can be in the form of popup or drop down questionnaires (Capgemini, 2010).	· SQ 1.1: Are feedback icons available? · SQ 1.2: Are surveys used regularly for rating the portal?
· **SG2: Getting offline feedback.** · SP 2.1: Use of offline surveys for people vising public offices or elsewhere (whether users or not) (Benchmark Measurement of European eGovernment Services, 2009), (Capgemini, 2010). This can be also in the form of interviews, telephone interviews, meetings or focus groups in order to give an idea to governments on the barriers that prevent the citizens from using the portal.	· SQ 2.1: Are offline surveys been used for people vising public offices or elsewhere (whether users or not)?
· **SG3: Allowing citizens to make complaints.** · SP 3.1: Presence of complaints or ideas box (Capgemini, 2010) that allows citizens to send complaints to government, which should have a way to manage them.	· SQ 3.1: Is there a possibility for users to complain?

Table 27. Service Customer Centricity

Specific Practices (SP) by Specific Goal (SG)	Assessment Questions (Yes/No Binary Scoring)
· **SG1: Assess the needs of citizens during implementation of services.** · SP 1.1: Using observatory and eye tracking approaches (Benchmark Measurement of European eGovernment Services, 2009). This is normally done during the build phase of services. The users are observed trying e-government services and the portal's functionalities during implementation in a living lab. This helps understand the citizens' needs to design the services accordingly.	· SQ 1.1: Are observatory methods or eye tracking approaches being used (i.e. observe users trying e-government services during implementation)?
· **SG2: Assess the needs of different groups of citizens during implementation of services.** · SP 2.1: Ethnographic approach; to group users according to some socio-cultural criteria like age and educational level (Benchmark Measurement of European eGovernment Services, 2009).The primary aim is to involve the groups in the design of the services or portals and assess the need of each group.	· SQ 2.1: Is ethnographic work being used (i.e. group users according to some socio-cultural criteria like age and educational level)?

Table 28. Service Help

Specific Practices (SP) by Specific Goal (SG)	Assessment Questions (Yes/No Binary Scoring)
· **SG1: Providing help to the citizens.** · SP 1.1: Providing live support like click to chat or ask question (Capgemini, 2010), (Tinholt & van der Linden, 2013) which can be used by citizens at any time, to chat with government officials to ask questions and receive answers.	· SQ 1.1: Is there a click to chat or ask question options available?
· **SG2: Helping citizens during the execution of services.** · SP 2.1: Providing help for citizens when filling the form or during the execution of online services (Makolm, 2002), (Rohleder, 2004). This can be in the form of unprompted help (Rohleder, 2004); that is, the citizen may not seek it or a popup that offers help during the execution of online services.	· SQ 1.2: Is there help provided for users when filling the form or during the execution of online services?
· **SG3: Answering the citizens' most common questions about services.** · SP 3.1: Providing FAQs sections for the services or for each service (Capgemini, 2010), (Tinholt & van der Linden, 2013). FAQs provides relief to the citizens and do answer all their common questions about the services.	· SQ 3.1: Is there FAQ sections for services?
· **SG4: Providing interactive help for the citizens on how to complete the services.** · SP 4.1: Providing a demo (video or downloadable manual) for the services (Capgemini, 2010), (Tinholt & van der Linden, 2013) that provides a step by step guide on how to complete the e-government services.	· SQ 4.1: Are there demos (videos or downloadable manuals) on how to use the services?

Table 29. Payments

Specific Practices (SP) by Specific Goal (SG)	Assessment Questions (Yes/No Binary Scoring)
· **SG1: Pay for the services that require payments online.** · SP 1.1: Offering payments via credit cards or electronic banking (Berntzen & Olsen, 2009), (Makolm, 2002). This means that the citizens can pay for the services online for services that require a fee, or pay fines or taxes online.	· SQ 1.1: Is there a possibility for payments via credit cards or electronic banking (if applicable)?

Table 30. Workflows

Specific Practices (SP) by Specific Goal (SG)	Assessment Questions (Yes/No Binary Scoring)
· **SG1: Informing the citizen of the series of process steps of the application (from initiation to completion).** · SP 1.1: Delivery notice of successful completion of the process step online (Tinholt & van der Linden, 2013). · SP 1.2: Offer possibility for users to track status of their application via dashboard or emails (UN, 2012). The citizens can know in which step their application is, or this information is sent to them by emails.	· SQ 1.1: Is there a delivery notice of successful completion of the process step online? · SQ 1.2: Is it possible for users to track the status of their application via dashboard or emails?
· **SG2: Transparency with the service delivery timelines.** · SP 2.1: Indication of delivery timelines of the service (Tinholt & van der Linden, 2013). · SP 2.2: Indication of the maximum time limit it would take to deliver the service (Tinholt & van der Linden, 2013).	· SQ 2.1: Is there an indication of delivery timelines of the services? · SQ 2.2: Is there an indication of maximum time limit it would take to deliver the services?

Table 31. Service Performance Ratings

Specific Practices (SP) by Specific Goal (SG)	Assessment Questions (Yes/No Binary Scoring)
· **SG1: Getting feedback from citizens on e-government services.** · SP 1.1: Possibility to rate or score services (Assar et al., 2011), (Benchmark Measurement of European eGovernment Services, 2009), during or after completion by the user. This can be in the form of a request of feedback, scoring the service using emoticons or even popup or drop down questionnaires (Capgemini, 2010)	· SQ 1.1: Is it possible to rate or score services (for instance, use of emoticons)?
· **SG2: Encouraging the users to make suggestions to improve services.** · SP 2.1: Publishing information about service performance like performance assessment to encourage users to make suggestions to improve the services or even to create new ones (Tinholt & van der Linden, 2013).	· SQ 2.1: Is there information about services performance like performance assessment?

Table 32. User Forms

Specific Practices (SP) by Specific Goal (SG)	Assessment Questions (Yes/No Binary Scoring)
· **SG1: Providing easy to use user forms.** · SP 1.1: Indicate the mandatory fields (Makolm, 2002) which saves time for citizens. If not indicated, the user will be informed of the missing mandatory fields only after submission of the form. This means that the citizen needs to resubmit the form. · SP 1.2: Use drop down menus when necessary (Makolm, 2002) which saves time for citizens, since they will only select values from the drop down menus, instead of typing them. · SP 1.3: Reuse of citizen information or automatic recalling of users' personal data to facilitate future interaction in forms and automatic calculation of forms (Papadomichelaki & Mentzas, 2012). This saves time for citizens, since they don't need to re-type some information such as: name, address, etc. · SP 1.4: Possibility to save earlier versions of user forms as a draft (Benchmark Measurement of European eGovernment Services, 2009), (Capgemini, 2010), (Tinholt & van der Linden, 2013). This way the citizen can save the work done as a draft, and complete it later on. · SP 1.5: Providing overview on data needed to fill the forms or complete the service (Tinholt & van der Linden, 2013).	· SQ 1.1: Are the mandatory fields indicated? · SQ 1.2: Are drop down menus being used when necessary? · SQ 1.3: Is there automatic calculation of forms based on citizens' information or automatic recalling of citizens' personal data? · SQ 1.4: Is it possible to save earlier versions as drafts during the execution of the services? · SQ 1.5: Is there an overview on data needed to complete the service?
· **SG2: Providing citizens with a global view on the form.** · SP 2.1: Provide the possibility to preview the form before submitting it (Makolm, 2002). This provides citizens with a global view of the form, which could be beneficial to correct errors or missing information if necessary. · SP 2.2: Provide a possibility to print the accepted form (Makolm, 2002). · SP 2.3: Presence of progress tracking (Capgemini, 2010), (Benchmark Measurement of European eGovernment Services, 2009), (Tinholt & van der Linden, 2013) or with information regarding the citizen's stage of completion of the form (Makolm, 2002). It should be clear to the citizen, how much progress is made and how much is left to be done. For example, step 3/7 means that the citizen is in step 3 and he is left with 4 steps. · SP 2.4: Showing expectations how the entire process to complete the form is estimated to take (Capgemini, 2010).	· SQ 2.1: Is it possible to provide a preview of the form before submitting it? · SQ 2.2: Is it possible to print the accepted form? · SQ 2.3: Is there information regarding the stage of completion of the form? · SQ 2.4: Are expectations how the entire process to complete the service been shown?
· **SG3: Providing citizens with help when filling the user forms.** · SP 3.1: Provide online help when filling the form (Makolm, 2002).	· SQ 3.1: Is there online help when filling the form?

Table 33. Service Responsiveness

Specific Practices (SP) by Specific Goal (SG)	Assessment Questions (Yes/No Binary Scoring)
· **SG1: Making sure the services response time is acceptable.** · SP 1.1: Testing services and making sure the response time is not too long. All the steps of the services need to be tested to make sure that service can load in an acceptable amount of time. Users with limitations in hardware and bandwidth can be also taken into consideration. · SP 1.2: Performance enhancements by inspecting and enhancing the code and algorithms for services that takes a high response time to load or execute.	· SQ 1.1: Is the response time of services being measured to make sure it is not too long? · SQ 1.2: Is the code and algorithms being analyzed for performance enhancements when necessary?
· **SG2: Making sure the services does not take a large amount of time to fill and submit.** · SP 2.1: Designing the service in an efficient way to help users complete it in the shortest amount of time (Tinholt & van der Linden, 2013). This is more related to the organization of the service and not to its response time.	· SQ 2.1: Are the services designed in an efficient way that helps the citizens execute them in the shortest amount of time?

Table 34. Interoperability

Specific Practices (SP) by Specific Goal (SG)	Assessment Questions (Yes/No Binary Scoring)
· **SG1: Sharing information between agencies.** · SP 1.1: Vertical interoperability - Sharing information between agencies belonging to the same jurisdiction in higher and lower levels. This integration is between different departments such as: federal, provincial, and municipal in the same jurisdiction. · SP 1.2: Horizontal interoperability - Sharing information between agencies belonging to different various jurisdictions. This means that an agency is able to coordinate and share information with other agencies belonging to different jurisdictions.	· SQ 1.1: Is information shared with higher and lower agencies (national, regional or federal, provincial, municipal) within the same jurisdictions? · SQ 1.2: Is information shared with agencies belonging to different jurisdictions?
· **SG2: Facilitating interoperability efforts.** · SP 2.1: Use of technologies and open source standards (Al-Khouri, 2011), (eGovernment: International Best Practices, 2008), (Posch et al., 2011) in order to facilitate interoperability between systems. · SP 2.2: Use of SOA or Web services (Klischewski & Askar, 2010) to help achieve interoperability. This is because the Web services can run in harmony in different servers and are platform independent (Fath-Allah et al., 2014b).	· SQ 2.1: Is there a use of standards or open source standards (such as: XML, JSON or Dublin core metadata)? · SQ 2.2: Is there a use of SOA or Web services?
· **SG3: Managing cross agency coordination.** · SP 3.1: Implementing a database coordinator (eGovernment: International Best Practices, 2008), (X-Road, 2014). This means that if an agency needs to get information from other agencies, it will not contact each agency at a time. However, it will just contact the coordinator which will be responsible of querying the multiple agencies' databases and provide the agency with the requested information. · SP 3.2: Presence of a chief Information Officer (CIO) role or similar to manage cross-agency coordination (Main Features Reviewed by the United Nations e-Government Survey, 2008), (Accenture, 2003).	· SQ 3.1: Is there a database coordinator that is responsible of querying different databases from different agencies? · SQ 3.2: Is there a Chief Information Officer responsible of managing cross-agency coordination?

Table 35. One Stop Shop

Specific Practices (SP) by Specific Goal (SG)	Assessment Questions (Yes/No Binary Scoring)
· **SG1: Covering a large number of services.** · SP 1.1: Providing a large number of services in the portal including national services (Georgescu & Georgescu, 2008), (UN, 2012).	· SQ 1.1: Are there national services available in the portal?
· **SG2: Covering services from different agencies and jurisdictions.** · SP 2.1: Providing services from different agencies and departments into one point of entry (Georgescu & Georgescu, 2008). · SP 2.2: Providing services that cover almost all ministries (applicable if the portal is national), such as providing a single access to most of the administrative procedures (UN, 2012).	· SQ 2.1: Are there services from different agencies and departments? · SQ 2.2: What are the covered jurisdictions in the e-government portal services (applicable if the portal is national)? (health, education, etc., equally weighted scoring)

Table 36. Mobile Applications

Specific Practices (SP) by Specific Goal (SG)	Assessment Questions (Yes/No Binary Scoring)
· **SG1: Offering mobile applications for citizens.** · SP 1.1: Offering mobile apps such as IOS and Android apps. The iOS and Android mobile applications can be downloaded from their e-government portal. This includes e-learning application that offers the possibility for students to learn from their mobile, and also a job related application that offers the possibility to obtain information on jobs' market (UN E-Government Survey 2012, 2012).	· SQ 1.1: Are mobile apps available in the portal? (iOS/Android, equally weighted scoring)

Table 37. Interactive Games

Specific Practices (SP) by Specific Goal (SG)	Assessment Questions (Yes/No Binary Scoring)
· **SG1: Offer games for the citizens.** · SP 1.1: Offering games for users, where they can play alone or with/against each other, such as interactive games and decision games to increase citizens' knowledge in economics (UN, 2012).	· SQ 1.1: Are games available for users where they can play alone or with/against each other?

Table 38. E-participation

Specific Practices (SP) by Specific Goal (SG)	Assessment Questions (Yes/ No Binary Scoring)
· **SG1: Presence of e-participation policy or mission statement.** · SP 1.1: Presence of e-participation policy (Main Features Reviewed by the United Nations e-Government Survey, 2008) which should state how the government will manage the citizens' comments, for instance removing inadequate comments that violate laws or discriminate people.	· SQ 1.1: Is there an e-participation policy in the portal?
· **SG2: Allowing citizens to communicate with government officials.** · SP 2.1: Presence of e-participation tools that allows citizens to communicate with government officials such chat rooms, comment forms, ask an expert section that allows to ask questions and receive answers (Berntzen & Olsen, 2009), (UN, 2012).	· SQ 2.1: Are there e-participation tools that allow citizens to communicate with government officials available in the portal, such as: chat rooms, comment forms or ask an expert section?
· **SG3: Allowing citizens to communicate with each other.** · SP 3.1: Presence of e-participation tools that allows citizens to communicate with each other such as: online forums where users can chat, share information and advice with each other (UN, 2012).	· SQ 3.1: Are there e-participation tools that allow citizens to communicate with each other available in the portal, such as: discussion forums, comment forms and blogs (allowing citizens to leave comments and make discussions with each other)?
· **SG4: Seeking feedbacks, opinions, and suggestions from the citizens.** · SP 4.1: Presence of feedback icons or 'have your say' icon. These way users can share their feedback with governments (Benchmark Measurement of European eGovernment Services, 2009). · SP 4.2: Presence of suggestion programs where citizens can make suggestions (UN E-Government Survey 2012, 2012), (Benchmark Measurement of European eGovernment Services, 2009). · SP 4.3: Use of Surveys is an efficient way to gather citizens' feedback (Oakland, 1991), (UN E-Government Survey 2012, 2012), (Benchmark Measurement of European eGovernment Services, 2009). · SP 4.4: Presence of e-voting or polls to gather citizens' opinions (Hiller & Bélanger, 2001), (Moon, 2002), (Shahkooh et al., 2008), (Lee & Kwak, 2012), (Siau & Long, 2005).	· SQ 4.1: Is there a way to gather citizens feedback (such as: feedback icon or 'have your say icon') available in the portal? · SQ 4.2: Are there suggestion programs or mail boxes for suggestions? · SQ 4.3: Are surveys being used to gather citizens' feedback? · SQ 4.4: Are e-voting and polls being used to gather citizens' opinions?
· **SG5: Integrating the users' feedbacks into the portal.** · SP 5.1: Making continuous improvement by integrating the users' feedbacks into the portal regularly (Iribarren et al., 2008), (Georgescu & Georgescu, 2008), (Kumar et al., 2007). Seeking feedback should be accompanied by a 'review or integration of this feedback into the portal'. Otherwise, getting feedback will be unnecessary.	· SQ 5.1: Are the users' feedback integrated into the portal regularly?
· **SG6: Involving the citizen in the policy making process.** · SP 6.1: Possibility for citizens to lodge e-petitions (Hiller & Bélanger, 2001), (Moon, 2002), (Shahkooh et al., 2008), (Lee & Kwak, 2012), (Siau & Long, 2005). These petitions can then be voted or signed by citizens. If the petition receives enough signatures, it can be sent to the respective authorities for proposals. · SP 6.2: Possibility for citizens to send comments on draft regulations or laws to the intended ministries (UN E-Government Survey 2012, 2012).	· SQ 6.1: Can the citizens lodge e-petitions in the portal that are sent to the respective authorities for proposals if they receive enough signatures? · SQ 6.2: Can the citizens send their comments on draft regulations?
· **SG7: Gathering feedbacks from non-users (the citizens who didn't use the e-government portal or one of its services).** · SP 7.1: Gathering feedback from non-users using tools such: Computer-Assisted Telephone Interview (CATI) survey, field survey for citizens visiting public offices and Email and postal surveys (Main Features Reviewed by the United Nations e-Government Survey, 2008), (Benchmark Measurement of European eGovernment Services, 2009).	· SQ 7.1: Is feedback gathered from non-users using tools?

Table 39. Sociability

Specific Practices (SP) by Specific Goal (SG)	Assessment Questions (Yes/No Binary Scoring)
· **SG1: Use of social media Websites.** · SP 1.1: Use of social media Websites (UN E-Government Survey 2012, 2012) such as Facebook, Twitter, WordPress, YouTube, and Flickr to strengthen the relationship with the citizen and enables the user to engage with the e-government (eGovernment: International Best Practices, 2008), (Fath-Allah et al., 2014b). For these reasons, governments should reinforce their presence in social media and take advantage of its potential which could yield to the improvement of e-service usage (UN E-Government Survey 2012, 2012). · SP 1.2: Presence of an active page in Facebook. This means that the Webpage should be active and governments should contribute in the page regularly. · SP 1.3: Presence of an active account in Twitter. This means that the account should be active; the government should contribute in the account regularly.	· SQ 1.1: Is there social media links available in the portal? (Facebook/Twitter/YouTube/Flickr equally weighted scoring) · SQ 1.2: Is there an active Facebook page of the portal? · SQ 1.3: Is there an active Twitter account of the portal?

Table 40. Data Analytics

Specific Practices (SP) by Specific Goal (SG)	Assessment Questions (Yes/No Binary Scoring)
· **SG1: Analyzing Website performance for insights.** · SP 1.1: Measuring Website performance using data analytics like number of visitors and pages viewed, number of published datasets, the number of data downloads, and the number of visitors (Lee & Kwak, 2012), (Layne & Lee, 2001), (Kim & Grant, 2010). Moreover, on-line activities can be tracked and used to improve the Web site, for example, the frequently accessed products, the length of time spent on each page, and the length of time spent searching (Layne & Lee, 2001).	· SQ 1.1: Is the portal's performance being measured (using tools such as google analytics) via data analytics like number of visitors and pages viewed, number of published datasets, the number of data downloads, and the number of visitors?
· **SG2: Analyzing citizens' feedback for insights.** · SP 2.1: Using citizens' feedback for data analytics to improve decision making (Lee & Kwak, 2012). For instance, governments can analyze regularly the suggestions, feedback, votes, petitions, comments and complaints of citizens to take reliable decisions, such as making new laws.	· SQ 2.1: Is the data or feedback being analyzed for improving decision making?
· **SG3: Analyzing the data in social Websites for insights.** · SP 3.1: Use of listening tools – including sentiment analysis, which allow the analysis of online conversations on websites such as Facebook, LinkedIn, Twitter, blogging sites and wikis (Andreasson et al., 2012).	· SQ 3.1: Are listening tools – including sentiment analysis, which allow the analysis of online conversations on websites such as Facebook, LinkedIn, Twitter, blogging sites and wikis being used?

Table 41. Service Interoperability

Specific Practices (SP) by Specific Goal (SG)	Assessment Questions (Yes/No Binary Scoring)
· **SG1: Making the services integrated with other services or systems from the same or different jurisdictions.** · SP 1.1: Making the services vertically integrated. An example is a birth registry service in a hospital that triggers other services such as a vaccination service or sends information automatically to other agencies in the same jurisdiction (Fath-Allah et al., 2014c). · SP 1.2: Making the services horizontally integrated. As an example, the same birth registry service in the hospital can trigger other services such as an insurance service or send information automatically to other agencies in different jurisdictions (Fath-Allah et al., 2014c).	· SQ 1.1: Are the services vertically integrated when necessary (i.e. the service can trigger other services or notify agencies with higher and lower levels: national, regional or federal, provincial, municipal within the same jurisdictions, an example is a birth registry in a hospital that can share this information with a service from national level for birth statistics)? · SQ 1.2: Are the services horizontally integrated when necessary (i.e. the service can trigger other services or notify agencies belonging to different jurisdictions, example is a birth registry that can triggers an insurance registry service)?

Chapter 3
Leveraging Data Spaces in Digital Twins for Smart Cities Deployment:
How Digital Twins Powered by Data Spaces Can Help Us Build Inclusive, Sustainable Smart Cities

Ana Turcan
Data Intelligence Offensive, Austria

ABSTRACT

The purpose of this chapter is to offer the readers an overview about how a data space is defined, how digital twins powered by data spaces can help us build inclusive sustainable smart cities, how emerging technologies can enhance data spaces, how data spaces impact digital twins and smart cities, and to share high level insights related to initiatives that have been adopted at the European Union level in terms of data spaces as business value creation and sustainability for a common path towards a sustainable data-driven ecosystem.

INTRODUCTION

Digital twins as digital representations of physical objects, systems, or processes - can be used to simulate the behavior of real-world objects, systems, or processes in a virtual environment. This can help us understand how these objects, systems, or process's function, and identify potential improvements or issues.

DOI: 10.4018/978-1-6684-3833-6.ch003

In the context of smart cities, digital twins can be used to create digital models of urban environments. These digital models can be powered by data from a variety of sources, including sensors, IoT devices, and other data sources. By integrating this data into a digital twin, city planners and decision makers can gain a better understanding of how the city is functioning and identify areas for improvement.

For example, a digital twin of a city's transportation system could be used to simulate the flow of traffic and identify bottlenecks or congestion. This information could be used to develop strategies for improving the efficiency of the transportation system, such as implementing new routing algorithms or building additional infrastructure.

In addition to improving the efficiency of city systems, digital twins can also be used to help create more inclusive and sustainable smart cities. By analyzing data from a variety of sources, city planners can identify areas where certain groups of people may be underserved or disadvantaged, and develop strategies for addressing these issues. For example, a digital twin of a city's energy system could be used to identify areas where access to clean, renewable energy is limited, and develop plans for expanding access to these resources.

Overall, digital twins powered by data spaces can be a powerful tool for helping us build more inclusive and sustainable smart cities. By providing a virtual representation of urban environments, and integrating data from a variety of sources, digital twins can help us understand how cities function, identify areas for improvement, and develop strategies for addressing those challenges (Gauss Centre for Supercomputing, 2020).

Digital twins technology is directly related to data spaces.

A dataspace refers to a type of data relationship between trusted partners where each apply the same standards and rules for storing and sharing their data. However, central to the concept of a dataspace is that the data is not stored centrally, but rather distributed at the respective source, and is therefore only shared as needed in the context of common use cases. Data spaces focus on domains (economic areas, industrial sectors or other specialist application fields) and make metadata available for potential innovative services while maintaining data sovereignty, i.e., the greatest possible control and dominion over their own data. Domain-specific data spaces can also connect (federate) with other data spaces, such as a mobility data space with a tourism data space (Prinz, Rose, and Urbach, 2022).

Building a digital twin requires coordinating many components into a coherent whole. Most of the data needed for the creation of the digital twin is stored somewhere within a company. Building complex digital twins representing smart cities, food supply chain systems, logistics, and energy networks requires interoperability and collaboration with many participants and competitors. The world is beginning to make progress in developing the data spaces, a concept that takes care of coordinating the technical, business, and legal aspects of data sharing.

The focus topics in the development of the Data Spaces are on the one hand the technical operations, such as the establishment of procedures, and the improvement of the user experience. On the other hand, the expansion of the community for the identification of innovative use cases and the development of business models is the second important building block for further development. Thirdly, internationalization is under consideration, so that initiatives are also being driven forward at global level (Grove, 2020).

BACKGROUND

How Can Emerging Technologies Enhance Data Spaces

Emerging technologies, such as artificial intelligence and machine learning, can enhance data spaces by allowing for more efficient and accurate analysis of large amounts of data. For example, using AI and machine learning algorithms, it is possible to quickly identify patterns and trends in data that may not be immediately apparent to humans. This can help organizations make more informed decisions and gain valuable insights from their data. Additionally, emerging technologies can also help to automate many of the tasks involved in managing and analyzing data, freeing up time and resources for other important tasks.

Blockchain technology can enhance data spaces in several ways. First, it can provide a secure and immutable platform for storing and managing data. Because blockchain networks are decentralized and use advanced cryptography to secure their transactions, they can offer a high level of security and protect against unauthorized access or tampering.

Second, blockchain technology can enable data sharing and collaboration in a secure and transparent manner. By using smart contracts and other blockchain-based tools, organizations can create data spaces that allow multiple parties to access and update data in a controlled and secure way, without the need for a central authority.

Third, blockchain technology can help to ensure the integrity and provenance of data in data spaces. By using cryptographic techniques and distributed ledger technology, blockchain networks can provide a tamper-evident record of data transactions, allowing organizations to track the history and provenance of data in a data space.

Overall, the use of blockchain technology in data spaces can help to improve security, transparency, and trust in data-driven applications and services.

How Data Spaces Impact Digital Twins and Smart Cities

A digital twin is a virtual representation of a physical object or system. It is a digital replica of a real-world entity, such as a building, a manufacturing process, or a supply chain, that can be used for various purposes, such as simulation, monitoring, and analysis. Digital twins are often used in the context of the Internet of Things (IoT), where physical objects are equipped with sensors that collect data, which is then used to create and update the digital twin in real time. This allows organizations to gain insight into the performance and behavior of their physical assets, and to optimize their operations and maintenance strategies.

A smart city is an urban area that uses various types of electronic data collection sensors to provide information and improve the management of urban services, such as transportation and energy. This information is used to manage a city's resources more efficiently and improve the quality of life for its residents. A smart city typically uses a combination of technologies, such as the Internet of Things (IoT), data analytics, and cloud computing, to collect and analyze data from a variety of sources, including traffic sensors, energy meters, and weather stations. The goal of a smart city is to create a more sustainable and livable urban environment by using technology to improve urban services and address challenges such as congestion, pollution, and resource scarcity (*What is a smart city*?, n.d.).

Therefore, digital twins for smart cities are those digital representations of physical systems, such as buildings, infrastructure, and other assets. There is so much knowledge that can be extracted from all these data collected in the creation of such digital twins and used for the development of Data Spaces.

The design of digital twins for smart cities is impacted by the data spaces in several ways. First, the data spaces provide the source of information for creating and maintaining the digital twin. This means that the quality and quantity of data available in the data spaces will directly impact the accuracy and completeness of the digital twin.

Second, the data spaces determine the type of data that can be used to create and maintain the digital twin. For example, if the data spaces only contain sensor data, the digital twin will only be able to represent the physical system in terms of the variables measured by the sensors.

Third, the data space determines how data from different sources can be integrated and used to create a coherent and comprehensive representation of the physical system. This requires the data spaces to support data fusion, which involves combining data from multiple sources and formats to create a single, consistent view of the system.

Overall, the data spaces are a critical component of the design of digital twins for smart cities, as it provides the information needed to create and maintain these digital representations of physical systems.

With the power of emerging technologies, data spaces can only be enhanced and further developed. Below are some examples of how we can get more value from data in a secure manner.

Cybersecurity and Data Spaces

Cybersecurity and data spaces are closely linked because data spaces are vulnerable to cyber attacks, and strong cybersecurity measures are essential for protecting the data stored in these spaces. As the data spaces are digital environments where data is collected, processed, and stored, it can be a target for cyber criminals who want to access or steal sensitive information. To protect data spaces from cyber-attacks, organizations need to implement robust cybersecurity measures, such as encryption, firewalls, and access controls. These measures help to prevent unauthorized access to data spaces and protect the data stored within them from being compromised. In turn, this helps to ensure the integrity and availability of the data, and to protect the organization and its users from the potentially damaging effects of a cyber-attack.

Quantum Computing and Data Spaces

Quantum computing has the potential to significantly enhance data spaces by providing new ways to process and analyze data. One of the key benefits of quantum computing is its ability to perform certain types of calculations much faster than classical computers. This could enable data analysts to quickly sift through large datasets and uncover patterns and insights that would otherwise be difficult or impossible to find. In addition, quantum computers can also be used to develop new algorithms and data analysis techniques that can help make data more useful and valuable. Overall, quantum computing has the potential to transform the way we work with data and open up new possibilities for data-driven research and decision-making.

MAIN FOCUS OF THE CHAPTER

How Could Global Smart Cities, Powered by Global Digital Twins Be Powered by Global Data Spaces?

Global smart cities powered by global digital twins could be powered by global data spaces in a number of ways. For example, global data spaces could be used to store and manage the vast amounts of data that are generated by the sensors, devices, and systems that are used in global smart cities. This data could then be used to create real-time, accurate digital representations of the physical world, known as digital twins. These digital twins could be used to simulate and analyze different scenarios, allowing cities to make more informed decisions about how to manage their resources and infrastructure. Additionally, global data spaces could be used to securely and efficiently share data between different smart cities, enabling them to learn from each other and improve their operations.

Our Human Common Path Towards Sustainable Data-Driven Innovations

We, humans, consciously and unconsciously produce data – with every click, every step, every pulse, that can be documented and recorded today. The sheer volume of data and the associated options for using it for most diverse purposes, is summarized in the buzzword "Big Data". Data has become the central commodity of our age and therefore has a unique value for economic compatibility in global contexts, as well as political dynamics and socio-ecological challenges. On one hand, data collections can be used to optimally control machines, processes, products, energy, and traffic flows in Industry 4.0. On the other hand, major risks concerning the security and safety of personal data come with the above-mentioned potentials.

The most promising approach is a functioning data services ecosystem for countries and continents. Data and the impactful handling of it are a core component for success and competitive advantage in many industrial sectors, value chains or other economic processes and thus a decisive factor for production, in addition to work and capital.

European Values and the Power of Networks

Data, as we all know, has no national boundaries – just as collaboration should not stop at borders to further advance data economy issues. Data service ecosystems are a crucial factor in shaping cities and countries digital transformation.

Europe is working on the development of a Federated and Secure Data Infrastructure to ensure European digital sovereignty called Gaia-X. The Gaia-X initiative has taken the legal form of AISBL, an international non-profit organization based in Belgium. The project, initiated by France and Germany, aims to create a next-generation data infrastructure proposal for Europe and promote digital sovereignty for European cloud service users (*What is Gaia-X?,* n.d.).

Gaia-X defines a trusted framework based on established standards and taking into account EU values such as transparency, openness, privacy and security. The trust framework solution supports privacy and self-accountability requirements and creates a chain of trust without the need for a global and traceable unique ID across the ecosystem (Zimmer, 2022).

An identity is defined in the Gaia-X architecture as a unique ID and a set of associated attributes that uniquely describe an entity in a given context. The lifetime of an ID is permanent; it can be used as a reference to an entity - beyond the lifetime of the entity it identifies. Reuse of an identity for another entity is prohibited.

A "secure digital identity" is a unique identity with additional data for robust, trustworthy authentication of the entity (i.e., with appropriate measures to prevent impersonation). This means that Gaia-X participants should assign and verify IDs for such identities themselves. It is the sole responsibility of the participants to determine the conditions under which the IDs are assigned.

In order to bring storage resources into the ecosystem, they must be actively integrated. This requires so-called interfaces or connectors (APIs - Application Programming Interfaces), which enable the technical connection of the (meta) data on the storage resources to an infrastructure (e.g., software solution, connectors, high-performance computing resources) in the ecosystem. In many cases, there are already well-established standards for interfaces so that systems can easily communicate with each other (interoperability).

The EU FAIR principles provide abstract and domain-independent guidelines for describing data and metadata. The abbreviation FAIR stands for Findability, Accessibility, Interoperability, and Reusability. Each of these terms consists of a category of concrete guidelines that can be used as rules for certification of data sets (*Fair principles*, 2022; Wilkinson et al., 2016).

In this framework, Gaia-X architecture allows organizations to securely share data and monitor compliance while maintaining complete control over their data. This is made possible by combining the GAIA-X infrastructure with the International Data Spaces (IDSA) standard. This format specifies exactly how to create and control a protected data space, who can access it, and how to ensure data security, especially regarding sensitive information (*International Data Spaces*, 2023).

Challenges like climate change, energy and mobility transition determine our present and shape our future. These challenges need to be mastered and their risks managed through allowing the development of such innovative grounds and usage of new tools like the data spaces. Data is part of the solutions we need for a sustainable future.

Data needs to be better and more standardized in its collection and exchange. Personal data must be managed more sovereignly and stored more securely. Machines Data needs to be transmitted more securely and shared in a way that adds value.

A key success factor is the high-caliber founding partners who are willing to actively drive forward data transformation ecosystems and its application. Another factor for success is the interoperability, which allows cooperations with other projects, such as Catena-X (Gras, 2022, Mobilithek (Digital Matters, 2022) and Caruso (*Mobility Data Space – ITS World Congress*, 2021). Data spaces should not be closed ecosystems and there should not be created data silos. The solution is to enable interoperability. IDSA is securing the federalization by following the 4 principles of Interoperability Governance: Legal Interoperability – ensuring that organizations under different legal frameworks can work together; Organizational Interoperability – let organizations align business processes, responsibilities, expectations, and goals; Semantic Interoperability – ensure that format and meaning of shared data is preserved and understood.

European Data Spaces for Business Value Creation and Sustainability

All European countries are actively supporting the such initiatives in the digitalization transformation ecosystem and are keen on staying informed and involved.

Austria can lead by example in connecting to the European infrastructure and staying upfront in the process of developing the Data Spaces and continuously driving these transformations. Austria's digital economy is an international pioneer in the intelligent use of data. It can handle large amounts of data for the benefit of all and form a new type of marketplace with data providers and data seekers while complying with data protection principles. The volume of data generated grows from year to year but is often not used for relevant decisions. Reasons for this include the heavy distribution and question segmentation of data, lack of resources, lack of interconnectivity of already established data infrastructures, or simply inexperience for this important topic.

Through organizations like Austria's Data Intelligence Initiative (DIO), national stakeholders receive support in transitioning to the data economy, developing Data Spaces and federating them regionally and internationally. DIO's key initiative, the Green Data Hub (GDH)[1], aims to create a sustainability-oriented European data service economy in line with the goals of the European Green Deal, the Paris climate goals as well as the UN SDGs. GDH is the platform for Austrian and European actors to build a data service ecosystem for Green and Sustainable Data to create a sustainable European data service economy that enables the implementation of the climate goals, the energy transition and the mobility transition. To this end, GDH's network of data-producing, data-processing and data-using stakeholders, focuses on the four domains of data spaces implementation: energy transition, mobility transition, circular economy and neutral climate cities (*Mission statement*, 2021).

Based on collaborative, sovereign, and secure data use in Data Spaces, the Green Data Hub develops data-driven use cases for contributing to climate goals. The experience of the Data Intelligence Initiative (DIO) suggests 3 steps to achieve this goal. Firstly, identification of real pain points; secondly, clarification of roles and responsibilities of each participant within a Data Space; and thirdly, clarification and implementation of values (*About us*, 2022).

Within the Data Spaces, specific use cases are identified through matchmaking of stakeholders and metadata of the individual actors. Use Cases are specific application projects within a Data Space. In a Use Case two or more stakeholders work together on a project where the aim is to increase value through data-driven collaboration. E.g., a telecommunication provider and a mobility provider share data about streams of people and available means of transport to improve services. Following this collaborative approach to data usage, sovereignty and security can be ensured. In the European Union, at the current stage only metadata-matching is possible in data spaces, the actual data stream/transfer is outside of the space, via the connectors.

Figure 1. Green Data Hub- Data Spaces and Use Cases

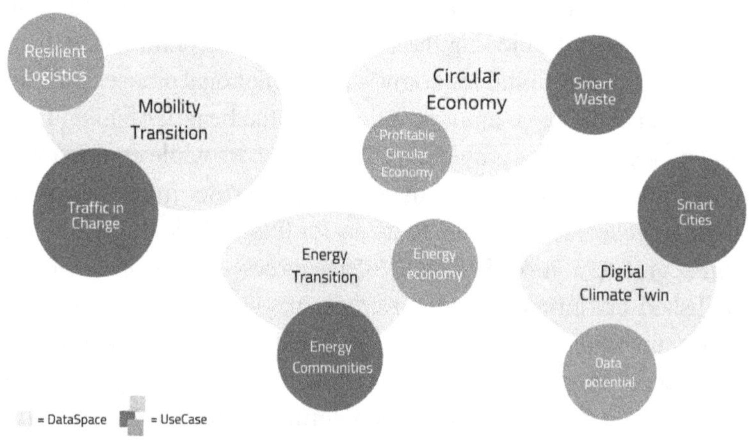

Figure 2. Green Data Hub-Use Case Diagram Example

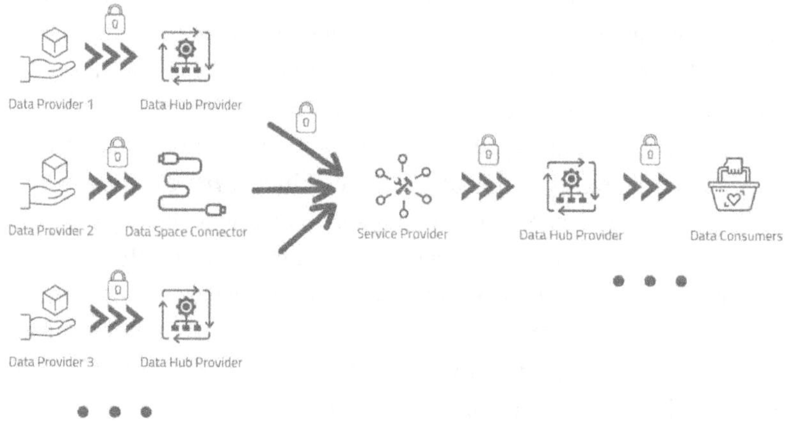

In addition to economic factors, the fulfilment of legal framework requirements at EU level (on security, data protection and trustworthiness), such as the "GDPR" or the "AI Act", plays an increasingly important role in the design of future-proof systems. As an official member of GAIA-X AISBL, DIO supports the GAIA-X project at national and international level to securing Europe's data sovereignty, to permanently building and maintaining its digital sovereignty. Together with the International Data Spaces Association (IDSA), DIO is collaborating on the growth of Data Spaces for sovereign and trustworthy data sharing, actively contributing to the principles of the Big Data Value Association (BDVA), who are setting the path in the digital transformation of economy and society on the European Level through Data and Artificial Intelligence by advancing in areas such as big data and AI technologies and services, data platforms and data spaces, Industrial AI, data-driven value creation, standardization, and skills. Together, all these national and international players are working toward cross-silo data networking in an open, sustainable, and secure digital ecosystem with common goals like increased efficiency, greater business agility and, ultimately, the unlocking of new values for governments, industry and civil society (*About EBDVF*, 2022).

Trust and Security by Design Through Data Spaces

The term data space refers to a data relationship between trusted partners who apply the same standards and rules for sharing their data. The decentral, federated technical architecture is of central importance. Data is not stored centrally in e.g. a Cloud – where stakeholders would lose the control over the use of their provided data – but remains at the source with the respective data owner and is only made available when required according to specified terms. This can be summarized as acting by choice.

'In practice, data exchange and sharing often happen via open or semi-open platforms, around which data and service ecosystems have evolved. Such platforms can be the intended side effect of main services such as search (e.g., Google), social networking (e.g., Facebook), entertainment (e.g., Netflix), or trade (e.g., eBay, Amazon, Otto), or much more specialized regional platforms such as the smart farming ecosystem around the farm equipment provider Claas in Germany' (Jarke, Otto, and Ram, 2019).

This dilemma is especially common in Europe, where there are many hidden champions and large market leaders. Data sovereignty refers to the self-determination of individuals and organizations regarding the use of their data. For example, in contrast to data protection as defined in the European General Data Protection Regulation (GDPR), data sovereignty plays a fairly passive role in protecting citizens from powers they cannot face on their equal footing. To do this, we need to obtain "data grounds" through clearly negotiated and closely monitored data usage agreements. IT security methods and tools play an important socio-technical support role in achieving data sovereignty.

Data Spaces are therefore associations of actors in superordinate domains (e.g., economic sector, industrial sector, etc.) with a decentralized (distributed) data infrastructure on which use cases are to be built. In a Data Space, data is made available for potential innovative services while maintaining data sovereignty, i.e., the greatest possible control and dominion over one's own data. The central point here is that different actors in a data service ecosystem access and use data in order to exploit the full innovation potential of data. Individual Data Spaces can also connect with other Data Spaces, such as the Mobility Data Space with the Tourism Data Space (keyword: interoperability). Due to the decentralized exchange of data in Data Spaces, the participants retain sovereignty and control over their own data.

The Data Spaces are becoming the new basis for a fruitful collaboration in the private sector as well as in the public sector.

In the private sector, key resource for the success is data. The volume of data generated grows from year to year, but is often not used for relevant decisions. Reasons for this include the high level of distribution and fragmentation of data, lack of resources, lack of interconnectivity of already established data infrastructures or simply inexperience or great respect for this important topic.

Data Spaces are pushing towards establishing a data service ecosystem in which a secure and sovereign cross-departmental, cross-organisational and/or cross-sectoral data exchange/usage takes place and the combination of relevant data along the data value chain is efficiently enabled.

Figure 3. Data Usage in Data Spaces

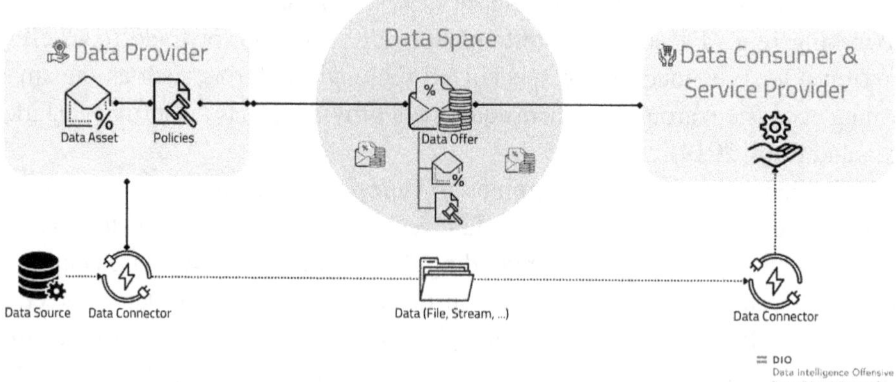

In a data service ecosystem, business models, products and services are developed and optimized through the networking and shared use of such decentralized data. The basis for this are the data spaces, as well as technical standards and basic legal definitions such as those developed by the Gaia-X Association, IDSA, BDVA, FIWARE etc. at the European level (data interfaces, governance, data sovereignty and security, and trust).

SOLUTIONS AND RECOMMENDATIONS

One can differentiate Data Spaces in all possible topics of activity: Data Space Energy transition, Mobility transition, Circular Economy, Neutral Climate Cities, Tourism, Media, Health, Government and so on. What makes a Data Space active, is the ultimate implementation of the Success Stories, the so-called Use Cases generation when a real data exchange is taking place and pain points within such a collaboration are solved or new business models are generated.

A Data Space demonstrates a simple and tightly controlled alternative for sharing data according to international standards: In a decentralized infrastructure, data is made available with the same standards and rules for sharing between trusted partners. Data is not stored centrally, but remains at the source and is only made available when needed and under certain rules. In a data service ecosystem, business models and services are developed and optimized through collaboration between data-producing, data-processing and data-using organizations. There are concrete use cases implemented by the Data Intelligence Initiative contributing to cross-company data-driven solutions. Talking about smart cities and optimization, for example within the DIO's Data Space Energy Transition, several use cases have been implemented for optimizing the energy efficiency of a district heating network.

Figure 4. Data Exchanges

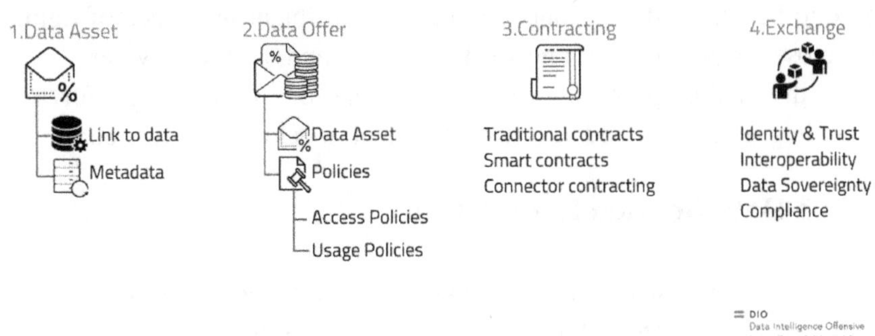

Use Case 1: Optimizing the Energy Efficiency of a District Heating Network

An alpine community covers up to 90 percent of its heating needs using district heating. Consumption is highly dependent on tourist frequency and capacity utilization. In the data service ecosystem, multiple data sources are sovereignly linked: Mobile frequency, energy consumption, calendar, weather and overnight stay data. This creates an information base that enables intelligent and optimized decision-making. This results in more efficient operation of the power plant, control of demand fluctuations, and a scalable solution for a more sustainable energy industry. In addition to increasing security of supply, a cost reduction for the energy company and a sustainable impact for the region is created.

Use Case 2: Thermal Earth Observation for Thermal Management

Heating and cooling of residential buildings and industrial processes causes significant greenhouse gas emissions. To detect unused temperature differences, various Earth observation data in the data service ecosystem are sovereignly linked and used for thermal monitoring and exploration. This enables monitoring and (predictive) maintenance of inefficient heat transitions, planning of investments for system stability, mapping of areas with available waste heat for increased cooling and heating demand, and for community policy planning of countermeasures against urban heat islands.

Another example of sustainable cooperation between public administration, companies and citizens is the energy community. Photovoltaic systems produce electricity on the roofs of public buildings, sports facilities, or private households, which can be made available to other actors if required or if there is a surplus. In order to enable an efficient operation of an energy community, a wide variety of data from different actors must be combined. These range from data on the degree of efficiency of a photovoltaic system, in order to enable the producers to plan with certainty, to weather data, in order to be able to predict when additional electricity will have to be supplied from outside due to insufficient incidence of light. This requires close cooperation with network operators and energy providers.

Below this author proposes a short summary of the learnings, for a better understanding of the new and yest fast approaching data sharing concepts.

What Is a Data Service Ecosystem?

In a data service ecosystem, business models, products and services are developed and optimized through the collaborative use of decentralized data.

The basis for this are so-called Data Spaces, as well as technical standards and basic legal definitions such as those developed by the Gaia-X association, IDSA, BDVA, FIWARE etc. at European level (data interfaces, governance, data sovereignty and security, trust).

What Is a Data Space?

A Data Space is an organizational and technical space where stakeholders and data meet and in which data is made available for joint use in use cases via a decentralized data infrastructure (federated architecture). There is no central data storage. The data stays where it originates and is exchanged as required (acting with choice). This forms the basis for data sovereignty. Data spaces usually focus on higher-level domains/business areas.

Who Meets in a Data Space?

Different stakeholders along the data value chain (data providers, data service providers, data consumers) meet in a Data Space. These are supported by pioneers from the fields of governance, standardization, law, business development and technology.

The aim is to jointly develop sustainable, value-added and technically feasible use cases that also comply with legal standards and governance obligations.

What Are the Technical Requirements for a Data Space?

International organizations, especially in Europe, are promoting the idea of a decentralized data service ecosystem. Leading initiatives such as IDSA, Gaia-X & GXFS, Ocean and the announced smart middleware SIMPL lead the way. The core idea is to create the basis and framework so that data can be shared with each other in a trusted way, even in decentralized systems. This requires solutions for finding the data (catalog / broker), for determining and defining the identities and access options (identity & access management), for traceability (logging) and for standardized data transfer (connectors such as the EDC) (*Introduction,* n.d.).

How Does Data Usage Work in a Data Space?

In data spaces, data is shared with one another via so-called decentralized data hubs and/or via data connectors. There are various providers on the market (nexyo, sovity, etc.) but also open source solutions that can be implemented (EDC etc.). Data Assets must be created in data hubs/connectors, and verifiable credentials and policies that regulate use and access must be maintained. The resulting Data Offers can be traded. The actual exchange of data takes place via the connectors peer-to-peer outside the data space only for the agreed purpose and based on the defined policies.

What Does a Data Exchange Include?

Data is considered as an asset of a company, which is why it is also referred to as so-called "Data Assets". Data Assets include a link to the data itself as well as metadata. Via "Policies", each "Data Asset" can be offered individually under one's own conditions by creating a "Data Offer". In public, private or restricted data spaces, as decentralized digital spaces, it is then possible for participants to see the "data offers" of other participants, to negotiate and to obtain them while preserving the underlying values.

Best Practices From Austria's Data Intelligence Initiative (DIO)

Smart Cities: Data Space

With many legal requirements for cities, Smart Cities should be created to facilitate the lives of citizens', affecting all areas of social interaction: from transport and healthcare to housing and security, environmental protection and infrastructure.

Output: The ultimate aim is to establish a data infrastructure or a data space, which can serve as a flexible basis to allow different smart city projects to emerge.

Digital Climate Twin: Data Space

A digital twin links multiple models of computable characteristics of an object and enables the simulation of scenarios at the same time. This can be used, for example, to demonstrate the impact of new technologies on local climate, air quality, and health.

Developing a digital twin (or digital shadow) requires access to multiple data sources, efficient data management, and data cleaning. It also requires resources for computationally intensive modeling as well as evaluation processes and (online) feedback on the simulated entity. The ability to share data in a decentralized manner while maintaining data sovereignty is a critical factor in this regard.

Output: By networking various actors along the data value chain and providing a data infrastructure, the creation of digital twins with sustainable impact should be enabled.

Energy Transition: Data Space

The Data Space Energy transition deals with all aspects around the transition from the non-sustainable use of fossil fuels and nuclear energy to a sustainable energy supply by means of renewable energies, this being an essential step to protect the environment. There are numerous activities in the field of research and development to develop efficient solar cells or to use wind energy efficiently. Besides the efficiency being optimized, the periodicity of renewable energies is another challenge. Furthermore, digitization plays a key role in the challenges of decentralization, flexibilization, efficient use of energy and stabilization of the networks. The success of digitization in the energy transition depends on the added value of the available data.

Output: The goal is to develop Use Cases that bring both economic and socio-ecological added value by networking a wide variety of actors along the data value chain and providing a data infrastructure.

Circular Economy: Data Space

The circular economy is a model of production and consumption in which existing materials and products are shared, leased, reused, repaired, refurbished and recycled for as long as possible. For this, an intra-company, but above all also an inter-company data exchange, which is based on principles such as data sovereignty and decentralization, is essential.

The Circular Economy Data Space links relevant actors along the value chain of various materials as well as waste management companies and research companies. Possible application scenarios include smart waste, optimized waste management, and recycling (e.g., increasing the reusability of materials in the textile industry, construction industry, semiconductors, and metal industry).

In order to achieve the greatest possible output from the meetings, with an appropriate number of stakeholders' for this purpose, the collaboration will be broken down according to different key topics (plastic packaging, textile industry, construction industry, etc.).

Output: The goal is to develop use cases that advance the circular economy by networking various stakeholders along the data value chain and providing a data infrastructure.

Tourism: Data Space

The vision of this Data Space is to rethink the guest experience in Austrian tourism to enable new business models, better services and seamless experiences. Sustainable design of the tourism landscape is essential for its long-term existence. Leveraging data enables tourism to scale with sustainability in mind. Sharing data within the tourism sector not only adds value, but also enables the management of one of the key issues of the 21st century - climate change. Pooling data from different organizations, such as transport associations, lift operators, booking platforms, catering companies, weather data providers, etc., enables the implementation of concrete use cases, such as visitor flow management, prevention of over tourism or hotspot analysis.

Output: Creation of neutral access to tourism-relevant data, applications and visualizations in order to drive innovations in digitalization faster and to offer third parties the opportunity to develop competitive business models along the value chain.

Mobility Transition: Data Space

Transport and mobility are to be converted to sustainable energy sources, soft mobility use and an interconnection of various forms of individual transport and local public transport. The mobility transition also includes a cultural change as well as a redistribution of public space. In addition, enormous added value can be created in the area of logistics through the networking of data.

The Data Space Mobility transition brings together the key players in the mobility sector. Enabling sovereign data exchange will enable a sustainable mobility transformation. Improved development of e-charging infrastructure, increased resilience to unexpected events in logistics, traffic management in cities, optimization and monitoring of existing infrastructure or demand-driven redistribution of public space are possible use cases.

It is important to integrate and learn from complex data in order to anticipate mobility needs when moving between locations. This is important in order to promote the most suitable environmentally friendly options where and when they are needed as alternative to private car use. The citizens therefore need to change their mobility behavior through appropriate incentives, which are developed in the Use Case ideation between stakeholders. The optimal mobility options can raise through innovative AI models backed by a knowledge graph that can predict future mobility choices based on different contexts and mobility path modelling.

Output: The goal is to develop use cases in the mobility sector that bring both economic and socio-ecological added value by networking a wide range of actors along the data value chain and providing a data infrastructure.

There are several lines of approaching the sustainable shift in mobility behavior while maintaining alignment with data sharing principles.

Citizens need to receive support and incentivize mobility behavior towards choosing more sustainable options, thus reducing carbon emissions from the use of private cars and petrol-/diesel-based means of transportation. Actionable datasets on

Figure 5. Mobility Transition in Data Spaces

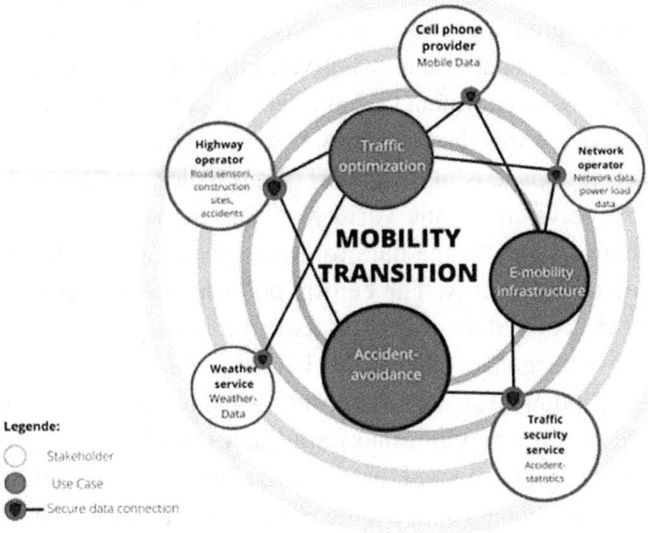

mobility choices and options are currently fragmented across data silos in different organizational networks. The sharing and merging of these datasets via a common data ecosystem and its processing by Intelligent Systems - allowing for data sovereignty, security and privacy - supports the training of AI models to explain how and why citizens make certain mobility choices, and to predict their future choices based on multidimensional context parameters such as the weather, the location and duration of upcoming events or the availability of environmentally friendly options. Customized incentives leveraging AI predictions aim to motivate citizens to adopt those new options and overcome remaining barriers to more sustainable behavior such as the need to sign up for a new service or the perceived convenience of travelling "as we have always done". To achieve this, there are needed AI-based approaches to predict complex mobility behavior and optimize incentives in a multidimensional manner, beyond currently available solutions.

There are several challenges (i) semantically integrating heterogeneous data from multiple sources into a dynamic mobility data ecosystem; (ii) understanding the evolving data ecosystem by means of a shared mobility knowledge graph; (iii) graph-based AI algorithms to learn from user mobility behavior and make predictions of future behavior and propose suitable incentives, and (iv) modelling different user mobility choices based on various incentive models in order to promote the most sustainable mobility behavior – that can be addressed to predict mobility choices.

There are needed explainable and understandable solutions so that national administrations can make informed decisions to promote and support more sustainable behavior in the future, thoroughly testing the results to verify and improve the approach.

Ultimate outcome is to enable and incentivize citizens to find more sustainable mobility choices, increasing awareness and affecting public opinion to develop a more positive attitude towards those choices.

The Mobility Data Space is just one part of a broader data ecosystem which supports interplay between a wide variety of actors from the data economy. The goal is to share and exchange data to drive innovation, under the highest preservation of data sovereignty. The central point here is that different actors in a data ecosystem access data to exploit its full innovation potential for their activities. To achieve the goal of sustainable behavior, it is imperative to ask what motivates every individual to change their established patterns. Domain-specific data spaces can also connect with other data spaces, such as the mobility data space with the tourism data space. Establishing a level playing field among all participants within such a data space is necessary to counteract a monopolistic structure of the data ecosystem. A level playing field for data exchange, data trading, and data use can be established when a data ecosystem is dominated against a background of collaborative cooperation rather than competition. This can be achieved through safe design and thorough maintenance of the federated infrastructure underlying data spaces. A sound data management approach is essential to the design, creation, and maintenance of ecosystems related to data-driven development. As a critical part of data governance, the roles and responsibilities of all stakeholders should be defined. These include companies, individuals and governments that act as data users or data providers, as well as their technology partners and IT experts. Data sovereignty refers to the greatest possible control over one's own data. The preservation of data sovereignty is a necessary condition for the competitive and secure development of a European data infrastructure - among others - within the framework of Gaia-X 27. The establishment of equivalent competitive conditions for all participants in data spaces is necessary to counteract potential monopolistic structures of a data ecosystem. Thus, new entrants do not face insurmountable obstacles when applying for admission to a data space. The transaction of data depends critically on accurate and error-free authentication, authorization, and guarantee of service levels. This concerns, for example, the security for data providers that users cannot circumvent their policies, as well as technically innovative solutions that allow data processes without direct access to original data sets. Access to data is critical for AI which should be trained to discover new insights and predict future states, such as for sustainable mobility.

FUTURE RESEARCH DIRECTIONS

It is difficult to predict the future of any field with certainty, especially when it comes to research and development. However, it is likely that data spaces will continue to be an important area of study and that advances in this field will lead to a better understanding of how to organize, manage, and analyze large amounts of data. This, in turn, could lead to a range of potential applications, such as improving decision-making, enhancing machine learning algorithms, and enabling more efficient data management. It is also possible that new technologies and techniques will be developed to facilitate the creation and use of data spaces, such as new ways to visualize and interact with data or more advanced methods for organizing and storing data. Ultimately, the future of data spaces research and development will depend on the continued efforts of researchers and developers in the field, as well as the direction of technological progress in general.

CONCLUSION

No matter what we do today, we consciously and unconsciously produce data - every click, every step, every pulse beat can be documented and recorded. The solution is a secure and established data market or a functioning data services ecosystem for sustainable social stability and prosperity. Environmental disasters and the consequences of climate change are shaping our present and dictate the imperative need to work together to counter these developments. Data is part of the solution we need for a sustainable future. All the communities shall actively participate in data sharing, take part in the Data Spaces with use cases that generate new business models for successful data strategies of your business.

Establishing a secure data market and a functioning data ecosystem requires careful planning and consideration of several key factors. Here are some steps you can take to create a secure data market and ecosystem:

- Identify the key stakeholders in the data market and ecosystem, such as data producers, data consumers, and data intermediaries, and understand their needs and motivations.
- Develop a clear and transparent governance framework that outlines the rules and regulations for the data market and ecosystem, including data privacy and security policies.
- Create a secure and scalable infrastructure for the data market and ecosystem, such as a blockchain platform or a distributed ledger technology, to enable secure and efficient data sharing and exchange.

- Implement robust security measures to protect the data and prevent unauthorized access or misuse, such as encryption and access controls.
- Foster a collaborative and open environment where stakeholders can freely participate and contribute to the data market and ecosystem, and incentivize their participation with rewards and recognition.
- Engage with the community and stakeholders to continually monitor and improve the data market and ecosystem, and address any challenges or issues that may arise.

Creating a secure data market and ecosystem requires a combination of technical, legal, and social measures, and the successful implementation of these measures will depend on the specific needs and circumstances of every market and ecosystem.

REFERENCES

About EBDVF. (2022, November 3). European Big Data Value Forum. Retrieved from https://european-big-data-value-forum.eu/2022-edition/about/

About Us. (2022, December 14). Green Data Hub. Retrieved from https://www.greendatahub.at/about-us/?lang=en

Digital Matters. (2022, January 7.). *Mobilithek - Germany's data plattform that gets you moving*. Federal Ministry for Digital and Transport. Retrieved from https://bmdv.bund.de/SharedDocs/EN/Articles/DG/mobilithek.html

Fair principles. (2022, January 21). GO FAIR. Retrieved from https://www.go-fair.org/fair-principles/

Gauss Centre for Supercomputing. (2020, May 7). *Using Digital Twins to design more sustainable cities*. Retrieved January 26, 2023, from https://phys.org/news/2020-05-digital-twins-sustainable-cities.html

Gras, N. (2022, September 8). *Catena-X: Building trust to ensure resilience*. International Data Spaces. https://internationaldataspaces.org/catena-x-building-trust-to-ensure-resilience/

Grove, J. (2020, October 15). *Second issue of data SPACES_NOW available*. International Data Spaces. https://internationaldataspaces.org/second-issue-of-data-spaces_now-available/

International Data Spaces. (2023, January 9). Gaia-X. Retrieved from https://internationaldataspaces.org/we/gaia-x/

Introduction. (n.d.). Catena-X. Retrieved from https://catena-x.net/en/angebote/edc-die-zentrale-komponente-fuer-die

Jarke, M., Otto, B., & Ram, S. (2019, August 19). *Data sovereignty and Data Space Ecosystems - Business & Information Systems Engineering.* SpringerLink. https://link.springer.com/article/10.1007/s12599-019-00614-2

Mission statement. (2021, July 6). DIO. Retrieved from https://www.dataintelligence.at/en/mission-statement/

Mobility Data Space - ITS World Congress. (2021, October 13). CARUSO. Retrieved from https://www.caruso-dataplace.com/mobility-data-space-its-world-congress/

Prinz, W., Rose, T., & Urbach, N. (1970, January 1). *Blockchain technology and International Data Spaces.* SpringerLink. https://link.springer.com/chapter/10.1007/978-3-030-93975-5_10

What is a smart city? Definition and Examples. (n.d.). TWI-Global. Retrieved from https://www.twi-global.com/technical-knowledge/faqs/what-is-a-smart-city

What is Gaia-X? (n.d.). Gaia-X. Retrieved from https://www.data-infrastructure.eu/GAIAX/Navigation/EN/Home/home.html

Wilkinson, M. D., Dumontier, M., Aalbersberg, I. J., Appleton, G., Axton, M., Baak, A., Blomberg, N., Boiten, J.-W., da Silva Santos, L. B., Bourne, P. E., Bouwman, J., Brookes, A. J., Clark, T., Crosas, M., Dillo, I., Dumon, O., Edmunds, S., Evelo, C. T., Finkers, R., ... Mons, B. (2016). The FAIR Guiding Principles for scientific data management and stewardship. *Scientific Data, 3*(1), 1–9. doi:10.1038data.2016.18 PMID:26978244

Zimmer, J. (2022, May 9). *Data spaces provide a new way to share information.* https://www.engineering.com/story/data-spaces-provide-a-new-way-to-share-information

ADDITIONAL READING

Shim, J. (2005). Roadmap for e-commerce standardization in Korea. *International Journal of IT Standards and Standardization Research*, *3*(2), 1–14. doi:10.4018/jitsr.2005070101

Section 2
Industry Perspectives

Chapter 4
Impact of Digital Twins on Smart Cities:
Healthtech and Fintech Perspectives - opportunities, Challenges, and Future Directions

Ingrid Vasiliu-Feltes
University of Miami, USA

ABSTRACT

There are several industries that have shown significant advancements in using digital twin capabilities. Healthcare, life sciences, and fintech constitute a powerful trifecta that can profoundly impact society at a global scale. The design and deployment of digital twins is complex and intimately connected to other digital technologies. However, in order to be successful, we will also need to overcome some of the challenges faced with any emerging technologies and mitigate against the risks of maleficent use. Specifically, ethics, cybersecurity concerns, and cross-industry standards will be on the agenda of most enterprises wishing to deploy digital twins. By leveraging a combined deployment of digital twins with other emerging technologies, we could potentially build a highly performant global digital twin ecosystem. Futurists would even venture to envision the existence of state-of-the-art smart cities where digital healthcare twins, digital life sciences twins, digital finance twins would be part of our daily lives.

DOI: 10.4018/978-1-6684-3833-6.ch004

INTRODUCTION

Digital twin technology is foundational to smart city deployments as they can serve a dual role. One role can be to create digital twins of smart cities to assist with the design, implementation and maintenance of the complex smart city strategic roadmap, as they are a powerful tool that can expedite the lifecycle of the smart city deployment. The other important role is their ability to offer key strategic intelligence for the adequate deployment of all other emerging technologies utilized in creating a fully operational smart city architecture and infrastructure. While there are numerous definitions of smart cities, one of the common denominators for all of them are:: leveraging technology and improving the quality of life for citizens. To be future-ready, we must strive to build inclusive, diverse, and equitable smart cities and digital twin technology can be a powerful catalyst in measuring our ESG impact. Given the global workforce trends predicted to occur by 2030, we must also strive to design and build smart communities for smart global citizens that will be connected by a smart global business ecosystem. This global ecosystem was valued at USD 1,226.9 billion in 2022 and is expected to register a compound annual growth rate (CAGR) of 25.8% from 2023 to 2030 in a recent report published by Grandview Research.

According to IBM, a digital twin is a virtual representation of an object or system that spans its lifecycle, is updated from real-time data, and uses simulation, machine learning, and reasoning to help decision-making. Their business impact is remarkable as the global digital twin market size was valued at USD 11.12 billion in 2022 and is projected to exhibit a compound annual growth rate (CAGR) of 37.5% from 2023 to 2030 as highlighted in a recent report by Market Research.

We are currently at the cusp of the 5th industrial revolution. Several emerging and frontier technologies are driving this new era and are foundational in building the smart urban metropolis of the future. Digital twin technology is one of the major technologies that can have a profound impact and act as a catalyst in designing safe, inclusive, diverse, and sustainable smart cities.

Other essential technologies include IoT, IIoT, 5G and 6G, satellite internet, biometrics, edge computing, cloud computing, spatial computing, distributed ledger technologies, quantum, brain-computer interfaces, AR, VR, XR, and the complex artificial intelligence portfolio. All these converging technologies can be embedded in complex smart technology architecture and can be enhanced by leveraging the data intelligence derived from digital twin technology either for safety, efficiency, research or to attain a higher degree of human-centricity. The literature reviews conducted and published by several authors (Filipe et al., 2021; Pintoa et al., 2022) provide an ample illustration of the combined power of distributed ledger technologies, the internet of things, and digital twin technologies. This review also showcases the people-centric attributes of DLT-IOT architectures.

For both healthcare and fintech, the powerful hybrid of digital ledger technologies and digital twin technology can offer increased operational efficiency, reduced overhead costs, reduced insurance premiums, enhanced auditing capabilities, optimized cyber-defense systems, as well as reduced fraud, waste, and abuse. When digital twin technology are coupled with federated learning capabilities there is an exponential benefit due to the ability to provide strategic intelligence for multiple stakeholders while maintaining data encryption.

Figure 1. Digital Twin Technology for Smart Cities -Domains
https://eartheon.digital/home/

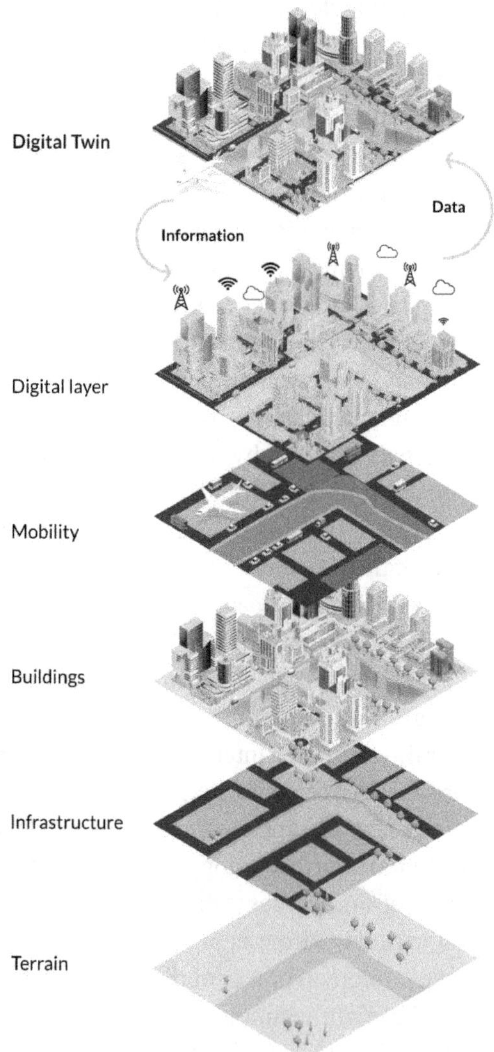

Figure 2.Digital Twin Technology for Smart Cities-Layers, Science Direct

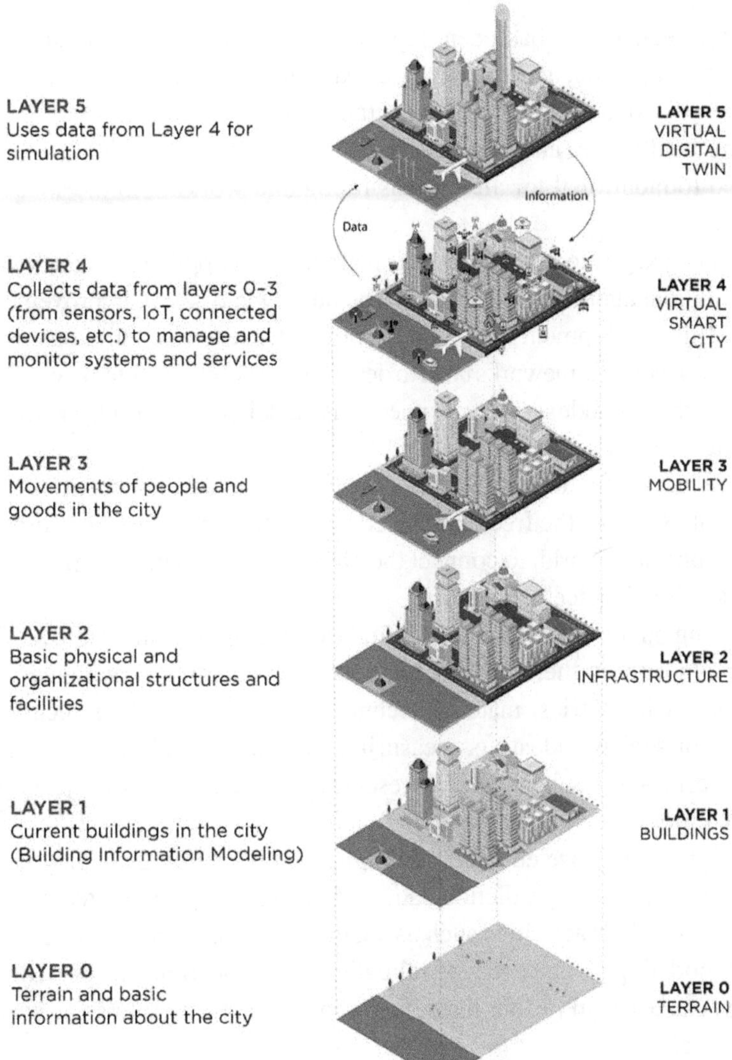

Furthermore, future-ready business leaders that wish to remain competitive or those that are aiming to gain a competitive edge have already begun leveraging digital twin technology technologies to accelerate their transition to the next iteration of the world wide web called web 3.0. Ledger technologies and digital twin technologies are poised to play a key role in providing the necessary strategic business intelligence required to develop the next generation of web 3.0-powered smart cities.

BACKGROUND

Digital twin technology has been deployed across numerous industries, such as automobile, aerospace, agriculture, construction, maritime, telecommunications, etc. While all industries play an important role in our society and the business ecosystem, healthcare, and finances deserve a close examination as they can be considered foundational for all citizens living and working in the smart cities of the future.

They play a key role in the production arena, with applications for logistics and supply chain management, as well as in the quality and safety improvement efforts across the full development life cycle for products, services, buildings, or cities.

There are several frameworks used to describe the complex digital twin technology portfolio and they include services, connections models, data,, and physical technology architecture.

They include tools for digital twin services, modeling, connections or for data management. Some of the frequently used connection tools are those used only for sensing the physical world, to connect the physical world with the digital, and those designed as digital space connectors.

Depending on the use case where digital twin technology are deployed we might encounter algorithms the, the internet, interfaces, artificial intelligence models, simulations, hydrometrics, material technology, etc. Digital twin technology can be utilized for sensing, ideating, measuring, controlling, refining, or dynamically adjusting various virtual models representing physical assets (objects, services, entities, or cities).

Perhaps the most coveted feature in digital twin technology deployments is the capacity for simulations, prediction,, and verification. This allows teams to perform a variety of modeling activities such as shape, behavior, performance, attribute, or constraint modeling. When used for validation or verification purposes, digital twin technology can perform various tasks, including correctness, tolerance, availability,, and result analysis.

Technology teams deploying digital twin technology for data management have a full arsenal at their fingertips such as collection, storage,, or fusion technologies.

The strategy for success in deploying digital twin technologies for any industry or use case includes a robust feasibility analysis and versatility in the tool selection process, as well as an in-depth understanding of the pros and cons of each digital twin tool. Furthermore, the deployment of digital twin technologies also demands the design and tracking of customized metrics and key performance indicators. Deploying digital twin tools also requires a commitment to an ongoing quality improvement and quality assurance process.

Additionally, adequate training and talent retention are equally crucial in attaining a sustainable digital twin deployment. Many pilots fail to scale due to a lack of a a skilled workforce that can adequately monitor and optimize existing applications.

Healthcare Technology Background (Healthtech)

Healthcare has traditionally lagged in deploying frontier technologies. The deployment of digital twin technology can break that pattern and revolutionize healthcare due to the impact on the full spectrum of healthcare domains, clinical care settings and specialties (James, 2021; Kobyakova et al., 2021; Laubenbacher et al., 2022). One of the major advantages is the deployment of digital twin technology to enhance safety in health care delivery, and medical and surgical devices, as well as the potential to accelerate the research and development lifecycle (Yuefei, 2021). Designing and building smart health cities powered by digital twins will likely become an expectation within the next few years.

Digital twin technology can have a significant impact on smart healthcare delivery by providing healthcare providers with a virtual representation of a patient, which can be used for a variety of purposes such as personalized treatment planning, remote monitoring, and virtual consultations. One of the key benefits of digital twin technology in healthcare is the ability to simulate and model a patient's condition, allowing healthcare providers to test different treatment options and predict potential outcomes before administering care. This can lead to more personalized and effective treatment plans, as well as improved patient outcomes. Another benefit of digital twin technology in healthcare is the ability to remotely monitor a patient's condition and vital signs. This can be particularly useful for patients with chronic conditions, such as heart disease or diabetes, who may require frequent monitoring but may not want or be able to come into the hospital, clinic or even critical care settings (Chase et al., 2021). Digital twin technology can also be used for virtual consultations, which can be especially beneficial for patients in rural or remote areas who may not have easy access to healthcare providers. This can be done by linking the digital twin with the patient's medical history, lab test results, and vital signs, allowing the healthcare provider to make an accurate diagnosis and treatment plan. Overall, the use of digital twin technology in healthcare can lead to improved patient outcomes, greater efficiency, and cost savings in healthcare delivery. Digital twin technology can be deployed for preventive and surveillance stages, the diagnosis and treatment of diseases. Additionally, they can be deployed to maintain wellness and better understand longevity. Several use cases have been described for leveraging digital twin technology in precision medicine, precision health, public health, mobile health, surgery, radiology, oncology (Pedersen, 2021), cardiology (Xing et al., 2022), etc. Researchers and clinicians around the world are continuously exploring novel applications and optimization of the currently deployed digital twin-powered solutions. Digital twin technology has also been deployed for the business of healthcare, as it can provide valuable strategic intelligence that can enhance operational efficiency and improve ROI. Examples include value-based care, revenue cycle management, actuarial and insurance services.

There are several review articles published over the past few years that eloquently highlight the benefits of digital twin technology in the business of healthcare. In a great article published in the Journal of Digital Economy (hi et al., 2022) the authors highlight several key models where digital twin technology deployments can be highly beneficial, such as the three C model, the six C model, and the four C model. These include context, configuration, constructs, coordination, capabilities, and change. They also call our attention to the holistic model of a business ecosystem with its four subsystems. Digital twin technology can be helpful in each of those subsystems in order to achieve optimal impact. In another publication by (Allam et al., 2020) the authors explore how 6G technologies will impact a digital future. Specifically, they highlight the role of digital twin technology in urban economies. They also emphasize the need for alignment with United Nations Sustainable Development Goals when deploying emerging technologies, as well as some of the unique 6 G-related challenges such as potency, interoperability, and lacking infrastructure in certain markets. Another Springer publication (Torky et al., 2022) shares the concerns about privacy and security in digital twin applications. The author (Cao, 2022) offers a rationale for blockchain technology as a potentially viable solution across numerous domains within the healthcare ecosystem, ranging from logistics, supply chain, utilities and care delivery. There are several published articles that illustrate the broad applications of digital twin technology in various medical disciplines (Boata et al., 2021), as well as for smart health devices, smart hospitals, and smart cities (Yang et al., 2021). Other articles also outline how digital twin technology could further amplify and augment various scientific advances in medicine (Wickramasinghe et al., 2021; Lauer-Schmaltz et al., 2022).

Other authors (Volkov et al., 2021) provide a detailed explanation of the important and impactful application of digital twin technology in mobile smart health (Yu, 2022), while a publication in the Journal of IEEE Internet Computing (Ricci et al in 2022) highlights the role of digital twin technology in shaping the broader vision for digital health. One of the published articles in the Journal of Engineering and Technology is in alignment with this author's view that digital will revolutionize healthcare (Benson, 2021). Several articles point out the benefit of digital twin technology in precision oncology (Wu, 2022), precision public health, and precision health (Kamel, 2021). Nutrition and the development of smart personalized healthcare (Meraghni et al., 2021; Hernandez-Boussard et al., 2021 ; Pedersen et al., 2021 ; Dhanapal et al., 2022). Other subject matter experts (Ferdousi et al., 2022) have taken a different preventive approach and emphasized how digital twin technology can be deployed for wellness.

Figure 3. Digital Twin for Healthcare-Future Directions. Source: IEEE, Roberto Saracco

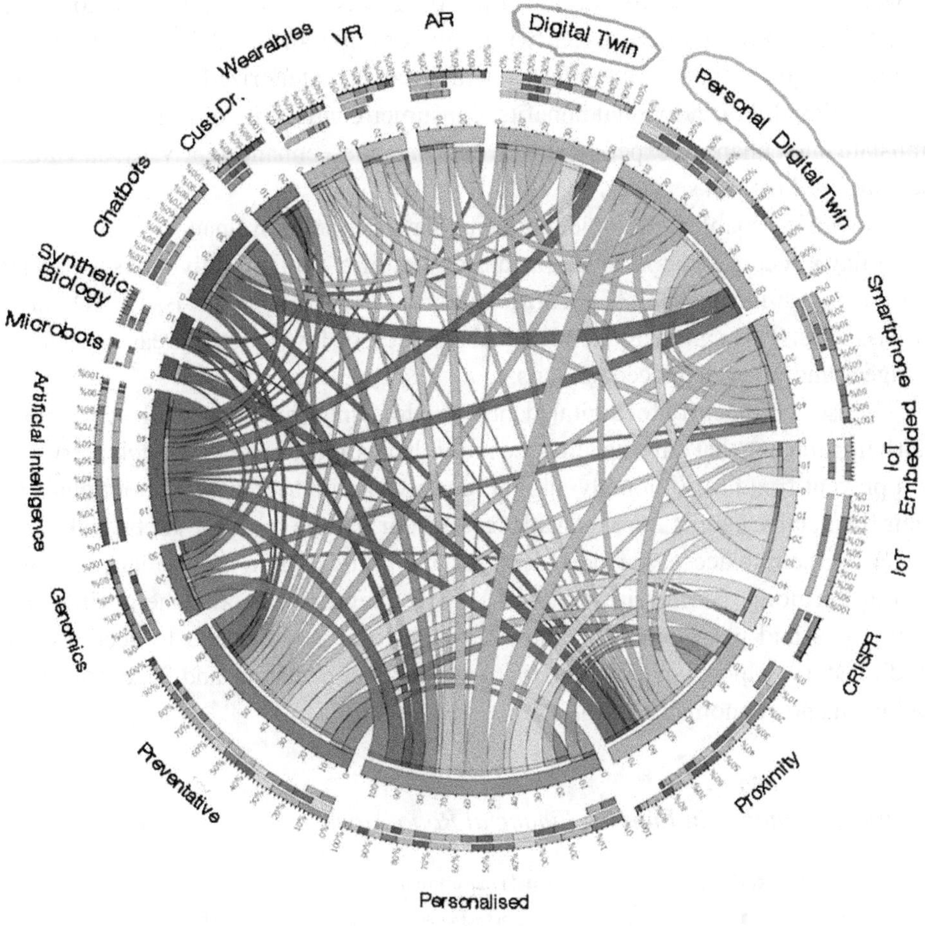

Financial Technology Background (Fintech)

Digital twin technology has the potential to impact all domains within the fintech industry, including mobile payments, neo banking, insurance, investments, capital raising, market provisioning, etc. There are broader impacts that are similar to other major industries, such as ability to perform a state of the art digital twin powered strengths, weaknesses and threats analysis, shorter development time for new applications or services, improve productivity, lower maintenance costs, and less business risk due to enhanced predictive strategic intelligence gathered. Additionally, financial organizations can deploy digital twin technology to perform cybersecurity vulnerability testing and optimize their cyber-resilience efforts. Digital twin technology can also enhance the quality improvement and quality assurance efforts, as well as augment proactive digital ethics programs via an auditing functionality.

Furthermore, customized digital twin technology offers a few additional advantages that are unique to the fintech industry, which can be categorized as individual and organizational. For individual benefits, consumer-centric digital fintech twins can translate into enhanced experience, less data breaches, enhanced privacy, and ultra-personalized fintech services.

Biometrics-enabled digital identity can play a key role in financial inclusion by providing a secure and convenient means of identification for individuals who may not have traditional forms of identification, such as government-issued ID cards or passports. This can include individuals in rural or low-income areas, as well as refugees and other displaced persons.

The use of biometrics for digital identity, such as fingerprints or facial recognition, can help to ensure that only the intended individual is able to access financial services and prevent fraud. Additionally, digital identity can be used to link individuals to their financial accounts, such as bank accounts or mobile money wallets, making it easier for them to access financial services and participate in the formal economy.

Furthermore, Biometrics-enabled digital identity can also help financial institutions and government to know their customers better and to access their credit history, which would help them to make better decisions and foster financial inclusion and economic growth.

Figure 4. Digital Twin Enabled Financial Robo-Advisor

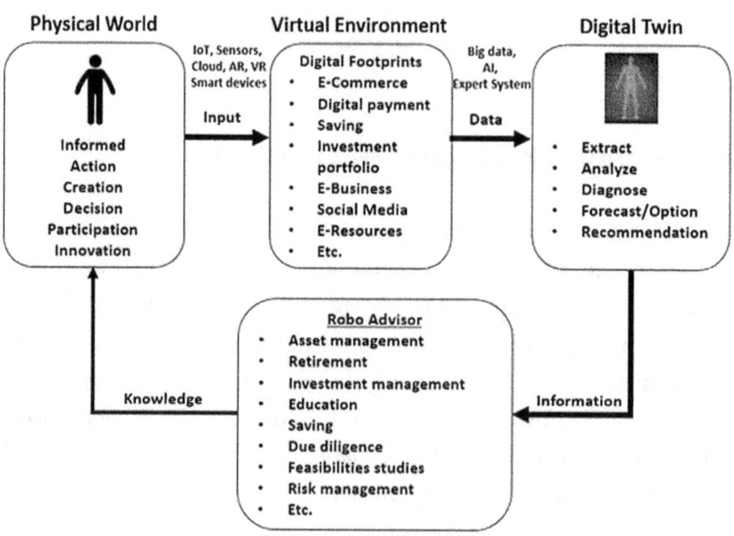

Overall, the use of biometrics-enabled digital identity can play an important role in increasing access to financial services for under-served populations, which in turn can contribute to overall economic development and financial inclusion.

Perhaps the most well-known application of digital twin technologies in the fintech industry are digital twin-powered robo-advisors for smart citizens living in smart cities. A recent review article (Anshari et al., 2022) emphasizes the role of Digital twin technology in fintech and considered it the next frontier as it will allow the financial advisory services to evolve from ad-hoc to a dynamic continuous model.

Financial organizations deploying digital twin technology can deploy digital twin technologies across all their service domains and throughout the life cycle of their products or services. They can significantly enhance the organizations financial returns either by realizing cost savings during new product development and rollout phases, or by generating additional revenue due to maximized operational efficiency, as well as the ability to roll out numerous products simultaneously. By using digital twin technology, they can also run their data centers more efficiently, achieve higher data management fidelity and save costs by having less overall downtime. Additionally, fintech organizations can engage in an ongoing infrastructure quality improvement process powered by digital twin technology, which allows them to improve efficiency and consumer satisfaction without any service disruptions to their operations. One additional major advantage is digital safety and ability manage cyber-risk more efficiently. The ability to incorporate digital twin technology to run

Figure 5. Digital Twin for Financial Business Operations

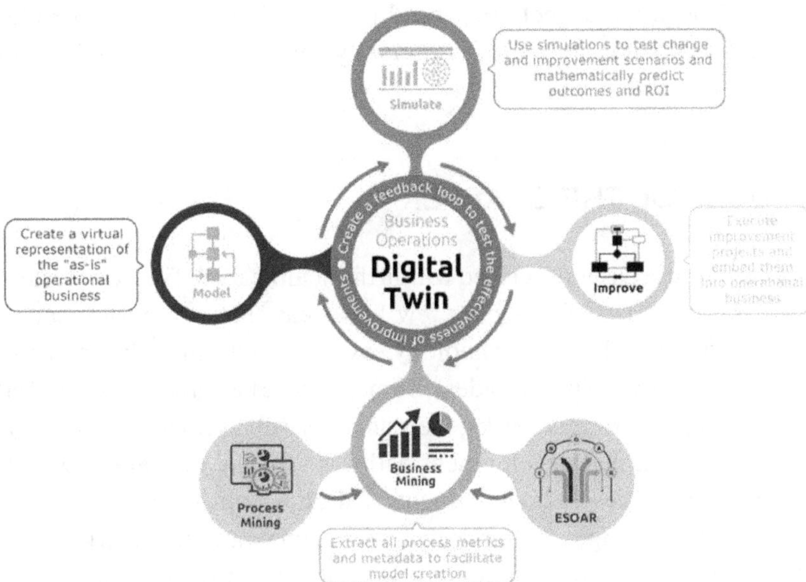

simulations that assist in developing robust cybersecurity defense systems will be paramount in protecting our digital identities and an ongoing process that can be fully automated.

Federated analytics processes have the potential of revolutionizing fintech and bridge some of the strategic intelligence gaps caused by an ongoing digital divide in certain regions around the world. In a recent article (Wang et al., 2022) outline the challenges and opportunities related to this new computing paradigm. Their article aims to be helpful resource for key stakeholders designing digital twin technology deployments for smart cities.

Digital twin technology can also be embraced to assist financial organizations in developing comprehensive solutions that facilitate attainment of the United Nations Sustainable Development Goals and meet the expectation to design, deploy and track ESG metrics. Experts also forecast that digital twin technologies will be leveraged in building the Fintech Metaverse for global digital citizens that will expect their unique Financial Identities in the Metaverse.

In summary, there are several pros and cons that need to be considered by business leaders when integrating digital twin technology in the broader enterprise strategy:

Some of the advantages are: improved risk management, increased efficiency and optimized customer experience. Given that digital twin technology can help financial institutions better understand and manage risks by providing real-time data and insights. They can facilitate task automation, reducing the need for manual input and increasing the speed of decision-making. They can be instrumental in optimizing customer experience and offer personalized and population demographic-tailored services (Verma, 2022). However, there are also potential negative aspects that need to be mitigated. These include high cost, as well as a high degree of complexity that requires special workforce technical skills and expertise.

MAIN FOCUS OF THE CHAPTER

Regardless of the use cases described within the healthcare or financial technology industries, we must first focus on a few key areas that can ensure success in deployments of digital twin technology, as well as ensure digital trust and increased adoption by business leaders. Some of these key areas are digital ethics, cyber-resilience, recalibrating risk governance, as well as revising our education, training and professional development pathways for the digital era workforce. To accomplish a successful deployment of digital twin technology for smart cities we must also change our mindset, embrace innovation, embark on a global collaborative digital transformation journey, adapt to the need for continuous quality improvement, and dedicate sufficient human resources. Furthermore, novel financing instruments and business models will be required to sustain large-scale deployments that can make an impact. Novel financing instruments will have to include public-private partnerships, as well as creative "digital twin technology" bonds similar to those customized for the green or blue economies. Business leaders will also have to disrupt their traditional business models to accommodate, as highlighted by several authors (Allam et al., 2021) emerging technologies such as digital twin technology, IoT and 6G will have a profound socio-economic impact on smart cities, and influence accelerate the attainment of United Nations Sustainable Development Goal 11. Similarly, a Springer article (Xu et al., 2022) showcases the important role digital twin technology in building a sustainable industry and powering the circular economy. The article by (Swarna et al., 2021) calls attention to the need to engage in ongoing quality improvement for smart city deployments and emphasizes the combined role of federated learning and digital twin technology to accomplish this essential task. Their article showcases several applications of federated learning-enabled digital twin technology and the positive secondary impact data governance and optimization privacy efforts (Gupta, 2021; Ramu, 2022). Other authors (Shia et al., 2022) eloquently point out the role of digital twin technology in the global digital economy. They argue that the industrialization of digital technologies can be foundational to novel business models, newly configured metaverse-like consumer experiences and give rise to a new digital industry characterized by a fusion between digitalization and industrialization.

Figure 6. Different levels of industrial systems and the shared core – value chain (Science Direct, 2021)

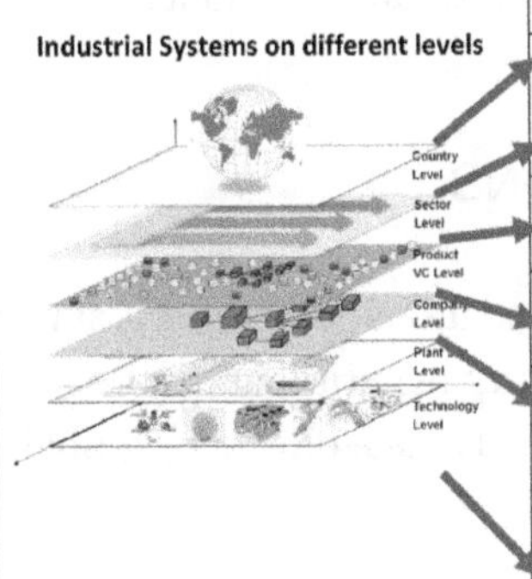

Cluster, Region, and Global Level	A geographic area with single or multiple industries and interactions
Industry / Sector Level	A similar product group with similar business processes
Product Supply Chain Level	An end-to-end product full life cycle related productive unit
Firm / Company Level	A product oriented business unit with comprehensive functions
Factory / Plant / Site Level	A product oriented productive unit defined by a group of buildings
Core Process Technology Level	A single productive unit defined by a core process technology.

Issues, Controversies, Problems

Several authors and global experts underscore the important role of digital twin technologies and cyber physical systems in the development of industry 5.0. For example (Taj et al., 2022) explore the role of data governance, data privacy and data security in building digital twins-powered smart cities of the future. Their article also shares the gaps that must be addressed to accomplish the transition from mass production to mass personalization. The digital twin technology and smart cities industries are both plagued by an inefficient approach with numerous disjointed pilots that cause lack of scalability, redundancies and waste in valuable resources. This is partly due to risk-adverse budgeting and financing, as well as trying to apply traditional tactics to modern technology deployments. Having the appropriate dynamic budgeting, as well as alternative financing mechanisms would be optimal and allow larger scale deployments that could truly change the paradigm and lead to enhanced quality of life for citizens.

Furthermore, we must call for global stakeholder collaboration to develop regulatory guidelines and frameworks that can accelerate digital twin-powered smart cities deployments. Lagging international standards, poor interoperability and portability, lack of adequate quality and safety metrics, deceased awareness about the need for proactive digital ethics and cyber-resilience programs are causing hesitation or resistance to large scale adoption among many business leaders. Given the advanced skills required for the design and deployment of digital twin-powered smart cities we must also devote adequate resources to educate the global workforce to be able to sustain large-scale deployments.

Last but not least, it is essential to address the financial and generational divides we must overcome of we wish to achieve exponential growth of smart cities worldwide.

When we examine the impact of these challenges and problems, we realize that not addressing them in a proactive way is particularly costly to our society. A few solutions have been considered by global thought leaders that involve ethics, cybersecurity, education, workforce management, financing and leadership.

SOLUTIONS AND RECOMMENDATIONS

While there are numerous solutions and recommendations possible, this author would like to emphasize the far-reaching and exponential benefit that can be attained by considering these 10 solutions that can be designed and deployed by global experts:

1. Proactive Digital Ethics Programs
2. Proactive Cyber Resilience Programs

3. Smart Cities Interoperability and Portability
4. Digital Twin Standards (International & Industry-specific)
5. Customized Digital Twin Powered Smart Cities KPIs
6. Education
7. Talent Development
8. Innovative Business Models
9. Alternative Financing Model
10. Leadership

Why do we need proactive digital ethics programs? We are currently witnessing a marked gap between the scale and pace of emerging technology adoption and deployment of resources to design and implement effective ethics programs that uphold moral principles (Braun, 2021; Popa, 2021). Few companies have any digital ethics programs in place and those few that do focus mostly on regulatory compliance with AI acts or set a low ethics bar fror themselves due to financial or operational pressures. Few digital era leaders are versatile in applied ethics and dedicated to building a proactive, sustainable, and impactful digital ethics program. Therefore, we continue to experience a high frequency of data privacy breaches, and upon retrospective analysis lack of fairness or even violations of human rights due to inappropriate deployment and monitoring of emerging technologies. While digital twin technology can be highly beneficial in our quest to deploy technology for good causes, if the data is used with malicious intent, it can cause severe negative individual and social consequences.

Why do we need proactive cyber-resilient programs? While governments and companies direct massive resources toward traditional cybersecurity measures and programs, there are several factors that would be required to building proactive, cyber-resilient enterprises: increased awareness about the need for zero-trust cyber security, customized workforce training, educating the workforce of the future and a leadership mindset change to accept the heightened cyber-risks of the digital era. Digital twin technology plays a dual role on the cybersecurity stage: on the one hand they demand heightened cyber-defense systems, while on the other they can actually be deployed to become cyber-agents. Digital twin deployments can facilitate the development of zero-trust cybersecurity environments in smart cities. This strategic approach to cyber defense adheres to the motto "never trust and always verify". It can protect smart cities from operational breakdowns related to cyberattacks and combined with digital twin tools can also expedite the forensic analysis post-cyberbreaches and prevent repeat cyber-attacks.

Why are interoperability and portability key in deploying digital twin technology for smart cities? Both are valuable lessons learned from past failed emerging technology deployments where business leaders failed to appreciate the need to think in advance about how to bridge physical and virtual boundaries. By focusing on interoperability and portability early on, governments and businesses can save costs and increase the pace of adoption, while avoiding digital divides.

Similarly, we must insist on developing digital twins and smart city standards that are internationally recognized and enforced, as well as key performance indicators to inform our continuous quality improvement efforts.

Educating young generations in the skills required for digital twin technology-powered smart cities, as well as upskilling and reskilling the current workforce will be crucial for the long-term sustainability and financial viability of smart cities. Redesigning the education and certification ecosystem and building a robust talent development pipeline are akin to moral imperatives if we wish to ensure future generation's long-term success in living and working in digital twin technology-powered smart cities.

Designing alternative global financing models will be essential as traditional finance instruments and budgeting mechanisms are not conducive to building a state-of-the-art, digital twin technology-powered global smart cities ecosystem. These alternative financing models must support a long-term strategy and be aligned with the 2030 United Nations Sustainable Development Agenda. Additionally, they must support emerging markets and developed markets in order to attain the economy of scale required for optimal impact.

Similarly, novel business models will need to be designed and validated in this fast paced, high-risk, high volatility economic environment, some of the innovative models that could be suitable for digital twin deployments in smart cities are freemium, e-commerce, marketplace, digital ecosystem, access over ownership, and even creative open-source models. For example, freemium and open-source models can increase adoption rates, allow customization, quality improvement and bridge the ongoing global financial divide. For technology vendors, the e-commerce and marketplace models can also be highly successful.

Digital era leaders will have to embrace innovation, recalibrate their risk-tolerance and develop digital ambidexterity to manage their business and digital citizen's expectations. Leaders that adapt their risk tolerance, embed innovation in the existing enterprise strategy, and deploy a complex portfolio of management styles are more likely to succeed in smart cities digital twin deployments.

FUTURE RESEARCH DIRECTIONS

Future research could be dedicated to exploring the impact of advanced computing techniques on amplifying digital twin technology applications (Yuefei, 2021). Quantum, edge-, cloud-, neuromorphic, and DNA-computing are all poised for use in both health tech and fintech. Given the transition to web 3.0 architecture, the higher degree of interoperability and converging technologies that will be possible over the next decade. Not only will they offer additional value in research and development, but also in safety, security, operational efficiency, customer experience, and ongoing quality improvement.

We are only beginning to grasp the magnitude of opportunities that are offered by the dual deployment of quantum computing and digital twin technologies for health tech, fintech and all other industries that would be represented in the smart urban metropolis of the future. The strategic intelligence derived from combining the advanced superior computing power offered by quantum computers and the predictive capabilities derived from digital twin deployments are poised to perhaps trigger the 6th industrial revolution.

Although we are currently preparing our transition to web 3.0, there are already forward-thinking companies that are laying the foundation for web 5.0. Future research will have to take into consideration lessons learned from the current business and digital transformation journeys experienced during the transition from web1.0 to web 2.0 or deployment of other frontier technologies.

A deeper dive into how the Metaverse and Omniverse ecosystems could accelerate or enrich our current health tech and fintech consumer engagement efforts would also be valuable given the massive investments in these domains by governments, venture capital and private equity firms.

Further studies will be required to better understand how digital twin technology and living in smart urban ecosystems will impact our specific health and finance digital identities, as well as the extent of the influence they can have on global diversity, equity and inclusion efforts. Optimized digital identity combined with powerful digital twin technology could change the paradigm in human resource management, as well as board room and C-suite governance. Future diversity, equity and inclusion officers or even ethics officers could perhaps be avatars powered by digital twin technology and truly disrupt existing workflows and traditional organizational structures.

Additional research efforts could also identify how to leverage human-brain interfaces safely and equitably to improve the outcomes or experiences in health tech and fintech. Digital twins could be creatively deployed to dynamically manage the newly created virtual human-computer workflows inherent to smart cities and escalate to appropriate human oversight as needed.

Digital twin technology, when combined with human-computer interfaces (HCIs), can create a powerful tool for human-in-the-loop workflows. Human-in-the-loop workflows refer to the integration of human decision-making into automated processes, allowing for a more collaborative and efficient approach to tasks.

Digital twin technology can be used to create virtual representations of physical systems or processes, such as manufacturing systems or power plants. These virtual representations can then be linked to human-computer interfaces, such as control panels or virtual reality interfaces, allowing human operators to interact with and control the physical systems in real-time. One of the key benefits of digital twin technology with human-computer interfaces is the ability to simulate and test different scenarios and control options before implementing them in the physical system. This can lead to improved decision-making, increased efficiency, and reduced downtime. Another benefit is that digital twin technology can be used to provide real-time monitoring and visualization of the physical system, allowing human operators to quickly identify and respond to any issues that may arise.

Furthermore, digital twin technology can also be used to improve training and education, allowing operators to practice and become proficient in operating complex systems in a safe and controlled virtual environment before working on the physical system.

Overall, the use of digital twin technology with human-computer interfaces can greatly enhance human-in-the-loop workflows by providing a more collaborative, efficient, and safe approach to controlling and monitoring physical systems.

Futurists would also like to see research dedicated to integration of digital twin technology with bioimplants, bionic limbs, and /or exoskeletons (healthtech) and with robo-advisors (fintech)

CONCLUSION

A cross-disciplinary and harmonized approach is not only recommended but imperative for the success of smart city digital twin technologies across all industries. However, health tech and fintech are great examples of the widespread negative consequences that can occur if we do not secure global, multi-stakeholder collaboration. They are both highly regulated industries, with broad societal and economic impact. State of the art successful digital twin deployments for these two industry sectors can serve as great blueprints for other industries and enhance digital trust. In the near future, we could envision smart communities powered by digital twin technology, where smart cities are interconnected and smart citizens have their unique cyber digital twin-secured digital identity, their unique digital health twin, and their unique fintech twin that optimize their quality of life.

Furthermore, these two industries can have an augmenting and amplifying effect on other industries. By facilitating optimization of public health and financial inclusion, we can expect a butterfly effect for all other domains included in the United Nations Sustainable Development Agenda for 2030. For instance, in addition to the direct correlation and impact on SDG #1,3,9, and 11 digital twin technology deployment could also generate key strategic intelligence to accelerate efforts to reduce poverty, reduce hunger, access to decent work, quality education and attain a net zero economy. Specifically, digital twin technology can help cities better understand and manage their infrastructure and resources, leading to more efficient and sustainable use (which are aligned with SDG #11. Building resilient infrastructure, promoting inclusive and sustainable industrialization, and fostering innovation, as well as more efficient and sustainable design, construction, and operation can contribute to SDG# 9. They can also have a major impact on SDG #7 by providing key strategic intelligence to improve availability of affordable and clean energy. Additionally, the technology can be used to optimize the performance of renewable energy systems, helping to increase the use of clean energy in cities. Digital twin technology can help cities better understand and manage their carbon footprint and make more informed decisions about how to reduce it and attain the net zero 2050 pledge. Overall, digital twin technology can help cities to become more efficient, sustainable, and resilient.

In conclusion, digital twin technology can facilitate building smart, inclusive, diverse cities for future generations.

REFERENCES

Allam, Z., & Jones, D. S. (2021). Future (post-COVID) digital, smart and sustainable cities in the wake of 6G: Digital twin technology, immersive realities, and new urban economies. *Land Use Policy*, *101*, 105201. doi:10.1016/j.landusepol.2020.105201

Anshari, M., Almunawar, M. N., & Masri, M. (2022). Digital Twin: Financial Technology's Next Frontier of Robo-Advisor. *Journal of Risk and Financial Management*, *15*(4), 163. doi:10.3390/jrfm15040163

Benson, M. (2021). Digital twin technology will revolutionize healthcare. *Engineering & Technology*, *16*(2), 50–53. doi:10.1049/et.2021.0210

Boată, A., Angelescu, R., & Dobrescu, R. (2021). Using digital twin technology in health care. *UPB Scientific Bulletin, Series C: Electrical Engineering and Computer Science*, 53–62.

Braun, M. (2021). Represent me: Please! towards an ethics of digital twin technology in medicine. *Journal of Medical Ethics*, *47*(6), 394–400. doi:10.1136/medethics-2020-106134

Cao, L. (2022). Decentralized ai: Edge intelligence and smart blockchain, metaverse, web3, and desci. *IEEE Intelligent Systems*, *37*(3), 6–19. doi:10.1109/MIS.2022.3181504

Chase, J. G., Zhou, C., Knopp, J. L., Shaw, G. M., Näswall, K., Wong, J. H. K., Malinen, S., Moeller, K., Benyo, B., Chiew, Y. S., & Desaive, T. (2021). Digital twin technology in critical care: What, when, how, where, why? *IFAC-PapersOnLine*, *54*(15), 310–315. doi:10.1016/j.ifacol.2021.10.274

Dhanapal, A. C. T. A., Subapriya, S., Subramaniam, K., & Appukutty, M. (2022). Virtual Digital twin technology: A Renaissance in Precision Nutrition Approach. In *Integrating AI in IoT Analytics on the Cloud for Healthcare Applications* (pp. 1–23). IGI Global. doi:10.4018/978-1-7998-9132-1.ch001

Digital Twin Market Size, Share, Growth & Trends Report, 2030. (n.d.). Retrieved January 10, 2023, from https://www.grandviewresearch.com/industry-analysis/digital-twin-market#:~:text=Report%20Overview

Ferdousi, R., Laamarti, F., Hossain, M. A., Yang, C., & El Saddik, A. (2022). Digital twin technology for well-being: An overview. *Digital Twin*, *1*(7), 7. doi:10.12688/digitaltwin.17475.2

Gupta, D., Kayode, O., Bhatt, S., Gupta, M., & Tosun, A. S. (2021). Hierarchical federated learning-based anomaly detection using digital twin technology for smart healthcare. *2021 IEEE 7th International Conference on Collaboration and Internet Computing (CIC)*, 16–25.

Hernandez-Boussard, T., Macklin, P., Greenspan, E. J., Gryshuk, A. L., Stahlberg, E., Syeda-Mahmood, T., & Shmulevich, I. (2021). Digital twin technology for predictive oncology will be a paradigm shift for precision cancer care. *Nature Medicine*, *27*(12), 2065–2066. doi:10.103841591-021-01558-5 PMID:34824458

James, L. (2021). Digital twin technology will revolutionize healthcare: Digital twin technology has the potential to transform healthcare in a variety of ways—improving the diagnosis and treatment of patients, streamlining preventative care, and facilitating new approaches for hospital planning. *Engineering & Technology*, *16*(2), 50–53.

Kamel Boulos, M. N., & Zhang, P. (2021). Digital twin technology: From personalized medicine to precision public health. *Journal of Personalized Medicine*, *11*(8), 745. doi:10.3390/jpm11080745 PMID:34442389

Kobyakova, O. S., Starodubov, V. I., Kurakova, N. G., & Tsvetkova, L. A. (2021). Digital twin technology in Healthcare: An Assessment of Technological and Practical Prospects. *Annals of the Russian Academy of Medical Sciences*, *76*(5), 476–487. doi:10.15690/vramn1717

Laubenbacher, R., Niarakis, A., Helikar, T., An, G., Shapiro, B., Malik-Sheriff, R. S., Sego, T. J., Knapp, A., Macklin, P., & Glazier, J. A. (2022). Building digital twin technology of the human immune system: Toward a roadmap. *NPJ Digital Medicine*, *5*(1), 1–5. doi:10.103841746-022-00610-z PMID:35013539

Lauer-Schmaltz, M. W., Cash, P., Hansen, J. P., & Maier, A. (2022). Designing human digital twin technology for behavior-changing therapy and rehabilitation: A systematic review. *Proceedings of the Design Society*, *2*, 1303–1312. doi:10.1017/pds.2022.132

Meraghni, S., Benaggoune, K., Masry, Z. A., Terrissa, L. S., Devalland, C., & Zerhouni, N. (2021). Towards digital twin technology driven breast cancer detection. In *Intelligent Computing* (pp. 87–99). Springer. doi:10.1007/978-3-030-80129-8_7

Pedersen, O. L. (2021). *A proof-of-concept Digital Twin implementation for monitoring patients through the Clinical Pathway for Prostate Cancer in the Norwegian Health and Care Service*. NTNU.

Pinto, F., da Silva, C. F., & Moro, S. (2022). People-Centered Distributed Ledger Technology-IoT Architectures: A Systematic Literature Review. *Telematics and Informatics*, *70*, 101812. doi:10.1016/j.tele.2022.101812

Popa, E. O., van Hilten, M., Oosterkamp, E., & Bogaardt, M.-J. (2021). The use of digital twin technology in healthcare: Socio-ethical benefits and socio-ethical risks. *Life Sciences, Society and Policy*, *17*(1), 1–25. doi:10.118640504-021-00113-x PMID:33397487

Ramu, S. P., Boopalan, P., Pham, Q.-V., Maddikunta, P. K. R., Huynh-The, T., Alazab, M., Nguyen, T. T., & Gadekallu, T. R. (2022). Federated learning enabled digital twin technology for smart cities: Concepts, recent advances, and future directions. *Sustainable Cities and Society*, *79*, 103663. doi:10.1016/j.scs.2021.103663

Ricci, A., Croatti, A., Mariani, S., Montagna, S., & Picone, M. (2022). Web of digital twin technology. *ACM Transactions on Internet Technology*.

Ricci, A., Croatti, A., & Montagna, S. (2021). Pervasive and Connected Digital twin technology—A Vision for Digital Health. *IEEE Internet Computing*.

Shi, Y., Gao, Y., Luo, Y., & Hu, J. (2022). Fusions of industrialization and digitalisation (FID) in the digital economy: Industrial system digitalization, digital technology industrialization, and beyond. *Journal of Digital Economy*.

Smart Cities Market Size, Share & Growth Report, 2030. (n.d.). Retrieved January 10, 2023, from https://grandviewresearch.com/industry-analysis/smart-cities-market#:~:text=The%20global%20smart%20cities%20market

Taj, I., & Zaman, N. (2022). Towards Industrial Revolution 5.0 and Explainable Artificial Intelligence: Challenges and Opportunities. *International Journal of Computing and Digital Systems*, *12*(1), 295–320. doi:10.12785/ijcds/120128

Torky, M., Darwish, A., & Hassanien, A. E. (2022). Integrated Digital twin technology and Blockchain Framework for Privacy and Security of Applications. In Digital twin technology for Digital Transformation: Innovation in Industry (pp. 99–112). Springer.

Verma, R., Patel, S., Minj, S. V., & Bhat, A. (2022). Healthcare 5.0: A Study on Improving Personalized Care. *2022 6th International Conference on Intelligent Computing and Control Systems (ICICCS)*, 1815–1818.

Volkov, I., Radchenko, G., & Tchernykh, A. (2021). Digital twin technology, Internet of Things and Mobile Medicine: A Review of Current Platforms to Support Smart Healthcare. *Programming and Computer Software*, *47*(8), 578–590. doi:10.1134/S0361768821080284

Wickramasinghe, N., Jayaraman, P. P., Forkan, A. R. M., Ulapane, N., Kaul, R., Vaughan, S., & Zelcer, J. (2021). A Vision for Leveraging the Concept of Digital twin technology to Support the Provision of Personalized Cancer Care. *IEEE Internet Computing*, *26*(5), 17–24. doi:10.1109/MIC.2021.3065381

Wu, C., Lorenzo, G., Hormuth, D. A. II, Lima, E. A., Slavkova, K. P., DiCarlo, J. C., Virostko, J., Phillips, C. M., Patt, D., Chung, C., & Yankeelov, T. E. (2022). Integrating mechanism-based modeling with biomedical imaging to build practical digital twin technology for clinical oncology. *Biophysical Reviews*, *3*(2), 021304. doi:10.1063/5.0086789 PMID:35602761

Xing, X., Del Ser, J., Wu, Y., Li, Y., Xia, J., Xu, L., . . . Yang, G. (2022). *HDL: Hybrid Deep Learning for the Synthesis of Myocardial Velocity Maps in Digital twin technology for Cardiac Analysis.* ArXiv Preprint ArXiv:2203.05564.

Xu, L., de Vrieze, P., Lu, X., & Wang, W. (2022). Digital twin technology Approach for Sustainable Industry. *International Conference on Advanced Information Systems Engineering*, 126–134.

Yang, D., Karimi, H. R., Kaynak, O., & Yin, S. (2021). Developments of digital twin technologies in industrial, smart city and healthcare sectors: A survey. *Complex Engineering Systems*, *1*(1), 3. doi:10.20517/ces.2021.06

Yu, Z., Wang, K., Wan, Z., Xie, S., & Lv, Z. (2022). *FMCPNN in Digital twin technology Smart Healthcare*. IEEE Consumer Electronics Magazine.

Yuefei, C., Sisi, W., Mingqi, T., & Chuan, C. (2021). Application of Digital twin technology in Medical and Health Fields and Related Research Progress. 计量科学与技术, *65*(10), 6–9.

ADDITIONAL READING

Allen, A., Siefkas, A., Pellegrini, E., Burdick, H., Barnes, G., Calvert, J., Mao, Q., & Das, R. (2021). A digital twin technology machine learning model for forecasting disease progression in stroke patients. *Applied Sciences (Basel, Switzerland)*, *11*(12), 5576. doi:10.3390/app11125576

Baranov, L. I., Tikhonova, O. A., Tsarev, A. N., Tverdynin, N. M., & Grishakina, N. I. (2022). Experience of Creating Digital twin technology as a Result of Monitoring the Health of Industrial Personnel. *International Scientific and Practical Conference Strategy of Development of Regional Ecosystems "Education-Science-Industry"(ISPCR 2021)*, 15–21.

Bisht, D., Singh, R., Gehlot, A., Akram, S. V., Singh, A., Montero, E. C., Priyadarshi, N., & Twala, B. (2022). Imperative Role of Integrating Digitalization in the Firms Finance: A Technological Perspective. *Electronics (Basel)*, *11*(19), 3252. doi:10.3390/electronics11193252

Digital Twin Technologies and Smart Cities. (n.d.). SpringerLink.

Erol, T., Mendi, A. F., & Doğan, D. (2020, October 1). *The Digital Twin Revolution in Healthcare*. IEEE Xplore. doi:10.1109/ISMSIT50672.2020.9255249

Khan, A., Milne-Ives, M., Meinert, E., Iyawa, G. E., Jones, R. B., & Josephraj, A. N. (2022). A scoping review of digital twin technology in the context of the Covid-19 pandemic. *Biomedical Engineering and Computational Biology*, *13*. doi:10.1177/11795972221102115

Luo, D., Thevenin, S., & Dolgui, A. (2022). A state-of-the-art on production planning in Industry 4.0. *International Journal of Production Research*, 1–31. doi:10.1080/00207543.2022.2122622

KEY TERMS AND DEFINITIONS

Digital Twin Technology: A virtual representation of a real-world object, process, entity or a city.

Distributed Ledger Technology: Technology infrastructure and protocols that are consensually shared, allowing simultaneous access and validation of transactions.

Federated Learning: One of the artificial intelligence methods that allows computing without decryption of data.

Human-Centricity: Focusing on self-sovereignty, self- determination, self-governance, autonomy, or agency.

Metaverse: A shared virtual environment that aims to replicate the physical world and offers individuals novel experiences related to agency, social presence and share spatial awareness.

Omniverse: Collective set of metaverses.

Personalized Medicine: Emerging practice of medicine that uses an individual's genetic data or other unique biomarkers to customize their treatment plan.

Precision Health: Medical discipline that leverages genetic, behavioral, and environmental data to generate customized treatment approaches.

Robo-Advisors: Online applications that provide automated advice and pre-determined services.

Smart City: A technology-enabled urban area that collects and processes data with the intent to improve the efficiency and quality of life for its citizens.

Chapter 5
Thinking Otherwise:
Integrating Existing Buildings in Smart Cities – Best Practice

Bianca Christina Weber-Lewerenz
https://orcid.org/0000-0002-8406-7119
RWTH Aachen University, Germany

ABSTRACT

Smart cities include the intelligent networking and smart operation of existing ones in accordance with the latest energy standards, SDGs, climate strategies, and common good. The building sector causes almost 40% of global CO_2 emissions and is therefore the key to central climate neutrality by 2050. Innovative building technologies in existing buildings enable intelligent ecosystems within the 5th industrial revolution. Experience from the practical application provides signposts for the use of the latest digital technologies for the cities of tomorrow, with more efficiency, comfort, and safety for everyone who lives and works in them.

INTRODUCTION

Smart Cities not only involve new construction of high-tech infrastructure and buildings but upgrading existing buildings to intelligent building automatization systems as per Smart Cities standards, SDGs, Climate and Ecological Strategies. The building sector causes almost 40 percent of global CO_2 - Emissions and is therefore key to achieving climate neutrality by 2050. Innovative building technologies in existing buildings support independent monitoring, control, analysis and optimization

DOI: 10.4018/978-1-6684-3833-6.ch005

of energy consumption. These technologies enable smart ecosystems ready for the 5th industrial revolution.

Experiences shared from user practice show how new intelligent technology such as digital twins, Artificial Intelligence (AI), Artificial Intelligence of Things (AIoT), Cloud Computing, Intelligent Building Operation and Networking increase efficiency, safety and sustainability.

They mean powerful tools for integrating existing buildings and upgrading to smart cities standards; they ease smart communication between new and existing buildings, new and existing infrastructure and facilities by digital planning, constructing and the overall holistically safe and economical operation. Digital twins create technical possibilities for significantly more efficient life cycles, for reducing operating costs, minimizing natural resources as to reach sustainability goals. Practical Use Cases refer to expert interviews from a larger research project (Weber-Lewerenz, 2021a).

Conventional cities highly consume non-renewable resources of energy, management for garbage and control pollution is poor. To gain economic and environmental sustainability as the cities grow, humans have to deal with diverse technical, social, societal, economic and organizational pressures. Mammoth tasks lie ahead in the construction sector. Hightech-aligned Smart Cities raise multiobjective problems.

Smart Design, Smart Construction and Smart Operation (BMWI, 2021) put focus on the Construction Branch. Smart city development bases not only on new construction but on how to integrate existing buildings, green buildings, electrical vehicles, roads, bridges and the overall infrastructure as to ensure smart, intelligent operation. The search for integrative options is part of smart cities modelling including Monument Preservation (Figure 1.-3.) and old buildings protection e.g.

Figure 1. Digital recording of existing Monastery Benediktbeuren, Germany.
Source: https://demonstratorenmanagement.gfe-net.de/demonstratoren/kloster-benediktbeuern-als-digitaler-gebaeudebestand-und-ki-demonstrator

Figure 2. Digital recording of Monument Preservation, Germany.
Source: www.metrik3d.com

New Harbour Village in the old town of Hamburg, The New Humboldt Forum in Berlin. Green energy technologies, green buildings are examples on how to upgrade existing buildings to modern standards of smart cities (Winkowska, 2019): urban flexibility, circular economy, and clean energy utilization.

This chapter presents up-to-date information and describes novel insights about the use of computational intelligence techniques applied to integrate existing buildings in smart cities. They consider urban requirements and services, societal and ethical responsibilities that belong to urban governance, and sustainability goals.

This chapter will surely be useful not only for the scientific and academic community but for the industry professionals, designers and planners, policy makers as the aspects comprises innovative and precise information regarding state-of-the-art applications and ideas for the future of smart cities – cities that are liveable for today`s society and the following generations.

The chapter leaves space for equally important areas such as urban mobility, urban transport and traffic systems, assistive smart cities, realistic vehicular networks simulations, social science behind the cities of the future, multi-criteria view about judicial and legislative decision making in digital cities and societies, Critical systems for Smart Cities: towards certifying software.

All these areas have one thing in common: the digital twin as the most inclusive, transparent, efficient method recognizing and embedding such multi-diverse aspects.

Buildings give a city its face, create identity. They add value, give home and attract people who like to live and work in them. They offer a sense of wellbeing. They are part of the value chain by merging the historical with the modern and preserving values (Figure 4.).

Existing buildings represent an important lever for reducing CO_2 considerably. They make up the vast majority of objects. In order to drop CO_2 emissions of real

Figure 3. Reconstruction Notre Dame Paris, France, via Digital Twin.
Source: AGP Art Graphique & Patrimoine

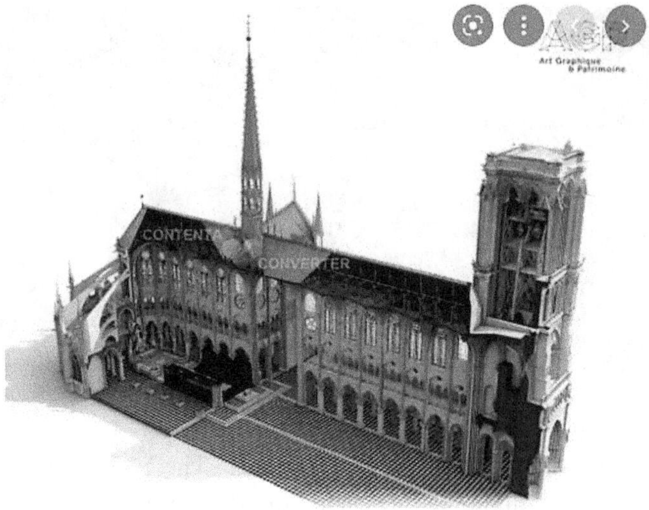

Figure 4. Merge existing and historic buildings and smart cities
Source: Smart Cities World, Leuven, NL

estate in Germany to zero by 2045, these must be almost halved by 2030. Smart cities come with the expectation that emissions will need to be reduced by at least two thirds by 2035 and by at least 90% by 2050 (Razmjoo, A. 2021). However, this goal can only be achieved if the potential of digital solutions is already being leveraged on a large scale today, e.g. air conditioning systems are regulated efficiently and according to needs by a cloud-based solution. But it is of highest important, that such application of technologies for smart cities` infrastructure does not have to be built upon efficiency, but on resilience. Thus, smart cities can adapt to changes in society.

Taking into account the multiple challenges to cope with climate change causing natural disasters, it is clear that cities and buildings will have to be planned and built even more "with the nature" in the future. Smart cities` and smart buildings´planning particularly requires to put focus on less surface sealing, intelligent smart wastewater management, warning systems e.g. in the event of a flood (e.g. Ahrweiler Germany in 2021), flood protection, stability and safety of tunnels, bridges and dams. The adaptation extends far beyond new necessary building materials and standards that are adapted to the requirements of climate change (e.g. acid rain, heat periods). Smart buildings and smart cities aim for an overall smart ecosystem, circular economy and resilience.

DIGITAL REVOLUTION IN CONSTRUCTION

The digital twin and AI applications implemented in practice contribute to an eye-opening effect. They represent signposts for the digital transformation in construction and the sustainable design of environment and society, living and urban worlds of tomorrow. They strengthen the citizens`, building and infrastructure users' and operators` trust. They offer full transparency in terms of a economic, society- and value-based design of living and urban worlds of tomorrow, future-proof urban infrastructures, space and society. Due to the numerous impacting factors creating a digital revolution in Construction, the holistic analysis makes sense (Weber-Lewerenz, B. (2022a).

Emerging technologies such as 10-D digital twins, AI, AIoT, BIM and Urban Information Modelling (UIM) (Standford University, 2016) enable the substantial increase of economic efficiency, transparency, confidence and the share of the value chain. UIM links BIM data to achieve a framework that allows the synthesis of the many data streams available in cities describing the spatial, temporal, and human dimensions of a city. This framework is an ideological extension of the BIM ontology at a larger urban scale. Such information modelling describes city life, since it is difficult to identify and understand the relevant multi-dimensional effects on urban environment that come from city-scale interventions. Such model enables decision-makers to understand impacts resulting from policy recommendations or new proposals for urban systems. On the practical user side, it especially enables contractors, city planners, architects, sociologists and policy makers to leverage integrated engineering tools for better design and analysis of city-scale interventions (Figure 5.).

Figure 5. Digital Twins to plan and operate smart cities like Singapore
Source: publish-industry Printing (https://www.industr.com/de/digital-twins-bilden-die-grundlage-fuer-smart-city-konzepte-2616454)

With Smart Service World Projects it will be possible in the future to exchange, store and trade decentralized and renewable electricity even more intelligently and efficiently - via Blockchain. The daily world of citizens is constantly changing: Starting with regional trading platforms, local delivery networks and logistics solutions, over sensor-networked city quarters and multimodal transport marketplaces, leading to an ecosystem for secure digital identities that, in the future, will allow simple registration by smartphone with citizen services, car sharing or online shopping. With the increase of data networking and data exchange, not only the capacity of technical development increases, but also the legal questions raise, e.g. related to data protection, data and IT security, the design of interoperability and patents as well as standards and norms. Last but not least, it is also about the development of suitable business models and exploitation channels (Figure 6.).

RETHINKING CITIES – BEST PRACTICES IN INTEGRATING EXISTING BUILDINGS AND PRESERVING MONUMENTS

Digital Twins combined with AI-based technologies facilitate routine processes. Automated Compliance Checking (ACC)(Amor, R. et al, 2021) serves as a good example: Compliance with the contract, Compliance with the legal and operational requirements for sustainable construction projects, Compliance in building permit procedures that are implemented with BIM, Automated Sustainability Compliance Checking Process (BIM conformity assessment with a focus on sustainability-

Figure 6. Application areas related to Smart Cities
Source: Lima, C.; Kim, K.-J.; Maglio, P.P. Smart cities with big data: Reference models, challenges, and considerations. Cities 2018, 82, 86–99 (https://doi.org/10.3390/su13147808)

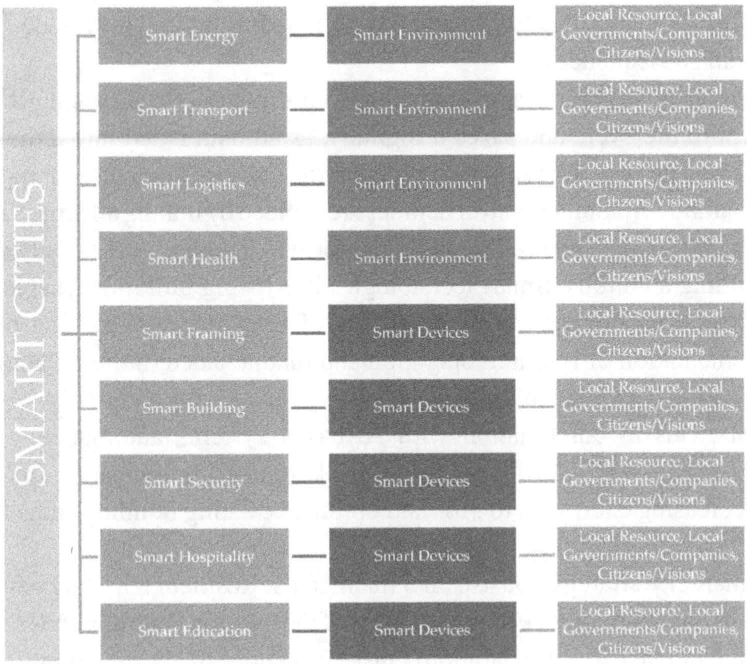

related standards, Proof of Concept) (Kasim, T. et al., 2013), Compliance with public concerns (e.g. the environment), Compliance with construction planning and building regulations (e.g. Evidence of the test engineers and experts), completeness of the documents (all documents necessary for a construction project, e.g. statics evidence, structural engineering evidence, building site stability, geotechnical evidence, official requirements such as fire and fire protection evidence), Compliance with standards and guidelines (through automated testing in BIM and AI-based processes), conformity tests (of building materials and components (product data sheets for safety, building material properties).

Additionally, digital twins enable to virtually adapt and measure buildings into the existing environment and simultaneously point out problem areas. Work and improvements can be carried out in a virtual state in which the building or the urban environment to be processed already exist (Figure 7.).

Multiple disciplines involved - no matter where they work from and in which country they are located - can contribute their expertise real time. Based on networking via clouds approvals and communication can be processed more efficiently. The effects of decisions can be visualized in real time and interactions can be traced at any time. Digital twins enable highly agile, interdisciplinary, diverse, inclusive and intercultural collaboration (Weber-Lewerenz, B., 2021b). Digital twins minimize the risk of loss of information by maintaining complex data in a structured, clear manner. The digital twin helps to simulate all possibilities and how to deal with its impacts and consequences.

BIM collects structural and operational information from existing buildings over its entire life cycle and saves it digitally, as all data is fed into a digital twin platform offering real-time monitoring, operation and ad-hoc corrections. Loss of information and communication difficulties are avoided by linking all project involved who work with the model on this single platform. All data is part of fully digitized smart building solutions. Smart technologies in existing buildings enable a much more efficient, economical and secure form of tests, decision-making processes and implementation of building operation and maintenance (building automation, communication between buildings).

Digital twins in combination with AI-based systems automatically learn a building`s and system`s behaviors and are able to provide forecasts. They reduce CO_2 by decreasing energy up to 30% and optimize existing buildings to get climate neutral.

The smart city strategy consists in 3 main goals: Resilient City, Transformation City and Circular Economy (Federal Ministry of the Interior, Building and Homeland, 2021). The Circular Cities Declaration (European Union, 2021) shows how the digital transformation in the municipalities can be made sustainable, and how the smart networking and intelligent operation of urban infrastructure as per the European

Figure 7. Working with Digital Twins within existing urban environment
Source: University of Lucerne

Green Deal increasingly find their way. Cities and their circular economy are keys for a prosperous society and climate neutrality, but require a big change of the citizens` behavior and urban design. The idea behind consists in decoupling resource use from economic activity by maintaining the value and utility of products, components, materials and nutrients for as long as possible in order to close material loops and minimize harmful resource use and waste generation. There is enormous potential as 75% of natural resource consumption occurs in cities and 50% of global waste comes from cities, in the EU, this number is estimated to be even higher. Signatory cities recognize their full commitment of transforming towards a circular economy and increase the share in the value chain. In-line with the Climate Strategy 2030 and UN`s Agenda 2030 some cities are role models e.g., the city of Aachen, Germany. Marzia Traverso with her Institute of Sustainability in Civil Engineering (INaB) at the University of Aachen (RWTH) add significant value by contributing their multiple expertise and supporting the Center for Circular Economy (CCE) at the RWTH Aachen on the city's transformation to a circular one. It is the full commitment for a more resilient future of urban planning.

The Governing Smart Cities Report (WEF, 2021) provides policy benchmarks for ethical and responsible smart city development. Such ethical responsible act offers ways to strengthen the trust of everyone: citizens as well as planners, manufacturers, users and operators of building and infrastructure.

Modeling city data is critical to optimizing according to economic, social and ecological criteria. It is also critical to supporting automated code review processes. Ideally, digitally modelled building regulations and urban planning data are retrieved from municipalities and the digital building model is automatically checked on the basis of the information collected. However, BIM tools do not include or allow management of geographic information. Therefore, a coupling between GIS

Figure 8. Improved city planning with holistic digital twins with GIS data.
Source: Autodesk

(geographic information system) and BIM environments (e.g. buildings, streets, bridges, railways, airports) is necessary (Demir, 2021). With the help of a proof-of-concept GIS application, geometric and semantic data can be stored and queried. Using this data model for automated code review is an example of how GIS data can seamlessly complement BIM data. It demonstrates how it supplements other important data in order to create an even more holistic digital twin (Figure 8.).

Monument Preservation, duplication of severely damaged parts of historic buildings within smart cities mean applying modern technologies, creating digital twins and connecting the concerned modernized buildings with surrounding intelligent buildings and infrastructure.

Here, designated digital technologies and AI methods apply, especially in measuring and reconstructing works, cuts and measurements of building component, fitting of re-manufactured components into the existing building, production of building materials, Forecast Modelling (investment planning and control) and simulation of construction models (3-10-D, e.g. with construction-accompanying continuous recording of the actual construction time, costs and quality compared to the original planning). Virtual Reality and Augmented Reality are used to create virtual models of churches, castles and historical buildings to "walk through". Object recognition using Machine Learning enables the identification of building types, classes, damage or anomalies in images. Defective areas, often in hard-to-reach locations, can be allocated, repairing and maintenance can be planned. Digital online site visits with VR glasses get routine, Simulation of Renovation, Acoustic and Fire Protection Concepts (Weber-Lewerenz, 2022b) can be realized. AI-based Sensors recognize and track defects and damages with changes of material, air humidity and temperature.

AI technologies compare new data with stored data to track compliance with structural requirements and regulations (monument protection, fire protection, environmental protection) when taking measures.

Via 3D measurements and using drones for UAV photogrammetry and UAV laser scanning in the area of monument protection digital twins can be created. It is possible to record the inventory and 3D documentation helps to digitize cultural assets, stone-based measurement of facades using UAV photogrammetry with high-resolution, calibrated camera systems or LiDAR scanners. To produce new parts, stucco, decorations, windows and doors, high-resolution 3D scans of relics for the digitization of cultural assets and high-resolution facade views in the form of Orthophotos (corrected measurement images) are basis for designing and for recording and documenting damages.

Architectural monuments, ground and area monuments can be used for digital tours in public relations with the help of digital twin models. In this way, the high accuracy requirements of monument preservation and monument protection can be met, (damage) documentation in the area and monument protection can be performed highly efficient, as well as the entire documentation, cataloguing and archiving, planning of renovation measures and reconstruction. Digital twins have become indispensable methods both for highly sensible restoration and their use in public relations.

Buildings represent about 40% of the total primary energy consumption and contribute approximately to one-third of the total greenhouse gas emissions. Green building standards and certification systems include requirements for both new and existing buildings. However, a vast part of the existing building stock in cities was built with limited consideration for green design, energy efficiency and low carbon emissions, yet its refurbishment can help reach national energy reduction goals, decreasing environmental impact. Furthermore, retrofitting existing buildings can often be more cost-effective than building new facilities. The potential of an existing building to achieve certification under a certain Green Standard, defined in this work as its Green Potential (GP), depends heavily on factors that were determined during the original design and construction process.

In the research alliance for cultural heritage, the competencies of the partners are bundled in order to jointly develop and test new technical, digital or AI-supported processes and methods for the restoration and conservation of cultural property, to further intensify the transfer of knowledge between research and restoration practice and to increase the importance of cultural heritage more firmly anchored in the public consciousness.

There is a strong social responsibility - not limited to the digital era - to consciously preserve the ephemeral cultural heritage - which is not a renewable resource.

Best Practice: Planning of Buildings and Cities With Virtual Reality (VR)

VR and digital twins deliver consistently positive empirical values to monument preservation building planning. VR glasses enable the spatial customer journey, in which the project performance, promised by the one who will realize it, is already made accessible and transparent for the client, customers and everyone involved in the project. By walking through buildings, city quarters and cities, the customer establishes his first emotional ties to the project. From early beginning, the client is involved in the design, development and all other construction and operation phases. This method helps to avoid misunderstandings; it demonstrates high safety and quality performance based on responsible handling of the project budget. One of the entrepreneurial advantages is that the customer can influence the performance and make changes in the earliest project phase in a cost- and time-efficient manner with an overall cost transparency. All change requests are visible to everyone, errors are recognized early, flow of communication is given between everyone involved and anytime.

Best Practice: Operating Existing Buildings – Smart Buildings by BOSCH (Bosch, 2021)

With networking solutions and new IoT services, such as "Bosch Energy and Building Solutions", commercial buildings are becoming smart, more efficient, sustainable, safe and comfortable environments adapting to and responding to people's needs.

But what are other advantages for owners, operators and users? And how does an existing building, which has been "silent" until now, become intelligent?

Advantages cover:

- More efficiency, comfort and safety
- Increase of efficiency and nee intelligent services via intelligent networks of the technical building equipment
- Construction sector causes almost 40% of global CO_2 emissions (= essential lever to achieve climate neutrality by 2050)
- Independent monitoring, regulation, analysis and optimization of energy requirements
- Buildings that "think for themselves"
- Identify and repair incidents quickly, identify upcoming maintenance early on and plan automatically, optimize the use of areas and rooms
- New dialogue between building and human = Changed processes in management, maintenance easier, extended maintenance cycles, remote maintenance are simply possible
- The building itself becomes the best partner for facility and energy managers, supports the daily work of all those involved, simplifies processes and helps to permanently reduce operating costs.

Building technology solutions can be:

- Building automation, light and energy management systems, security technology
- Integration of building technology equipment and systems with fire alarm systems, access and intrusion alarm systems, building automation and video surveillance systems and IoT components: connection with cloud-based platforms
- Digitized frequency of use: utilization of areas in the building known and thus solutions for space management
- Basic requirement: digital building twin which generates the holistic digital image of the building.

Technologies, software, systems and sensors are integrated with the knowledge about the diverse processes and procedures within the digital environment of a building. Challenges consist in connectivity, secure data storage and processing as well as visualization.

Best Practice: Monument Preservation by VINCI Construction

Europe is characterized by a rich building culture. Maintenance with the help of reproductions that are as identical as possible, and the restoration of existing buildings are performed by using AI methods beyond the digital twin:

Cathedral Notre Dame Paris, France

After the fire in 2019, 3D scans were carried out to create the digital twin. The Point Cloud-based model captures the original structures and digitally preserves them for future generations.

Sixense-Software

VINCI uses „Sixense" Software created by VINCI Construction Group and its ScanLab. High-precision 3D and photogrammetric measurements enable automated risk detection. The company's own AI program is called "Leonard VINCI". Recordings of the façade automatically map the technical floor plans. The advantages and experience consist in the intervention times; scaffolding construction working period are drastically reduced. This results in a lot more efficient and shorter restoration time, so that tourist masterpieces can be accessed within short period of time. Classical example from practice is Mont Saint Michel in Normandy, France.

VINCI Chronsite Program (www.leonard.vinci.com)

The program involves the automated consolidation of all construction property data and information obtained by computer vision and deep learning technology. This AI method offers an automated complete analysis of the construction site: Tasks are identified, their duration, as well as the associated work capacities by camera monitoring. This method derives precise data recorded onto online interfaces. The practical experience lies in the fact that the person who applies the program can analyze and differ, what human work can do best, at what point and in which manner it can be optimized. This AI method can be transferred to technical building automation, which means that energy consumption can be efficiently controlled and reduced.

SOLUTIONS AND RECOMMENDATIONS

Catalyst for Innovation: Education and Legal Regulation

Education

Digital technologies and AI mean a milestone as German location of research and education. Digital transformation requires resources of knowledge, education, new digital qualifications and skills. Education in the Digital Era means Compass for "Building Forward Better" (Weber-Lewerenz, 2022c).

Digitization, BIM and AI are opening up both new fields of education - requiring new curricula - and new professional opportunities that could revive the lack of attractiveness of the entire construction industry for years. New technologies and the need of smart buildings and smart cities create new job profiles, e.g. Digital Construction und Digital Engineering, and services (Smart Services). They require new skills and diverse, inclusive environments (Weber-Lewerenz et al., 2022d). BuildingSMART Germany as part of buildingSMART International designed a program to train certified BIM Professionals, customized to both corporate and application needs of BIM (buildingSMART, 2021). This program offers a global quality standard for the evaluation and comparability of BIM knowledge and skills.

Education is the driver of success; it requires the adaption of teacher`s qualification equally. Currently advertised professorship positions call for skills in digital urban designing and planning, concepts and methods of virtual cities and smart cities, embedding societal requirements, climate engineering, optimizing sustainable building life cycles and construction processes - based on applying BIM and AI and the overall work on the technical interfaces.

Resource conservation in the building sector needs to be pushed also in energy consulting by advancing appropriate qualification of consultants, e. g. through an additional training module as a prerequisite for the list of energy efficiency experts. Digital transformation and intelligent design and operation of smart cities require the adaption of organizational structures in the municipalities (Smart City Board, Chief Technology Office) (Figure 9.).

What does sustainable building life cycle management as part of the new urban agenda (Council of the EU, 2016) (UN, 2016) mean? There are two goals: develop liveable cities and empower designers and operators of urban development (Federal Institute for Building, Urban and Spatial Research, 2017). Sustainability, as international pioneer cities such as Amsterdam, Copenhagen, Vienna, Barcelona and Singapore demonstrate, goes further in many steps.

Figure 9. Chief Technology Office City of Amsterdam, NL
Source: https://amsterdamsmartcity.com/network/chief-technology-office

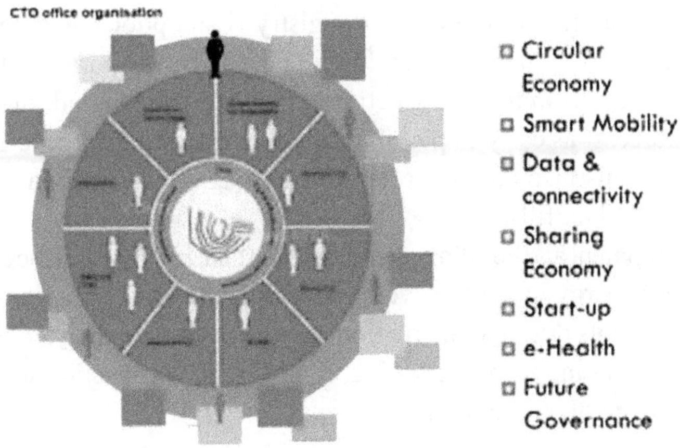

Legal Regulation

The European Union previously compiled Regulations for facilitating transformation of existing cities, starting from existing buildings, into smart buildings.

The Internet of Energy (IoE) impacts on smart cities' power sector. IoE is an implementation of the Internet of Things technology (IoT) into distributed energy systems and aims to achieve energy efficiency, to avoid energy wasting, and improve environmental conditions. IoE technology includes, among others, utilizing smart sensors and renewable energy integration. Therefore, the IoE is becoming a legal science tool to serve the purpose of a smart city.

A smart building template can manage the performance of all technical systems through IoT technology towards achieving energy efficiency. There are methods that minimize time consuming procedures and store, on a cloud platform the energy performance of each building, for the purpose of drawing conclusion and applying measures. An automated remote, control method supported by cloud interface helps significantly to improve the certification of existing buildings, as for energy performance.

Sustainability also means integrating the principle of Cradle-to-Cradle: utilizing the environment instead of only reducing damaging effects. The concept is based on three principles: the understanding of waste as food, the use of renewable energies and the support of diversity. Practical example: Digital twins - both of buildings and infrastructure - generate automatically real time types and quantities of materials, the data is stored and can be called up again when the building is due to be dismantled and the recycling material can be used again for the new construction. Cradle-to-Cradle means full efficiency of resources and energy. It means building in the material cycle in compliance with the sustainability goals.

Setting up digital urban-mining-land registry is an option - with an evaluation scheme standardized by means of building and goods passports. In this way the stock of durable goods in the construction sector can be recorded, future material flows can be forecast as early as possible, and the best possible recycling channels can be derived. This also applies to landfill mining, the extraction of recyclable materials from old landfills.

Urban development and building planning could take into account the consumption of materials and energy in the life cycle and drastically reduce the consumption of mineral raw materials during construction and the ongoing energy consumption of buildings, by combining resource-saving, lightweight construction with the use of reusable and recyclable building materials.

FUTURE RESEARCH DIRECTIONS

With increasing technical feasibility, social responsibility the need to specify and tighten the applicable legal framework to maintain data sovereignty increase (Weber-Lewerenz, 2021c).

On the construction site, quantity of data, its origin, transparency and control, data maintenance, use and processing, exchange and data interfaces are extremely complex, sometimes completely confusing and they constantly increase. Compliance with the principles and accountability of the General Data Protection Regulation, the processing of personal and project-related data and compliance with the responsibility of third parties are hardly controlled and even less prosecuted. Who controls and prosecutes that personal data may only be processed under official supervision or - if provided for - by law? And who has the earmarked user rights accessing data for the planning, execution and management of buildings? Who controls data clouds, without which BIM and digital twins do not work? Who will ensure and protect ownership of data? The trade in personal data has become a money machine; Al Gore calls such business model the "stalker economy".

Who checks, during the daily construction progress documentation, that all persons visible on photo documentation have given their declarations of consent beforehand and that the data are forwarded only to the addressees specified earlier? How does the handling of data work outside of companies, e.g., at home office workplaces as during the times of pandemic, and who ensures that data is controlled and protected?

Citizens of a smart city are connected to intelligent urban ecosystems in a variety of ways, e.g. by using smartphones and mobile devices as well as connected cars and apartments. Coupling devices and data with a city's physical infrastructure and services can reduce costs and improve sustainability. Thus, it is important to protect this highly sensitive data, its use and communication.

Digital self-determination is increasingly seen as a task for the entire legal system in order to ensure, among other things, the protection of personal and project-related data, protection against cyber- attacks and data misuse. The handling of data, such as the digital twin, the user data of smart technical building automatization, smart buildings, smart cities, and the training of AI algorithms require clarification and criminally enforceable regulations as well, who, where, for what and to what extent has rights and obligations and is liable for disregard.

In Germany there are no uniform standards with regard to BIM, customers and contractors apply different methods and they vary from project to project. The inequality of the standards means that no faster solutions to the problems can be obtained. An adapted legislation is important to overcome these obstacles, speed up digital transformation, create more jobs in innovative technology areas, to more cost and resource efficiency in the construction life cycle. Legal security in construction drives innovation and sustainability equally.

According to Paul Nemitz, the main advisor for judicial policy to the European Commission, the focus is on ensuring that EU companies "[...] will have a level playing field. It doesn't matter whether the AI comes from outside or inside the [EU] comes". He affirmed, "[...] that the AI-Act does not hinder or slow down the technology, but rather makes this technology even safer and more trustworthy." (Nemitz, 2021). Thus, these who act constructively with these rules have a more sustainable profit perspective than those who act against any principles of responsible innovation and responsible engineering.

Surveys` evaluation shows clearly that statutory requirements by the state are absolutely necessary for the secure handling of data. In its "White Paper on Artificial Intelligence" in 2020 (European Commission, 2021), the EU Commission called for the consideration and protection of values and fundamental rights, the careful demarcation between machine and human intelligence and careful handling. In July 2018, the German government called for the development of a transparent and ethical AI in its AI strategy paper and sends an important signal with its € 3 billion plan in November 2018. The Data Ethics Commission DEK (Data Ethic Commission of German Government, 2019) sees the state as having a special responsibility to formulate ethical standards for the digital space. In order to be able to keep this guarantee towards the citizens, an international position of political and economic strength is required. Efforts to secure the digital sovereignty of Germany and Europe in the long term are therefore not only a requirement of political foresight, but also an expression of ethical responsibility. The General Data Protection Regulation (GDPR) came into force on May 25, 2018 (European Union, 2018).

Protecting human and democratic values in the sense of CDR means designing, maintaining and protecting trustworthy technologies; it builds up trust, increases the acceptance of data-driven technologies and promotes innovative ecosystems. Engineers, architects, designers and craftsmen are more than just designers of living and working environments. They shape technical, social, societal and human changes in the construction industry 4.0 (Weber-Lewerenz, 2021d; Weber-Lewerenz, 2021e). Having more technological possibilities in life and at work as a person means at the same time "a greater responsibility on the part of people for technology-based innovations ... and for the sustainable preservation of the earth ..." (Kirchschläger, 2021).

The Step-by-step Plan Digital Construction by BMVI (Bodden, et al., 2020) is the first significant step into the complete digital project life cycle in the public sector for all traffic and infrastructure projects.

On the other hand, the implementation and application of new technologies also demonstrate the critical side: to ensure compliance of data exchange processes and information transfers in project communication and in the cooperation process, to protect and secure personal and project-related data, ensure cybersecurity, when using BIM and AI in construction, there is a urgent need for legal regulations.

In China e.g., the law on the protection of personal data (People's Republic of China, 2021) came into force on January 1st, 2022; some contents of the GDPR have been adopted, many other regulations do not (yet) exist in Europe.

CONCLUSION

The key challenge of the hour is balancing the potential of smart cities, the increased risk from extensive cyber-attacks, and preserving values and cultivating culture. The consideration of buildings worth preserving by integrating and networking and upgrading facilities to smart cities` standards prove, that smart city development has highly positive impacts on future cities. Smart buildings and smart cities are resident-oriented and environmentally sustainable.

Integrating existing buildings into smart cities means giving existing and historical buildings a future and equipping them with the most modern technologies for today's needs.

Though strong research in the fields of Smart Cities Development and new technologies applied, the key question still remains unanswered: What measures have to be taken to adapt existing buildings and cities to the new climatic conditions? Humans face completely new climatic conditions caused by climate changes - climatic conditions which have a very serious influence on the human organism, the animal and plant world. Speaking of today`s sustainability, it usually occurs in the course of speaking about new buildings, new construction. But what will happen to the existing buildings and the urban space?

When reflecting about sustainability it is essential to understand what the cost implications are and the impact on the resources themselves - the ones we want to protect but, on the other hand, need for such adjustments.

Innovative digital methods, e.g., digital twins, and AI technologies represent cutting-edge technologies offering high efficient methods to make full use of key competencies, demonstrated bybest practices in Smart Buildings and Smart Cities. Designing, realizing and operating smart cities means turn-around, radical re-thinking to reach a sustainable technical, societal and human change of the construction culture 4.0.

ACKNOWLEDGMENT

I would like to thank all of the experts interviewed in this research. They not only shared valuable insights but share the vision of the “Excellence Initiative for Sustainable, human-led AI in Construction”. I would like to thank the European Association for REsearch on SERVices (RESER) and the buildingSMART Network for the great cooperation and support addressing the diverse ethics debate for moving digital innovations forward.

In particular I would like to thank my supervisor, Prof. Marzia Traverso (PhD), Head of Chair of the Institute of Sustainability in Civil Engineering, Faculty of Civil Engineering at RWTH Aachen University, Germany.

Statistics and reports were of great support with data and facts to underlay the statements and recommendations made in this paper. Some public statements, which come from internet, literature and archive research, also underline the quality and statistical values of the expertise and survey values obtained, as well as limitations and urgently necessary measures. The author of this study conducts external research, is company-independent and is not financially supported by third-party funds, companies or other institutions. She is free in her research and shares her findings at the interface of “application practice - applied technical research - economic and social transfer”. In this way, the author is researching the responsible use of digitization and AI -neutrally, critically and inclusive - and promoting the ethical debate about the AI technologies.

This research received no specific grant from any funding agency in the public, commercial, or not-for-profit sectors.

REFERENCES

Amor, R., & Dimyadi, J. (2021). The promise of automated compliance checking. *Journal of Developments in the Built Environment*, *5*, 100039. Advance online publication. doi:10.1016/j.dibe.2020.100039

BMVI. (2021). *Smart Design, Smart Construction, Smart Operation*. BMWI Publication. https://www.digitale-technologien.de/DT/Redaktion/DE/Downloads/Publikation/SSW/2021/SSW_Baupublikation.pdf?__blob=publicationFile&v=3

Bodden, J., & Dittschar, F. (2020). *Umsetzung des Stufenplans Digitales Bauen* [Implementation of the step-by-step Plan for Digital Construction]. BMVI Publishing, Working Paper 3.1 – Klärung von Rechtsfragen.

Bosch. (2021). *Fire Protection*. White Paper by Bosch Energy and Building Solutions. https://media.boschsecurity.com/fs/media/ib/media_1/solutions/Bosch-Whitepaper-Brandschutz_1.pdf

buildingSMART. (2021). *BIM Professional Certification Program*. https://education.buildingsmart.org/

Council of the European Union. (2016). *Urban Agenda for the EU, Pact of Amsterdam*. https://e3p.jrc.ec.europa.eu/articles/pact-amsterdam-urban-agenda-eu-european-cities-get-their-say-eu-policy-making

Data Ethic Commission of German Government. (2019) Assessment on Data Ethic. Brandenburgische Universitätsdruckerei und Verlagsgesellschaft Potsdam mbH (bud).

Demir Altıntaş, A. (2021). Loose coupling of GIS and BIM data models for automated compliance checking against zoning codes. *Journal of Automation in Construction*. DOI: doi:10.1016/j.autcon.2021.103743

European Commission. (2020). *White Paper On Artificial Intelligence - A European approach to excellence and trust*. European Commission.

European Union. (2018). *General Data Protection Regulation*. https://www.datenschutz-grundverordnung.eu/

European Union. (2021). *The Circular Cities Declaration*. https://circularcitiesdeclaration.eu/

Federal Institute for Building, Urban and Spatial Research (BBSR) in the Federal Office for Building and Regional Planning (BBR), Bonn. (2017). *Smart City Charta - The sustainable design of Digital transformation.* https://www.bmi.bund.de/SharedDocs/downloads/DE/veroeffentlichungen/themen/bauen/wohnen/smart-city-charta-kurzfassung-de-und-en.pdf?__blob=publicationFile&v=4#:~:text=Die%20Smart%20City%20Charta%20richtet,Bund%20und%20die%20L%C3%A4nder%20angesprochen

Federal Ministry of the Interior, Building, and Homeland. (2021). *Model projects Smart Cities.* https://www.bmi.bund.de/SharedDocs/kurzmeldungen/DE/2021/07/smart-city.html

Kasim, T. (2013). Automated sustainability compliance checking process: proof of concept. In *Proceedings of the 13th International Conference on Construction Applications of Virtual Reality* (pp. 30-31). School of Engineering, Cardiff University Press.

Kirchschläger, P. G. (2021). Ethische Verantwortung in der digitalen Transformation [Ethical Responsibility in Digital Transformation]. *Journal ZRFC, 6*, 253.

Nemitz, P. (2021). *Executive Interview.* Aitrends. https://www.aitrends.com/executive-interview/executive-interview-paul-nemitz-principal-adviser-on-justice-policy-for-the-european-commission-brussels/

People's Republic of China. (2021). *Personal Information Protection PIP Law.* https://usercentrics.com/knowledge-hub/china-personal-information-protection-law/

Razmjoo, A. (2021). Investigating Smart City Development Based on Green Buildings, Electrical Vehicles and Feasible Indicators. *Sustainability Journal 2021, 1*(14), 7808. doi:10.3390/su13147808

Stanford University. (2016). *Urban Information Modelling.* https://cife.stanford.edu/Seed2016_UrbanInfoModeling

United Nations. (2016). *New Urban Agenda. United Nations Conference on Housing and Sustainable Urban Development (Habitat III).* https://www.urbanagendaplatform.org/about

Weber-Lewerenz, B. (2021a). CDR and Ethical Aspects in AI in Construction. The Disruptive Role of Data, AI and Ecosystems in Services. *Proceedings of the 31st RESER Conference.* https://publica.fraunhofer.de/dokumente/N-642928.html

Weber-Lewerenz, B. (2021b). Gender Innovation in Construction in Digital Era. The Disruptive Role of Data, AI and Ecosystems in Services. *Proceedings of the 31st RESER Conference.* https://publica.fraunhofer.de/dokumente/N-642928.html

Weber-Lewerenz, B. (2021c). Corporate digital responsibility (CDR) in construction engineering - Ethical guidelines for the application of digital transformation and artificial intelligence (AI) in user practice. *Springer Nature SN Applied Sciences, 3*(10). DOI: doi:10.1007/s42452-021-04776-1

Weber-Lewerenz, B. (2021d). Die unternehmerisch verantwortungsvolle Digitalisierung im Bauwesen. *Journal Der Bauingenieur, 96*(1-2), 19-25. DOI: doi:10.37544/0005-6650-2021-01-02-45

Weber-Lewerenz, B. (2021e). Corporate Digital Responsibility in Construction Engineering. *International Journal of Responsible Leadership and Ethical Decision-Making, 2*(1). Advance online publication. doi:10.4018/ijrledm.2020010103

Weber-Lewerenz, B. (2022a). *Accents of Added Value in Construction 4.0 – Ethical Observations in Digitization and AI.* Springer Nature Publishing.

Weber-Lewerenz, B. (2022b). Technological Dream or Safety Traumata? Fire Protection in Smart Cities-Digitization and AI ensure burning ideas and a New Culture of Thinking in Construction 4.0. In I. Vasiliu-Feltes (Ed.), *Impact of Digital Twins in Smart Cities Development.* IGI Global Publishing. doi:10.4018/978-1-6684-3833-6

Weber-Lewerenz, B. (2022c). Impact of education and training for diversity, inclusion and sustainable digital innovations. In I. Vasiliu-Feltes (Ed.), *Impact of Women Empowerment on SDGs in the Digital Era.* IGI Global Publishing. doi:10.4018/978-1-6684-3637-0

Weber-Lewerenz, B., & Vasiliu-Feltus, I. (2022d). Empowering Digital Innovation by Diverse Leadership in ICT. In I. Vasiliu-Feltes (Ed.), *Impact of Women Empowerment on SDGs in the Digital Era.* IGI Global Publishing. doi:10.4018/978-1-6684-3637-0

Winkowska, J. (2019). Smart city concept in the light of the literature review. *Journal of Engineering Management in Production and Services., 11*(2), 70–86.

World Economic Forum. (2021). *Governing smart cities: policy benchmarks for ethical and responsible smart city development* [Paper presentation]. WEF White Paper. https://www.weforum.org/whitepapers/governing-smart-cities-policy-benchmarks-for-ethical-and-responsible-smart-city-development

KEY TERMS AND DEFINITIONS

AI: Artificial Intelligence. AI is the so-called umbrella term for applications in which machines provide human-like intelligence such as learning, judgment and problem solving. The development, application and maintenance of such data-driven technologies requires orientation in order to be able to fulfill the associated digital responsibilities. AI is understood as the ability of machines to emulate human intelligence and to learn independently, without being dependent on instructions from a human developer. For this purpose, software uses machine learning to evaluate large amounts of data for matches and draws conclusions from them.

BIM: Building information modeling.

CDR: Corporate digital responsibility.

Digitization: There is no clear definition for the term. In the original sense, digitization refers to converting analog data values to digital formats. This data can be processed by information technology. Another meaning of digitization is the digital revolution, also known as digital change or digital transformation.

Ethics: Ethics deals with the prerequisites and the evaluation of human action and is the methodical reflection on morality. At the center of ethics is specifically moral action, especially with regard to its justification and reflection (ethics describes and judges morality critically). The focus of ethics is on the three questions of the "highest good", the right action in certain situations and freedom of will.

IoE: Internet of Energy. Not to be confused with Internet of Everything (IoE). It is a concept that goes beyond the confines of the Internet of Things (IoT). IoE is an implementation of Internet of Things (IoT) technology in distributed energy systems. IoE is a technology term that refers to the modernization and automation of power infrastructure for energy producers and manufacturers. As a result, energy production can be driven more efficiently and cleanly with the least amount of waste.

IoT: Internet of Things. The term describes that the computer in the digital world is increasingly being supplemented by "intelligent objects" to "AI", artificial intelligence. Instead of being the object of human attention itself, as is currently the case, the "Internet of Things" is intended to support people in their activities.

Chapter 6
The Workforce Needed to Support Digital Twins in the Global Space Ecosystem:
Advocating for Innovation, Bettering Life on Earth

Shelli Brunswick
Space Foundation, USA

ABSTRACT

Ever since the first spaceflight, humanity around the world has changed for the better as the reality of what's "up there" and here on Earth become inextricably linked. Today, every continent, country, community, and citizen are impacted by innovations that originated from space exploration. This includes everything from the soles of athletic shoes to developing groundbreaking technologies like remote sensing, broadband connectivity, and high-performance computing used by modern enterprises all over the world. In truth, space supplies the critical infrastructure upon which every other infrastructure is dependent. Once considered the final frontier and only accessible to a select few, space is now home to the most dynamic and innovative environment on the planet and beyond. Cultivating the talent to fuel and drive bold innovations like digital twins will be one of the greatest challenges we face. Workforce shortages and skills deficits across geographies, disciplines and demographics threaten not just future advancement but also expanding economic opportunity.

DOI: 10.4018/978-1-6684-3833-6.ch006

INTRODUCTION

Space supplies the critical infrastructure upon which every other infrastructure is dependent. A day without space access and space-inspired technologies would be catastrophic to the world's security, economies and ways of life. Putting this into perspective, the global space economy is estimated at $469 billion and forecast to grow to $1.4 trillion by 2030 (Space Foundation Editorial Team, 2022).

Once considered the final frontier and only accessible to a select few — space is now home to the most dynamic and innovative environment on the planet and beyond.

As a collection of integrated systems and diverse stakeholders, this environment is best viewed as a global space ecosystem — and every ecosystem must be managed and balanced to ensure its health, sustainability, and long-term impact and viability. To get there will require us to diversify the participants, investments and partnership models that enable information, education and collaboration across industry, finance, national security, research and other communities. We also must grow our workforce for both operations in space and on Earth.

Cultivating the talent to fuel and drive bold innovations like digital twins will be one of the greatest challenges we face. Workforce shortages and skills deficits across geographies, disciplines and demographics threaten not just future advancement but also expanding economic opportunity.

Recognizing all of the different ways people are tied to space, and space back to Earth, the following questions are raised: How can we use these connections for the greater good, how do we maximize our impact across all age groups and demographics all over the world, and what role can technology play?

This chapter explores the answers in depth in the pages that follow, with the understanding that there is nothing we cannot achieve without knowledge, dedication, courage and collaboration. Together, we will ensure that the global space ecosystem is managed and balanced for ongoing sustainability and impact.

People will power the innovations of the future. We will apply remarkable tools of transformation, such as satellite imaging, remote sensing, and digital twin technology, to create more efficient, human-centric cities and a better, healthier, more secure world — we just need to follow the right roadmap to get us there.

BACKGROUND

Era of Access and Opportunity

For as different as people and cultures may be around the world, they do share a common thread of history. Whenever access is provided to more people, it opens new opportunities to change life as we know it. From the Appian Way of the Roman Empire, to the boundless digital frontier of the Internet, history consistently details a blueprint for possibilities. Such is the *new era of access and opportunity* that is unfolding before us.

In their paper, *The evolution of mission-oriented policies: Exploring changing market creating policies in the US and European space sector*, Robinson and Mazzucato (2018) point out that over six decades, the handful of countries involved in space has grown to over 60 nations. By the time I participated in Expo 2020 in Dubai, more than 90 nations had become active in space, and a total of 178 countries visited its Expo City. This rapid growth underscores just how essential space technologies are becoming to the world at large. It's not just about looking at what's in outer space.

For example, tremendous work is being done globally in understanding and applying space technologies to help create smarter, safer cities. Li et al. (2022) tracked the work published on the use of satellites for emergency management services, finding nearly 3,000 academic articles stemming out of 128 countries and regions, on this one application of the technology. Whether it's determining how to improve traffic flow, manage natural resources, or better assist in a disaster, space technologies like satellites will play a vital role — and the world is watching. This much is clear by the level of interest in how Dubai is now transforming Expo City into the emblem of a clean, green, tech-enabled, human-centric city of the future.

And the same satellites and remote sensing technologies used in the development of smart cities are being used for so much more. As Zhao et al. (2022) note, the Earth observation they enable,

provides an effective way of exploring the physical, chemical and biological information related to the Earth and of supporting the sustainable development of modern human civilization and the global environment by monitoring and assessing natural and artificial environmental conditions and changes.

That's a pretty big deal!

Organizations are also adopting digital twins to monitor, diagnose, predict and optimize systems and processes, as noted in Qi et al. (2021). Researchers found that by merging our virtual and physical worlds, digital tools will drive innovation across all sectors. Additional research shows that digital twins can be used for everything from protecting satellite systems from cyberthreats (Hóu et al., 2022) to fault diagnosis and health monitoring (Shangguan, Chen, and Ding, 2020) that can prevent "considerable loss or catastrophic impact on people, equipment and the environment."

Born of ingenuity, demand and competition, the space sector now demands attention beyond that which the government has historically provided. There are more companies and citizens looking to build a future beyond the bounds of Earth, while still bringing solutions and rewards to the lives back on it.

MAIN FOCUS OF THE CHAPTER

Access to space is rapidly increasing as new avenues to reach space have opened. The recent commercial crew flights by SpaceX, followed by the two suborbital flights by Virgin Galactic and Blue Origin, not only put more space vehicles into operation but also powered new investors, customers and market entrants to become part of this expanding economic sector. Weinzierl (2018) noted that if space-economy visions like those put forth by Jeff Bezos, Elon Musk, Neil deGrasse Tyson and Peter Diamandis "are even partially realized, the implications for society — and economists — will be enormous."

That new type of access creates opportunity for those investors, customers and market entrants — the cutting-edge pioneers who are willing to challenge the status quo, bring new ideas to the forefront, and revolutionize a community that is always pushing into that next bold frontier.

Space technologies are experiencing a boom — the result of public and private companies joining with governments to open opportunities, advance commercial products, and deploy services with far greater speed and scale. In fact, Rathje and Katila (2020) developed an enabling technology index examining the impacts of private-firm relationships with the public sector. Across a 21-year period, they found that agreements and grants with mission agencies, such as NASA and the Department of Defense, can positively impact growth trajectories.

Currently, the space economy equates to approximately $469 billion, while it is expected to reach $1 trillion before 2030. This wave of massive market growth was seeded back in 2006, the first performance year following the passage of the 2004 Commercial Space Launch Amendments Act, which led to the NASA contract for the Crew Exploration Vehicle and the first contract awarded to SpaceX. A period of space innovation, development and investment not seen since the 1960s was born.

According to The Space Report, space-related mergers, buyouts, and equity financings totaled $26.2 billion in the U.S. in 2021, with total transactions climbing to 316 from the 228 finalized in 2020. A number of disruptions, including supply chain issues, the war in Ukraine, and a cooling toward SPAC-based financing have pushed investors toward more traditional investment cycles, which has ensured continued investor support for larger, defense-related space companies, such as Lockheed Martin, L3Harris Technologies, Northrop Grumman, and Raytheon Technologies. All headed into Q4 of 2022 with strong performance compared to many newer space company entrants.

This level of commercial activity will only increase as established and emerging companies offer goods, services, and space-inspired technologies that fuel space operations, enrich space supply chains, develop smart cities, and add value to lives everywhere.

But in their work studying smart cities from a systems perspective, Ammara et al. (2022) wrote:

A smart city can be defined not by its degree of adaptability to the most cutting-edge technologies currently available on the planet; rather, it is defined by the ability of urban inhabitants — whether they are leaders or members of their communities, or even both — to use ingenious solutions to solve lingering problems.

The takeaway? Even with the best technologies available, people have to know how to apply them.

Space-related organizations employ hundreds of thousands of highly skilled individuals to design, produce, commercialize and operate cutting-edge technologies. Its talent pool spans from highly specialized experts required to operate remote sensing technologies and applications to much-needed trades like welding and wiring. There is even a need for urban planners. Among the biggest areas of space growth are telecommunications, manufacturing, engineering and high-performance computing.

This wide-ranging workforce has expanded, despite the global pandemic. According to data from The Space Report, U.S. space-related labor increased 2.6% in the U.S. in 2021 and 5.4% in Europe over the year prior.

Fields associated with space technology continue to broaden and intersect in exciting ways, and traditional barriers to entry are falling away. Access is no longer limited to those with advanced degrees, Ph.D.s and rocket scientists, nor are the jobs solely based in tech hubs or government centers. These are well-paying jobs that can be found in nearly every community.

For example, in 2020, the most recent year for which salary data is available in the U.S., The Space Report found:

The average private sector space salary was $125,214, more than double the average salary for all U.S. private sector jobs of $62,247. It is also well above the average annual salary for STEM occupations, which was $98,340 in 2020. Broadcast and wireless communications equipment manufacturing and satellite telecommunications, guided missile and space vehicle manufacturing, and space vehicle propulsion units and parts manufacturing all experienced growth in real wages from 2014 to 2020 (Bolton and Bauerlein, 2020).

By offering financial security and a greater quality of life to a wide range of workers, there is no better time to be a part of the collaborative space ecosystem.

Issues, Problems and Controversies

A Workforce Shortage and Skills Deficit Conundrum

Even as the space technology landscape expands and an abundance of new opportunities surface, the candidate pool of qualified workers is shrinking dramatically. Statistics and analysis from the U.S. Department of Labor's Bureau of Labor Statistics (BLS), the Commerce Department's Bureau of Economic Analysis (BEA), the U.S. Census Bureau, the Organisation for Economic Co-operation and Development (OECD), and the American Action Forum all forecast significant shortages in workers and needed skills.

The situation has not changed much. Executive search agencies (Baker, 2021) are finding talent demands to be as high as at any point in recent history. Without enough qualified workers to fill space technology jobs, the space ecosystem will stall — space technology patents will remain undeveloped, and the pace of space technology innovation and advancement will be stunted.

This workforce shortage is happening for several reasons. Prior to the pandemic, the number of available jobs was outpacing the number of people willing and available to work — and not just in space-related fields. Data showed that the U.S. set new records for the number of open jobs exceeding the number of unemployed workers and shrinking applicant pools for space-related positions. Space-related fields were less affected than many others by COVID-19 shutdowns and layoffs, enabling the industry to start to address deficits through greater access to workers. The industry is also successfully weathering an unstable post-pandemic economy, but employment gains are still not enough to fill the pipeline of workers required for the thriving space ecosystem that's both possible and necessary.

Additionally, many of the most specialized workers in space technology (the Baby Boomer generation) are aging out of their fields and have not kept pace with changes in technology or their respective industries. A new crop of workers has not been trained to replace them. As a result, the number of eligible workers with the STEM-related skills necessary for many space technology jobs is nearing crisis levels.

Consider NASA's staffing, a 2019 report from the Office of the Inspector General found that 28% of the total science and engineering employees, or roughly 3,000, are currently eligible to retire, with an additional 2,000 employees eligible within the next five years.

On a wider workforce front, 25% of the manufacturing workforce is over 55 years of age. A survey by the Manufacturing Institute's Center for Manufacturing Research with partner Alfred P. Sloan Foundation found that 97% feared the loss of institutional knowledge in the manufacturing workforce when aging workers retire. In addition to educating and training the next generation of space workers and professionals, programs are needed to extend the productivity of aging workers and create ways to transfer their knowledge.

"Manufacturers across the country are seeking creators — those individuals who want to help solve tomorrow's challenges today," said National Association of Manufacturers

President and CEO Jay Timmons.

Our industry is growing, and we need passionate, driven individuals to join our ranks. As the skills gap widens, 2.4 million positions across the manufacturing industry could go unfilled by 2028, so we must work to attract and develop new talent (Manufacturing Institute, NAM and Samsung Seek Creators at MFG Day Event in South Carolina, 2019).

Contributing to the skills shortage, 50% of U.S. jobs will require some level of STEM training by 2026, but schools — historically, the primary workforce training ground — are only graduating about 15% of students in STEM subjects and just one-tenth of the computer science students needed to fill those positions. These numbers will likely be further impacted by educational lapses caused by the pandemic.

And while vocational school training is on the rise, it is still recovering from huge drop-offs in enrollment in the 1980s and 1990s. Technical schools also struggle to combat the perception that their programs are subpar, a catch-all for the students who couldn't get into traditional college.

These workforce issues pose a significant threat to future space innovation. Even with a push to leverage more machines, such as using AI models and predictive analytics to handle some of the work traditionally done by humans, people are still needed to build the models, interpret or assess data, and take appropriate action with insights.

And this problem is not just isolated to the U.S. A study conducted in conjunction with the Australian Space Agency (SmartSat Cooperative Research Centre, 2021) shows that while they hope to add 20,000 space-related jobs, requiring 319 identified hard and soft skills, by 2030, there are pervasive skill shortages, as well as gaps in training providers. Even countries as small as Indonesia are seeking to solve for building and training a new space workforce. Researchers Agustan et al. (2018) have looked at how space technologies like satellites and remote sensors will be key to supporting national development initiatives across the political, economic, defense and security spectrum for the "world's largest archipelagic country with terrestrial and marine terrain."

Obstacles to Developing a Modern Space Workforce

As outlined, the future of the space ecosystem depends on a workforce of people at every level — laborers to suppliers to highly skilled space technology experts — to develop, commercialize and support the next generation of technologies for space-reliant industries. Equally imperative is a trained workforce, nurtured to ensure long-term, rewarding careers that enable entrepreneurship and progress.

Yet, there are obstacles standing in the way of building and sustaining a robust space technology workforce, including:

Limitations of the Current Educational System

Traditional Pre-K–20 education alone has not moved the needle in driving new generations of students to pursue careers in space. One reason is that space events haven't risen in popularity since the 1960s (with the exception of modest bumps around SpaceX launches), so the concept of space is often taught in terms of history — Sputnik, Apollo 11 Moon landing, the Challenger and Columbia accidents — as much as it is in science. Space-based applications across a broad range of industries are not highlighted in modern curriculums, which take a very narrow view of space.

Magnifying this issue, it was revealed that students associate astronauts, engineers and programmers with space exploration but not careers in farming, and only 25% recognized the connection between the products they use in their daily lives and space technology innovations. Discussions and activities are often limited to space exploration or studying the names and locations of planets and constellations rather than teaching that space is a fertile ground for problem-solving and entrepreneurship.

To learn more about space and engage with content beyond the basics, space education is often relegated to extracurricular activities and outside agencies. Formal programming that engages students and highlights space innovations and opportunities is severely lacking. Furthermore, what little does exist is constantly threatened by budget constraints.

Beyond STEM and space education, a growing perception exists that schools generally fail to prepare current and future generations for the workplace. *The Global Learner Survey* (2019) indicates that students feel like they are not ready for the modern workforce and that necessary higher education and/or job training is cost-prohibitive or viewed as insufficient. As a result, students are forced to look for new pathways to enter the workforce. Unfortunately, even when they find and pursue alternative routes, the opportunities to break into their chosen professions remains elusive.

Lack of Minority Representation and Missing Demographics

As in many industries, the lack of diversity in the space technology workforce has become a systemic problem. Women, minorities, veterans and other underrepresented groups are an untapped demographic that can positively and productively help address the void of space entrepreneurs as well as workforce shortages.

Women, for example, make up nearly half the U.S. labor force, but currently represent only 27% of employees (Martinez and Christnacht, 2021) in STEM-related fields, even as research across sectors proves that the most innovative and productive companies are those that include women in leadership roles and maintain gender balance. A Bayer-commissioned study (*Making Science Make Sense,* n.d.) that surveyed STEM department chairs at 200 top research institutions found that women and other underrepresented groups drop out of their respective degree programs at a disproportionate rate to white males, even when they are well prepared. They are also routinely steered in other directions by professors and advisors. According to a 2019 report from Boston Consulting Group (BCG), startups led by women are less likely to survive, despite being very successful as individuals. Analysis suggests that global GDP might rise by 3-6% and the global economy by $2.5-$5 trillion if men and women participated equally as entrepreneurs (Unnikrishnan and Blair, 2019).

Another comprehensive study (Eagan et al., 2014) revealed that only 29% of Latino students, 25% of Native American students, and 22% of Black students finished their STEM degree program within six years, while their white and Asian counterparts fared better — with a completion rate of 43% and 52%, respectively. Minimal support mechanisms are accessible to help close these gaps, resulting in the frequent elimination of space technology career trajectories before they start.

Compounding the barrier to entry, the companies hiring space technology workers tend to place an undue emphasis on traditional, fixed skills and degrees that align with outdated job descriptions. Instead of placing value on soft skills, proven problem-solving abilities, and unconventional yet effective professional or educational backgrounds, companies miss out on potential candidates with much to offer.

Military veterans are an example. A veteran who joined the military straight out of high school, led teams in combat or reconnaissance missions, is highly proficient with GPS technologies and has practical experience with materials used to make uniforms for a variety of climates might not seem like a natural fit for a space tech startup based on a resume that does not include a college degree in STEM or product design. Nevertheless, his or her skills, perspective and experiences could prove invaluable to the company if given a chance. Yet, the doors of opportunity currently remain closed to some of the best candidates for space tech jobs.

In her work, House Committee on Science, Space and Technology Chairwoman, Rep. Eddie Bernice Johnson (D-TX) shared the view that it is important "to develop and implement evidence-based policies to promote the progress of women, minorities, and other groups underrepresented in STEM studies and research careers." She went on to say:

The need for full engagement in STEM by women and underrepresented groups goes beyond enabling individuals to fulfill their dreams of becoming a scientist. Our economic future relies on what we do now to nurture the STEM talent that will be necessary to meet the demands of an increasingly technological and knowledge-based economy (Chairwoman Johnson and Ranking Member Lucas Introduce STEM Opportunities Act, 2019)

Insufficient Lifelong Learning Opportunities for Students to Professionals

Practices established during the Industrial Revolution assumed that once workers entered a field, that is where they would stay for the rest of their professional lives, and no further training would be needed. Technology has changed this expectation and mindset significantly, but the way companies consider their workforce has not.

Surveys show that 64% of workers (*Does Job Hopping Help Or Hurt Your Career?*, 2018) are in favor of job-hopping, often to pursue new challenges and higher salaries which are commensurate with skill level. The average employee tenure (*Employee Tenure Summary*, 2022) is 4.2 years. This number drops to just 2.8 years for employees ages 25-34. It is also anticipated that as many as 40% of U.S. workers will quit their jobs post-pandemic, causing a dramatic reshuffling of the labor force (Walsh, Pandey, and Muller, 2021).

Prudential Financial's *Pulse of the American Worker Survey: Post-Pandemic Work & Life – Expectations from the American Workforce* (n.d.) found that 48% of workers are completely reconsidering the type of job or work they want. Of those expecting to leave, 34% of workers are seeking professional growth opportunities and 24% are sick of working on the same things.

A Pearson survey (*The Global Learner Survey*, 2020) further supported the need for career upskilling with 24-76% of global learners, depending on region, needing further education to do their job in the last two years because their jobs have changed. The problem is that workers are not receiving the kind of continuous job training required to advance knowledge, research and innovation in light of the influence of evolving technology and automation. There is a growing need for lifelong learning from career entry to exit.

Additionally, despite research showing that less than 5% of occupations (Manyika et al., 2017) can be fully automated, there is an enormous fear that workers in space technology will be replaced by machines. McKinsey & Company (Lund et al., 2019) found that at least 30% of tasks could be automated in 60% of jobs by adapting technologies. Putting this in perspective, the World Economic Forum projects (Russo, 2020) that the division of labor between human and machine will shift dramatically, "leading to 85 million jobs being displaced and 97 million new ones created by 2025."

This highlights the urgent need for reskilling as technologies give rise to new positions for human workers with skill sets that can leverage big data, artificial intelligence, machine learning and more. Although future jobs may look different, there will be greater demand for augmenting and enhancing human capabilities in both soft and hard skills — for nurturing space entrepreneurship and jobs that haven't even been thought of yet. To secure their place in the workforce, individuals and employers will need new tools and solutions for continuous job training.

But, returning to school to prepare for a new, more rewarding career, augment an existing career or acquire the requisite skills for a post-machine, post-COVID world is not always possible. Between the expense of tuition and the time involved in completing traditional coursework, many workers find adult education incredibly difficult to fit into their lives, particularly as family responsibilities increase.

Approximately 74% of the 11,000 respondents to Pearson's survey of global learners indicated that colleges and universities offer only limited options for working adults. Instead, the workforce is turning to the internet for DIY online self-teaching and short courses offered by employers, associations or boot camps. The companies that succeed in the future will be those that step up to support ongoing upskilling and reskilling programs that benefit both workers and the organizations that benefit from employees' acquisition of essential skills.

Absence of Mentoring and Role Models

There was an unprecedented level of excitement and national recognition of space programs in the 1960s, prompting many to pursue their dreams of working in space-related fields, but that level of "space as my chosen career path" excitement has waned over the years as people mistakenly thought that only a select few can make the cut and that space careers were limited to space exploration. Exposure and popularity have played a crucial role. In a study by Lego and The Harris Poll (PRNewswire, 2019), 86% of polled students ages 8-12 have an interest in space exploration, but perhaps not surprisingly, youth in the Western Hemisphere chose being a YouTuber as their first career choice over being an astronaut, their last choice.

Part of the problem? From Pre-K–12 classrooms all the way up to the top space professionals, people fail to see others with whom they can identify, whether that is due to dissimilar race, gender, economic background, educational background or another classification. Without an example of someone "like me" in a space technology career, many abandon ambitions of space entrepreneurship and revert to fixed, exclusionary notions of who space workers are and what they do.

Dr. Mae Jemison, the first African American woman in space (#TeamEBONY, 2019) highlighted how the problem of exclusion has been evident as far back as the height of space enthusiasm, with many people failing to identify with those involved in space exploration, "even though we were there." Jemison noted this problem is not isolated to space. "Matthew Henson [Arctic explorer] was there. Lewis Latimer [inventor/draftsman] was there," she pointed out, emphasizing how minorities have long been underrepresented and unheralded in exploration and scientific achievement.

Even when role models are present, those who want to break into space technology have limited opportunities to engage with mentors. Doing so would allow them to learn more about potential careers, the technologies being utilized, and the coursework required to understand them, as well as how to persevere in their programs. They could also learn how to make the necessary business connections to enter space-related fields or pursue their entrepreneurial dreams of turning an idea into a commercially viable solution.

There remains little visibility into the expansive space ecosystem despite the removal of barriers to entry. People simply don't have access to robust, mentor-driven programs and resources that can inspire, guide and bridge them over to the space movement.

FUTURE RESEARCH

Commercializing Space Technologies to Improve Life in Space and on Earth

Virtually every industry on the planet has advanced as a result of tech transfer – the transfer of knowledge that enables tech advancements to reach other users, organizations or industries who can further develop or co-opt it. In this case, technologies designed for space have been adopted and commercialized for a variety of purposes for life on Earth. Such products and services can then be inserted -into established enterprises. This is the true return on investment of every space program.

Healthcare, transportation, communications, agriculture, energy, security, business and more have been transformed by space, and that transformation is only going to continue through the support of space policy, funding, engagement and new applications. Companies like Microsoft, Google and Garmin are using space technologies to provide navigation and tracking services, while real-time data is being used to track and monitor endangered marine species, locate survivors trapped in buildings after natural disasters, and find sources of safe drinking water. At the same time, digital twins are taking off to simulate and track environments and how equipment functions at unprecedented depth and scale, in ways previously impossible to mimic. The real-time insights digital twins produce have the potential to save lives (as well as billions of dollars in investments) and solve some of the most pressing issues of our time — and they can be deployed across industries in a seemingly infinite number of use cases.

But it is not just about what we can apply back on Earth today. There remain some great mysteries to uncover "out there" as well. With three missions to Mars over the last year, we are learning about our planetary systems in ways that were not previously possible. With the information supplied via digital twins, we can safely explore uncharted territory, not to mention successfully navigate around systems of satellites and reduce space debris. The implications of what is yet to be discovered could be enormous for tourism, human settlements beyond our home planet, and sustainability strategies for Earth that preserve and enhance life for future generations.

With luminaries like Elon Musk, Jeff Bezos, Richard Branson and others making space cool again and scientists like Jill Tarter studying extraterrestrial intelligence or Louis Allamandola searching for the chemical origins of life, new questions are being asked and answered. New seeds of excitement in space are being planted — now we just have to nurture them.

Not every person will get to ride on a rocket, see one lift off, or build a satellite or payload, but they can contribute to delivering space technology solutions that we depend on today and tomorrow.

In particular, NASA's patent portfolio offers thousands of technologies for commercialization. It, as well as the Pentagon and other government and research organizations, support aggressive tech transfer programs that are at risk without the vision and talent of entrepreneurs to craft, refine and apply these technologies to real problems in space as a new frontier as well as on planet Earth — that is if we can build a broad-reaching space ecosystem that truly unites us all to further innovation.

SOLUTIONS AND RECOMMENDATIONS

Space has never been more important. It unites and inspires people around the globe, and many public service and non-profit organizations like Space Foundation are committed to helping people find their place in the global space ecosystem. If we are truly focused on extending the reach and rewards of this new era of access and opportunity, now is the time for our extensive partner ecosystem to step up and support our mission to be the preeminent advocate and gateway for trusted information, lifelong education, and seamless collaboration for all people and organizations that engage in space exploration and space-inspired industries.

This is a tall order, but it is possible. Space Foundation's Workforce Development Roadmap provides fresh pathways for space awareness and inspiration at every stage, as well as creating scalable support structures — for access, networking and mentors — to cultivate a robust workforce candidate pool and turn ideas into viable contributions. Once candidates are trained, strapped in and careers are launched, workers require easily accessible tools, venues and collaborations to hone their skills and advance their careers.

To bring this approach full circle, a vibrant and capable workforce community will need to cultivate vital role models who can spark interest in future generations of entrepreneurs and lifelong learners. Their resulting innovations will carry the responsibility for retaining space as a critical infrastructure for the betterment of life on Earth for all global citizens.

For the entire workforce life cycle, lifelong learning is essential for students through senior professionals with a continuum of education and training at every stage, from career entry to exit. By offering end-to-end workforce development programs, research and education conducted throughout the industry and supported by a vibrant ecosystem of partnerships and mentors, the global space ecosystem will be able to engage and retain a qualified space technology workforce for the future.

Five Core Disciplines

To fully serve a deeply interconnected web of lifelong learners and satisfy the priorities outlined here, Space Foundation follows and advocates across five core disciplines.

1. Awareness of Space Technology's Impact and Opportunities: Raising space awareness has been the lifeblood of Space Foundation since its inception, but education and awareness of space technology must go beyond traditional education — and beyond the narrow views of space exploration — to secure a viable future workforce for the space ecosystem.

Commercializing technologies designed for space exploration but promising a better life on Earth is big business for aerospace as well as virtually every industry. There is a tremendous opportunity in the market to fill the innovation gap and help bring space technologies to market. Thousands of patents at NASA alone are waiting for commercialization, and enterprises across a broad spectrum of industries are actively seeking tech insertion of new commercialized products and services.

By highlighting the wide range of work conducted and the diverse backgrounds of the people involved, Space Foundation showcases the breadth and depth of the modern space workforce through its Symposium 365 Digital Platform. Serving commercial, government and education sectors, Symposium 365 produces signature programming — live and online — for professionals, entrepreneurs, educators and students to learn, connect and conduct business 365 days a year.

To further educate and inform a global audience of space enthusiasts, Space Foundation's research and analysis team publishes TechTrack Papers, Capitol Hill Updates, and The Space Report, a quarterly online publication that serves as an unbiased, authoritative guide to global space activities. Space Foundation also conducts a broad range of programs_to educate policymakers about the complex space-related topics that must be addressed to ensure that the space industry is well-represented on critical space policy issues.

Another way Space Foundation raises awareness is by profiling real-world use cases in a broad range of industries in its Space4U podcast series as well as leveraging firsthand accounts from young leaders and professionals at regional workshops, webinars and conferences to make space technology relatable to peers and aspiring space contributors.

Space Foundation also provides much-needed validation for the products and services developed out of space technologies through its Space Certification Program. Space Foundation acknowledges and celebrates the individuals and companies making the biggest differences in the industry by honoring Space Technology Hall of Fame inductees at its annual Space Symposium.

2. Access That Opens Doors to Space Industry Jobs, Careers and Contracts: The space workforce spans from STEM experts to laborers, and the future of the space ecosystem is a collaboration of communities, public and private companies, government entities, colleges and individuals — dynamics that open new pathways to all interested parties. Space Foundation's Center for Innovation and Education brings together these constituencies through partnership, sponsorship, fundraising and grants to serve the entire life cycle of the workforce. Its programs are designed to enhance the outlook and opportunities for new jobs, careers and new business ventures in the space ecosystem.

For example, through its Space in the Community program, Space Foundation endeavors to demystify access to space industries so that everyone can be a part of the space ecosystem at any stage in their career.

The Space Commerce Entrepreneurship team at Space Foundation engages with communities, entrepreneurs, small-to-medium businesses, educational institutions and incubators to determine needs and curate custom-tailored programs, workshops and events that further space technology innovation and open doors for participation and contributions.

For example, Space Foundation partnered with an incubator and state university in Youngstown, Ohio, to conduct a well-attended space commerce workshop. Businesses, entrepreneurs, innovators and students spent the day addressing the most common challenges companies face in growing their space- based technology businesses.

Space Foundation also works diligently to welcome underserved groups such as minorities, women and veteran-owned businesses to the space technology community. Through a grant from the U.S. Department of Commerce's Minority Business Development Agency (MBDA), Space Foundation provides an online, self-driven webinar series as well as regional training workshops through its Space Commerce Entrepreneurship program.

More than 275 minority enterprises have participated in this program within the first nine months of its inception. Small businesses gain the necessary insights and skills to open new revenue streams. They could become a supplier to public and private enterprises or government agencies in the space industry,_expanding into space-based technology industries, such as agriculture, telecom, finance and more. Individuals are also likely to find more rewarding career options.

The outlook for identifying additional demographics includes people in traditional roles displaced by automation in changing work paradigms, aging workers or retirees. Space Foundation is committed to offering professional support to empower these often-forgotten demographics through reskilling in fields such as artificial intelligence, machine learning and big data analysis for space technology industries.

3. Training for Lifelong Learning at Every Stage: Careers are no longer linear, and technology is transforming workplace dynamics at a rapid clip. Compounding this uncertainty, traditional education cannot keep pace with the requirements of a quickly evolving job market. Recognizing these changes, Space Foundation is a strong advocate for lifelong learning as an investment in continuing education throughout the career life cycle.

For Pre-K-12 students, Space Foundation delivers space education beyond the textbook by developing immersive experiences that make space technology come alive to students through innovative curricula in Discover the Universe field trips, Explore Space Experience summer camps, and an extensive outreach program to regional school districts.

Space Foundation Discovery Center in Colorado Springs, Colorado, is one of the most advanced space-based educational centers in the world, featuring immersive, hands-on exhibits through the El Pomar Space Gallery, Northrop Grumman Science Center, and the Lockheed Martin Space Education Center.

Space Across the Curriculum prepares teachers and educators to integrate space principles throughout their daily core curriculum. Through extensive partnerships with aerospace leaders like Northrop Grumman and Lockheed Martin, these programs inspire, educate and connect students to the broad range of space-based careers and bolster not only pursuit of STEM education but also build essential business and life skills such as creativity, critical thinking and problem-solving.

In fact, Lockheed Martin recently sponsored Space Foundation's Junior Space Entrepreneurs Program (JSEP), which won an honorable mention in Fast Company's World Changing Ideas Awards. JSEP was designed with the goal of building the future workforce of space professionals. The weeklong program inspires youth to secure STEM positions within the space industry through real-world simulations, which teach and connect critical 21st century skills to real-world work environments.

On the post-secondary level, Space Foundation is actively establishing partnerships with corporations, technical schools, community colleges and research universities to develop curricula and credentialing programs. A Space Commerce Webinar and Speaker Series provides preliminary self-service tutorials on preparing for business engagements.

Through collaboration with communities and incubators, the Space Commerce Entrepreneurship Program delivers one-day workshops that bring lessons to life with subject matter experts preparing not only a new crop of workers for the job market but also reskilling and upskilling those who are already a part of the workforce.

Movement through career paths — whether from job to job or up the corporate ladder — requires new and updated skills to be done effectively. Space Foundation is dedicated to opening pathways through its Space Commerce initiatives that make learning new skills a lifelong endeavor. Through the support of grants and sponsorships, these business development programs assist in the persistent training of the space technology workforce.

Collaborations facilitated by Space Foundation keep skills fresh without upending people's lives or contributing to greater student debt. By making ongoing reskilling and upskilling easy, affordable and accessible, Space Foundation helps workers meet modern demands and continue to serve as active members of the space technology workforce.

4. Connections to People Through a Vast Network: Gaining entry to the space community, like many fields, is hastened by connections. As one of the most respected authorities on space since 1983, Space Foundation inherently has one of the deepest worldwide networks for connecting workforce candidates to space professionals and industry opportunities. The massive reach of its Global Alliance division spans the world, while also working at the local, regional and national levels in order to have the maximum impact.

For example, Space Foundation's annual Space Symposium is the leading international event for the global space community. Attracting more than 15,000 representatives from around the world from the military, civil and commercial space sectors, Space Symposium is where they all gather to examine space issues from multiple perspectives, promote dialogue, conduct new business, and build relationships to advance shared interests.

On a regional basis, Space Foundation brings together communities, government entities and school districts through events, workshops and virtual programs. It understands the needs of each of these constituents and how they all must come together to achieve future workforce development and space innovation goals.

A number of Space Foundation mentorship initiatives connect students and young professionals to space entrepreneurs and space professionals for inspiration and guidance. Other programs ensure that entrepreneurs and small businesses meet with the right people to take their ideas and solutions forward to commercialization and business opportunities.

The New Generation Leadership program connects promising young professionals (ages 35 and younger) to space professionals that can provide real-world career advice, guidance and job roadmaps. Additionally, the new Swigert Society Young Leaders program connects tomorrow's leaders with philanthropists who want to make substantial innovations a reality by providing funds that will jumpstart promising efforts.

5. Mentorship of Young Leaders to Develop Role Models for the Next Generation: Recognizing that people are the most powerful assets, Space Foundation is focused heavily on expanding the role of mentorship at every stage. This does not only refer to mentorship in the traditional sense, but also cultivating relationships with those who can serve as coaches and champions as well.

Mentors may focus on very specific areas in which they can help an individual learn or improve, such as public speaking or budgeting, while coaches are concerned with how to make one a better leader overall. Champions tend to be those who rank several levels above the people they seek to help; they are the ones who inspire and open the right doors to enable individuals to reach the next level of an organization.

The remarkable people involved with the Space Foundation global ecosystem recognize that they would not be where they are today without contributions from mentors, coaches and champions who share everything from invaluable tiny nuggets of advice to career-defining moments. They look for ways to give back to others in an effort to ensure the future health of the space community.

Space Foundation International Teacher Liaisons, for example, comprises an extraordinary group of educators from nearly 15 countries across the globe who use space-related education programs and principles to inspire the next generation of world changers, as well as advocate for space-inspired education in their schools and communities. Similarly, Space Foundation's NewGen Ambassadors mentor middle and high school students, and its Senior Leader Mentors also guide young leaders, entrepreneurs and small businesses.

These mentors, coaches and champions not only aid in skill development and career advancement but give the curious an inside look into professions, thereby helping to drive and motivate them to pursue passions and continue developing talents throughout their careers. Integrating contributors with experienced leaders in the workplace will allow for active engagement, collaboration and the development of a thriving space industry.

CONCLUSION

We have a vision of a whole connected web of people engaged in space, and by their acts of engagement, we will satisfy core economic needs while still making the world a better place through collaboration. It is a vision shared by many and that will benefit millions of people worldwide.

The work being conducted by the Digital Twin Earth project, which has created a dynamic, digital replica of Earth that accurately mimics the planet's behavior is remarkable. One can imagine how teams can use the insights gleaned from such an effort to solve problems from climate change to food insecurity. We will need people to take those insights and apply them across all of the industries they touch, with workers at every level required at each step to generate change, which will in turn drive economies.

It is also worth considering all the people who come together for a space flight, one made possible by the efforts of countless members of the space community. Because of their work, we can now leverage things like digital twins to prepare for every conceivable variable. This preparation and knowledge will potentially prevent a catastrophic scenario by dramatically reducing risks and determining just the right adaptations. Instead of tragedy, we will walk away with even more knowledge that can be shared with an entire ecosystem for greater advances, jobs and discoveries.

Space has always been at the frontier of innovation, and it has never been more important for us to explore and understand how space can be used to better life on Earth. To do that, we need new kinds of tools and approaches, such as digital twins. These tools are imperative to our future — but they don't operate themselves. We have to have people to run them, people who understand them and know how to get the most out of technologies, and people who can continue pushing forward by applying key learnings in the real world. Billions of dollars — and people's lives — are at stake.

A better future awaits, thanks to advances brought forth by the global space ecosystem — but to get there, we need a fully trained and engaged workforce.

Fortunately, in this new era of access and opportunity, there is tremendous upside to space careers with room for anyone interested to advance in an industry that will surpass the trillion-dollar mark within the next 20 years. What's so exciting about this environment is that it provides the space community not only the chance to offer lucrative opportunities to workers displaced by the pandemic but also welcome new faces that historically have been missing from the space ecosystem and industry at large. That is only part of the movement.

It is also vital that we engage people at every stage of life — whether this means getting students excited about space-related careers at a young age, upskilling and reskilling the existing workforce, or establishing relationships between senior professionals and up-and-comers.

By inviting a whole spectrum of people into the space community, we can ensure that innovation continues well into the future so that our cities, and rural areas, grow smarter, safer and more plentiful — and so that life on Earth becomes truly sustainable.

REFERENCES

Agustan, S. M., Yulianto, S., Sutrisno, D., Bintoro, O. B., & Alhasanah, F. (2018). *Participatory Scenario Planning for Indonesian Space Industry in 2025.* IEEE International Conference on Aerospace Electronics and Remote Sensing Technology (ICARES), Bali, Indonesia.

Ammara, Ul., Rasheed, K., Mansoor, A., Al-Fuqaha, A., & Qadir, J. (2022). *Smart Cities from the Perspective of Systems.* Systems.

Baker, P. (2021). Upskilling and reskilling employees for a competitive edge. *SearchHRSoftware.* https://www.techtarget.com/searchhrsoftware/feature/Upskilling-and-reskilling-employees-for-a-competitive-edge

Bayer. (n.d.). *Making Science Make Sense.* https://www.bayer.com/en/us/makingsciencemakesense/home

Bolton, E., & Bauerlein, A. (2020). *Space Workforce 2023: Executive Leadership Pledges.* National Academies. Retrieved from https://www.nationalacademies.org/documents/embed/link/LF2255DA3DD1C41C0A42D3BEF0989ACAECE3053A6A9B/file/D3063C38C8FAB019DB855143D5B047EB77160FD3E974?noSaveAs=1

Committee on Science, Space, and Technology. (2019 May 17). *Chairwoman Johnson and Ranking Member Lucas Introduce STEM Opportunities Act*. https://science.house.gov/news/press-releases/chairwoman-johnson-and-ranking-member-lucas-introduce-stem-opportunities-act

Does Job Hopping Help Or Hurt Your Career? (2018 April 5). Robert Half. Retrieved from https://press.roberthalf.com/2018-04-05-Does-Job-Hopping-Help-Or-Hurt-Your-Career

Eagan, K., Hurtado, S., Figueroa, T., & Hughes, B. E. (2014). *Examining STEM pathways among students who begin college at four-year institutions*. National Academy of Sciences.

Employee Tenure Summary. (2022 September 22). U.S. Bureau of Labor Statistics. Retrieved from https://www.bls.gov/news.release/tenure.nr0.htm

Expo 2020 Dubai. (2020). *The human-centric city of the future*. https://www.expo2020dubai.com/

Hóu, Z., Li, Q., Foo, E., Dong, J. S., & de Souza, P. (2022, March). A digital twin runtime verification framework for protecting satellites systems from cyber attacks. In *2022 26th International Conference on Engineering of Complex Computer Systems (ICECCS)* (pp. 117-122). IEEE. 10.1016/j.jmsy.2019.10.001

Li, D., Jiao, J., Wang, S., & Zhou, G. (2022). *Satellites in Emergency Management: A Bibliometric Analysis*. Research Square., doi:10.21203/rs.3.rs-1533351/v1

Lund, S., Manyika, J., Hilton Segel, L., Dua, A., Hancock, B., Rutherford, S., & Macon, B. (2019 July 11). *The future of work in America: People and places, today and tomorrow*. McKinsey & Company. https://www.mckinsey.com/featured-insights/future-of-work/the-future-of-work-in-america-people-and-places-today-and-tomorrow

Manufacturing Institute. (2021). *The Aging of the Manufacturing Workforce*. Manufacturing Institute. https://www.themanufacturinginstitute.org/research/the-aging-of-the-manufacturing-workforce/

Manufacturing Institute, NAM and Samsung Seek Creators at MFG Day Event in South Carolina. (2019, October 4). National Association of Manufacturers.

Manyika, J., Lund, S., Chui, M., Bughin, J., Woetzel, J., Batra, P., Ko, R., & Sanghvi, S. (2017 November 28). *Jobs lost, jobs gained: What the future of work will mean for jobs, skills, and wages*. McKinsey & Company. https://www.mckinsey.com/featured-insights/future-of-work/jobs-lost-jobs-gained-what-the-future-of-work-will-mean-for-jobs-skills-and-wages

Martinez, A., & Christnacht, C. (2021). *Women Are Nearly Half of U.S. Workforce but Only 27% of STEM Workers.* United States Census Bureau. https://www.census.gov/library/stories/2021/01/women-making-gains-in-stem-occupations-but-still-underrepresented.html

NASA. (n.d.). *Patent Portfolio*. https://technology.nasa.gov/patents

National Association of Manufacturers. (2021). *Manufacturing Institute, NAM and Samsung Seek Creators at MFG Day Event in South Carolina.* https://www.nam.org/manufacturing-institute-nam-and-samsung-seek-creators-at-mfg-day-event-in-south-carolina-6047/

PR Newswire. (2019 July 16). *LEGO Group Kicks Off Global Program To Inspire The Next Generation Of Space Explorers As NASA Celebrates 50 Years Of Moon Landing*. Author.

Pearson. (2020, August). *The Global Learner Survey*. https://news.prudential.com/presskits/pulse-american-worker-survey-post-pandemic-work-life.htm

Pearson. (2021). *The Global Learner Survey*. https://www.pearson.com/content/dam/one-dot-com/one-dot-com/global/Files/news/gls/Pearson_Global_Learner_Survey_2019.pdf

Prudential. (n.d.). *Pulse of the American Worker Survey: Post-Pandemic Work & Life*. https://news.prudential.com/presskits/pulse-american-worker-survey-post-pandemic-work-life.htm

Qi, Q., Tao, F., Hu, T., Anwer, N., Liu, A., Wei, Y., Wang, L., & Nee, A. Y. C. (2021). Enabling technologies and tools for digital twin. *Journal of Manufacturing Systems*, 3-21.10.1016/j.jmsy.2019.10.001

Rathje, J. M., & Katila, R. (2020). Enabling Technologies and the Role of Private Firms: A Machine Learning Matching Analysis. *Strategy Science*.

Robinson, D. K. R., & Mazzucato, M. (2018). The evolution of mission-oriented policies: Exploring changing market creating policies in the US and European space sector. *Research Policy*, 936–948.

Russo, A. (2020 October 20). *Recession and Automation Changes Our Future of Work, But There are Jobs Coming, Report Says.* World Economic Forum. Retrieved fromhttps://www.weforum.org/press/2020/10/recession-and-automation-changes-our-future-of-work-but-there-are-jobs-coming-report-says-52c5162fce/

Shangguan, D., Chen, L., Ding, J. (2020). A Digital Twin-Based Approach for the Fault Diagnosis and Health Monitoring of a Complex Satellite System. *Symmetry in Mechanical Engineering, 2.*

SmartSat Cooperative Research Centre. (2021). *Space industry skills gap analysis.* Author.

Space Foundation Editorial Team. (2022, July 27). *Space Foundation Releases The Space Report 2022 Q2 Showing Growth Of Global Space Economy.* Space Foundation. Retrieved from https://www.spacefoundation.org/2022/07/27/the-space-report-2022-q2/

TeamEbony. (2019 September 12). *Mae Jemison Fights For Diversity In Space And In The Classroom.* Ebony. https://www.ebony.com/mae-jemison-fights-for-diversity-in-space-377/

The LEGO Group. (n.d.). https://www.prnewswire.com/news-releases/lego-group-kicks-off-global-program-to-inspire-the-next-generation-of-space-explorers-as-nasa-celebrates-50-years-of-moon-landing-300885423.html

Unnikrishnan, S., & Blair, C. (2019). *Want to Boost the Global Economy by $5 Trillion? Support Women as Entrepreneurs.* Boston Consulting Group. https://www.bcg.com/publications/2019/boost-global-economy-5-trillion-dollar-support-women-entrepreneurs

Walsh, B., Pandey, E., & Muller, J. (2021 June 21). *Axios What's Next? The age of the career-jumpers.* https://www.axios.com/newsletters/axios-whats-next-24766a4c-92cf-43cf-8ae9-540980b5f3f4.html?chunk=1&utm_term=emshare#story1

Weinzierl, M. (2018). Space, the Final Economic Frontier. *The Journal of Economic Perspectives*, *32*(2), 173–192. doi:10.1257/jep.32.2.173

Zhao, Q., Yu, L., Du, Z., Peng, D., Hao, P., Zhang, Y., & Gong, P. (2022). An Overview of the Applications of Earth Observation Satellite Data: Impacts and Future Trends. *Remote Sensing*, *14*(8), 1863. doi:10.3390/rs14081863

KEY TERMS AND DEFINITIONS

STEM: Science, technology, engineering, and mathematics.

Tech Transfer: The transfer of knowledge that enables tech advancements to reach other users, organizations or industries who can further develop or co-opt it.

Chapter 7
Technological Dream or Safety Traumata?
Fire Protection in Smart Cities – Digitization and AI Ensure Burning Ideas and a New Culture of Thinking in Construction 4.0

Bianca Christina Weber-Lewerenz
https://orcid.org/0000-0002-8406-7119
RWTH Aachen University, Germany

ABSTRACT

The new living and urban worlds of tomorrow and the sustainable design of space and society require even more efficiency, comfort, and security. Technological progress and safe design of smart cities can only be achieved with the help of a will to innovate and technical know-how of the partnering network constituting the building project team. Digital twins, AI, machine learning, deep learning, intelligent sensor systems, intelligent networking between buildings, artificial intelligence of things, clouds represent highly efficient technologies to both optimize fire safety and fire protection and increase human and building operation safety as part of the implementation of a holistic and comprehensive building protection solution in smart cities. A model says more than a thousand words: state-of-the-art fire protection systems, a customized automatized building solution, can be designed, planned, tested, individually adapted, and adjusted with the help of digital twins and AI (e.g., to optimize early fire detection, targeted alarms, and evacuation and to reduce costly false alarms).

DOI: 10.4018/978-1-6684-3833-6.ch007

INTRODUCTION

Architects, civil engineers, urban planners have a particularly high responsibility: to create living space with the highest level of security and comfort for people in Smart Cities environments and to use technologies such as BIM and Digital Twins in a economic efficient, resource-saving manner. On the one hand to protect resources, climate and environment and to design optimized efficient and safe building operations. On the other hand, to develop attractive buildings and smart cities that can be used for generations. They create places to live, work and meet but also predictive basis for decision-making, supported by constantly newly developed technologies (Sheikhnejad et al., 2020). Singapore, Smart Dubai, Songdo in South Korea and Shenzhen in China serve as samples of Smart Cities development.

The misuse of data and cyberattacks represent some of the diverse risks and challenges. It remains high priority to ensure security and safety of information technology, Cybersecurity of servers by supervising weak points and immediate mitigation of security vulnerabilities and the support of highly skilled partner maximizing safety. In order to ensure the efficient operation and safe functioning of networked buildings and to protect them from crises that threaten the existence of the company, sophisticated, highly modern fire protection is required (Liu, Zhang, and Wang, 2020; Fan et al., 2019).

The digital twin seems to be the all-purpose tool of the moment. In the Construction Industry in particular, BIM is the most efficient method to date, to merge technical interfaces on a single data platform and digitally map a building or an urban environment. The digital twin created in such process not only visualizes the building`s shell and interior (Figure 1.), but also all of the technical lines that

Figure 1. Digital twin of building and building operation
Source: https://new.siemens.com/de/de/unternehmen/stories/forschung-technologien/digitaler-zwilling/digital-twin.html

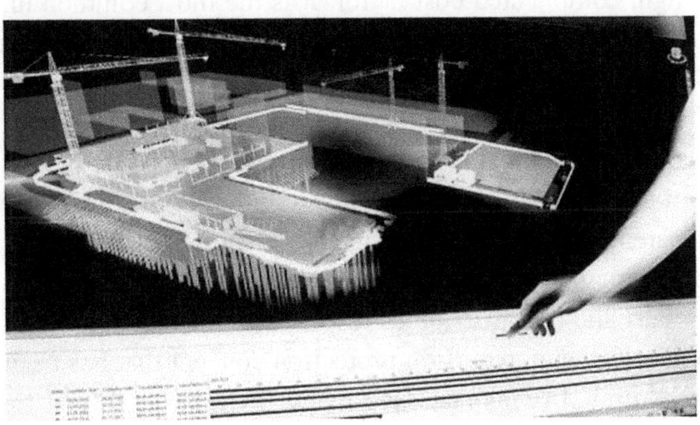

Figure 2. Digital twin of fire protection lines in a building
Source: www.cadsys.de

run through the building and penetrate parts of the building (Figure 2.). Siemens declared ‚Simulation and Digit Twin‘to its “Company`s Core Technology” (Van der Auweraer, 2021).

The digital twin offers to everyone involved one common exchange platform, in real time and without data loss. With increasing technical feasibility, the human need for security and the desire to handle and use new technologies responsibly increases (Weber-Lewerenz, B., 2021a). The research work is dedicated to CDR in Construction (Weber-Lewerenz, 2021b; Weber-Lewerenz, 2021c; Weber-Lewerenz, 2021d, Weber-Lewerenz, 2022a).

Fire safety and fire protection does not only mean safety and protection of human life. Intelligent fire protection, which is modelled using the digital twin on the projected building (Cheng, 2017; Li, 2014), means protection against a significantly high, complicated cost factor. It is the most common misconception to hold plant investors and operators, architects, product manufacturers and many other people responsible for fire protection. There must be a joint effort to engage highly qualified fire protection experts who design a smart and safe fire protection concept that is tailored to local and national requirements. On this foundation and with the help of the digital twin, the building solution can be presented in a visible and cost-transparent manner. The investment is definitely worth it in the event of a fire. But much more important: the building solution involves all those responsible right from the start and thus distributes responsibility evenly.

A Smart Cities fire and explosion protection concept that has been worked out with the help of digital twins can save many investors high construction costs, shorten construction times, enable more efficient plant processes, promote ecological developments, avoid major damage and ultimately save lives. It all starts with the right fire protection concept and only ends with active support in site supervision (the so-called fire protection specialist site management), inventory checks or demolition work.

As part of CDR, this chapter focuses on the changed security needs by investigating if new digital technologies and AI represent a technological dream or safety traumata when it comes to ensuring the overall fire protection in smart cities. Digitization and AI ensure burning ideas and a New Culture of Thinking in Construction 4.0. There is a significant need of new fire safety models in construction to build highly safe, liveable Smart Cities. However, the adoption of smart technologies offers new innovative ways.

FIRE PROTECTION IN SMART CITIES

Our environment gets more and more smart, faster and faster, but especially more complex. From Smart Building to Connected Building to Smart Cities: New modern cities require a new culture of thinking: Smart Cities – the global smart cities market will be valued at $3.6 trillion by 2025 (Iltisberger, 2021). Also, the retrofit of existing buildings (Weber-Lewerenz, 2022b), its upgrade to Smart Cities` fire safety standards and integration of existing buildings in Smart Cities by intelligent connection fully use the potential by digital twins (Bosch, 2021). This is where digitization and AI meets Fire Protection in Smart Cities.

Smart Cities consist in high-tech intelligently connected technologies for building and infrastructure operations. These are routinized highly automatized operations and require a broad knowledge on its functions, need for corrections, monitoring, maintaining and evaluating to meet highest safety standards and sustainability goals (Sovacool et al., 2020; Metallidou et al., 2020). Ideally, buildings networked in smart cities ensure that a building "warns" the neighboring building in the event of a fire, based on IoT (Minarčík et al., 2021; Minoli, 2017).

Due to the generated data volumina and exploded Big Data over the recent years, digitization and AI require high speed data transfer. Servers require extensions to provide increasing data and saving capacities. These servers create heat, the air is continuously purified. A customized fire protection concept detecting the most minor fumes and facilitating high speed action is the answer (Wrick, 2020).

E-mobility, electric cars and especially its batteries require adaption of fire protective measures to stop fires and at the same time professionally handle the highly explosive materials e.g., in a car battery.

Fire and explosion protection gain a whole new level of safety; digital twins help to simulate worst case scenarios to set up customized concepts to protect citizens and residents. Digital twins enable the simulation of lithium-ion batteries in the event of a fire in order to get to the bottom of possible causes of fire and fire behavior in advance. Error rate in fire protection planning can be reduced with the digital twin (DMT Group, 2021). The digital twin helps to play through virtually all questions on fire protection (buildingSMART, 2021) and simulate hybrid fire testing (Schulthess et al., 2020).

The results of the "Fire Protection Industry Barometer" study (Müller, R., 2019) document the unanimity of the experts:

- Approximately 88% of those surveyed are of the opinion that digitization will lead to an increase in efficiency.
- Over 80% of those surveyed are of the opinion that BIM increases planning reliability and productivity.
- Around 60% are of the opinion that BIM will largely replace other planning methods.

The potential of digitization in fire protection is high with increasing tendency (Figure 3. + Figure 4.). The German Fire Protection Branch Statistics in 2019 reflected that digitization is the driver of efficiency: almost 90% of the interviewees confirmed the high potential for efficiency in fire protection. Compared to 2018 (86,4%) 100% confirmed that digital twins are essential for the construction phase. 94,1% consider a high degree of digitization in building operations as high potential. 92,9% estimate high efficiency in permit application processes, 92% for approval and survey phases.

Figure 3. Digital Twin Building Operational System
Source: SIEMENS

Figure 4. High efficiency potentials of digitization in fire protection
Source: Branchenbarometer Brandschutz 2019

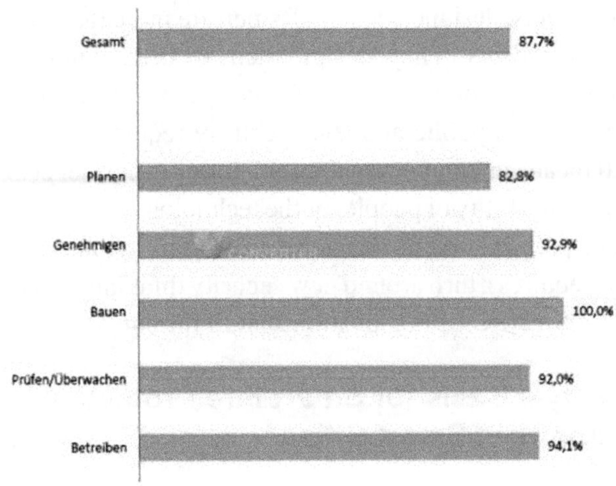

What role can innovative technologies such as digital technologies and AI play in meeting fire protection and fire safety compliance in smart cities? How can these technologies be used in a targeted manner in structural fire protection for more transparency, trust, added value and sustainability?

New innovative technologies for planning Smart Cities equally require responsibility for designing safe, highly efficient fire protection and fire safety, and meeting the need for concrete smart, intelligent but all the more secure building solutions. The degree of digitization of fire protection planning must increase significantly, because the contribution of the fire protection planning software must be firmly integrated into the BIM-based planning (Max, D., 2021). For example, initial walk-throughs or follow-up checks in remote mode are already part of everyday life of best practices. Digital twins integrate recommended hazard-based fire protection concepts into projects at an early stage; not only can optimally fire protection be achieved, but an economically attractive added value, for example by reducing operating costs. However, this could simply be a reminder that e.g. not necessarily 360 doors have to replaced, that were installed years ago, but the test of one of these types could accelerate the approval process to receive the fire authorities` confirmation that they meet today's requirements. The fire protection basic document No. 2 is an indication of the European Commission, defining protection goals of the Fire protection (Europe-wide) (European Commission, 2010).

AI-based technologies take a further step forward. Machine Learning Algorithm (MLA) e.g., can help to rapidly ascertain the optimal dynamics of fire stopping measures to maximize the fire-retardant release effectiveness (released material usage and target impact) (Zohdi, 2021).

Engineers, architects, designers and craftsmen are more than designers of living and working environments. They design their security, intelligent networking and communication - not only in case of emergency - and specifically address the associated changes in public and legal security requirements and regulations. Construction 4.0 means gaining new safety thinking (Agnusdei et al., 2021) and to fulfil "a greater responsibility of people for the technology-based innovations....and for the sustainable preservation of the world ….." (Kirchschläger, 2021).

However, this requires a turn around new security thinking with a new awareness of the dangers that Smart Cities innovations entail for humans at the same time.

Data Complexity – Basis for Smart Fire Protection Building Solutions in Smart Cities

Digitization and AI in construction require data protection that is appropriately adapted to the technologies. In construction, data quantity, data origin, transparency and control, maintenance and updating, data use and processing, data exchange and data interfaces are extremely complex (Ludwig, 2021), sometimes completely confusing and constantly increasing. Compliance with the principles and accountability of the General Data Protection Regulation, the processing of personal and project-related data and compliance with third-party accountability are rarely controlled and even less prosecuted.

Clouds, without which BIM does not work, form a uniform data platform for a construction project. Therefore, clouds require the secure design of access rights, secure data exchange and protection of personal and project-related data.

Close communication and exchange of data with fire authorities reflects best in the "Extinguishing according to building requirements", which means that the fire brigade has access to construction plans and building materials and can therefore act quickly in an emergency. Smart planning of Smart Cities is key as to reach the strived goal, and not use the most modern, intelligent technologies at any price.

Data governance and compliance (Weber-Lewerenz, 2021e) play an important role in fire protection in everyday construction: compliance with the contract, compliance with the legal and building authority requirements and regulations, compliance in building permit procedures (fire protection concept), which are implemented with the help of digital twins, evidence from test engineers and fire protection experts, whose profession is legally regulated (European Commission, 2021), Completeness of the documents (all documents required for a construction project, e.g. official specifications such as fire and fire protection certificates), compliance with standards and guidelines (through automated testing in BIM and

AI-based processes), conformity tests (of building materials and components that require special Fire protection and safety requirements must be met (product and safety data sheets, building material properties).

In practice, compliance control systems are essential, especially in building automation, as well as methods supporting the operator and user in the automated check. Here, Digital Twins offer real-time analysis and control of diverse characteristic measures of both the building, its technical equipment and its intelligent communication with the surrounding infrastructure. BIM and AI technologies offer important support for structuring complex data and interfaces merging to a single data platform for efficient collaboration of all participants (Weber-Lewerenz et al., 2021f). In Construction practice this includes approval and permit processes with the authorities, so that fast, smooth, structured and quality-enhancing building permit procedures (Ponnewitz, 2019) succeed and strengthen sustainability.

Also compliance with occupational health and safety on construction sites can be checked using BIM: by integrating crane information models into BIM, compliance with lifting plan requirements can be checked. Or via BIM-based compliance testing of performance-based parking lot ventilation concepts or via BIM-based automated code compliance checking systems.

The conquest of the hybrid world means that new work innovation hubs are formed: Leadership in Hybridity, Retention and Employee Experience, Innovation and Learning, Working worlds. The hybrid way of working is sustainably supported by AI technologies, especially in fire protection: remote diagnostics, remote monitoring and monitoring, tracking performance, intelligent networking of people and technology.

Intelligent smart Fire Protection solutions for buildings and cities strongly need a thorough critical reflection on data safety. Digital self-determination is increasingly considered as a task for the entire legal system in order to ensure the protection of personal and project-related data, protection against Cyberattacks and data misuse. The handling of data, e.g. feeding data for working with the digital twin, user data for Smart Technical Building Automatization, Smart Buildings, Smart Cities, and training AI-Algorithms equally require clarification and criminally prosecutable regulations as to who, where, for what and to what extent has rights and obligations as well as liability in the event of disregard.

Technological Dream or Safety Traumata? Best Practices Show the Sovereign Path

Utility-optimizing AI is an exploding business with high demand. Gartner (Wiggers, 2022) predicts that at least 50% of manufacturers with annual revenues in excess of $5 billion have at least one digital twin initiative launched for either products or assets. Analysts at the firm predict, moreover, that the number of organizations using digital twins will triple by the end of 2022.

Still, the number of BIM applicants in Germany is low. Thus, numerous associations and companies have founded the Planen Bauen 4.0 – Group to strengthen the implementation of BIM in the country.

The construction labor shortage only threatens to execrate the challenges, with the U.S. Chamber of Commerce reporting that 70% of contractors are struggling to meet project deadlines. Best Practices show how to cope with these challenges best by applying innovative technologies. BIM allows users to build models by drawing schematics and then generates a physics-based digital twin. Additionally, BIM and AI-based technologies require new skills and qualifications. Modern innovative companies may use these opportunities to create new working fields and offer the necessary adequate job positions. In the digital era, this tops overall the company`s unique selling point for job applicants. It also offers new niches for innovative universities that may fill the gap between required skills and education and academic training curricula (Weber-Lewerenz, 2022c; Weber-Lewerenz, 2021g). Best Practices also show how to make best use of these potentials for high quality projects by combining fire safety management and building information. Projects that satisfy both customers and corporate success; projects that lead to highly efficient project timelines and operation phases (Gartner, Inc., 2018) and add value in sustainability value chains.

Further Challenges in Shaping Sustainable Smart Cities

When it comes to designing intelligent cities of tomorrow, the question raises immediately on how cities shall be created to achieve smart digital standards. Furthermore, how can future Smart Cities not only be smart - but also sustainable and resilient? In this respect, Construction 4.0 can certainly be regarded both as a catalyst for digital transformation and as representing key competences: it carries high potential for an economic more efficient construction life cycle, while requiring responsible handling with innovative technologies. The building sector causes almost 40% of global CO_2 emissions. Buildings are responsible for around 40% of energy consumption and 36% of CO_2 emissions in the European Union (European Commission, 2020). The construction industry is the largest branch of the economy. This "sexy" sector not only has great potential but is also the key to centralized climate neutrality by 2050. Innovation Champions and Best Practices represent successful implementation, provide orientation and encourage the strengthening of the will to innovate. The exchange of knowledge between research and practice creates a practical approach, but also shows where exactly the adjustment screws for digitization and AI have to be turned in Construction. Corporate digital responsibility not only includes courage and awareness of the sensible use of innovative technologies, but also value-conscious consideration, the overall more efficient design of the construction life cycle from project idea to dismantling, resource-saving construction, more protection and more security of personal and data rights. The goal is the increase of added value and the achievement of sustainability goals.

Experiences in user practice range from structural and civil engineering, Technical Building Equipment and Operation, real estate management, monument preservation, formwork technology, tunnel technology, timber construction, to fire protection, intelligent buildings, Circular Economy, Cradle-to-Cradle, Smart Cities. There are no limits to the variety of fields of application, as surveys of building contractors and research institutions show. The crux lies in the development of a digital agenda that suits the company (Abdul et al., 2021; Ernstsen et al., 2021).

Innovative technologies enable a more structured, more transparent basis for decision-making and for forecasting increases in efficiency by visualizing the planned end product with its technical operating equipment, building usage data and operating data. Malfunctions, risk hazards, environmental impacts, user behavior, energy consumption and a holistic life cycle of the building can be simulated. This is done with a focus on sustainable economic development, SDGs and ISO/AWI 14075. BIM and AI lay the essential foundation for making decisions in agile environments using consolidated planning results and a common data environment. As consolidated coordination modeling in which all relevant data and data interfaces are brought together centrally, BIM brings considerable added value (Lo, C.K., 2021). BIM is rising towards the next consolidated methodology: Building Lifecycle Management. In this way, the digital transformation can be designed collaboratively and environmental protection-related challenges can be consciously tackled (Ye, Z. et al., 2020).

Technologies, software, systems and sensors are integrated with the knowledge of the diverse processes and procedures in the digital environment of a building. The challenges are connectivity, secure data storage and processing, and visualization. Building 4.0 means gaining a new sense of security (Agnusdei et al., 2021). However, such complex transformation requires a new safety mindset and a new culture of thinking in Construction 4.0, as well as a new awareness of the dangers that technological innovations entail for people at the same time in order to build safe, livable Smart Cities. Strong negative impacts of digital methods and AI are often referred to as the "dark side" of digitization (Pilgrim, 2017). They have social and ecological impacts on the raw materials sector. Economic growth is difficult to decouple from resource consumption. Resource consumption already exceeds the limits of our planet. The future technologies of the 4th Industrial Revolution require high quantities of resources and raw materials with low, particularly critical recycling potential and a lack of a return strategy. Four times current lithium production, a three-fold increase in heavy rare earths, and a 1.5-fold increase in light rare earths and tantalum. Due to the increased use of electronics, the global demand for copper will grow between 231% and 341% by 2050 (Elshaki et al., 2016). According to the DERA study, in 2035 up to 34 percent of the global indium production may be used exclusively for the production of displays (DERA, 2016). Digitization requires a significantly higher speed of data (fiber optic cables, routers, high-performance microchips, sensors, data transfer infrastructure) and data storage with larger volumes, i.e. significantly higher data capacity (data clouds, IoT, AIoT) for real-time transmission for error-free network communication. The devices must be continuously cooled. Overall, this results in higher additional energy consumption. More base stations need to be built. The permanent electromagnetic radiation caused by these new technologies also has severe health effects on humans and animals that have not yet been conclusively assessed. This leads to the conclusion, that digitization can become an energy guzzler, because in 2025 data centers will account for around 4-11% of global energy consumption; at the same time, high energy saving and the potential of waste heat utilization are forecast and localized (Höfer, Bierwirth, and Madlener, 2020).

The digital value chain is being taken ad absurdum. Where is the economic benefit? How do we want to reduce our ecological footprint and operate sustainably if we are increasingly consuming raw materials, accepting inhumane mining work, abuse of people and the environment and long-term damage to health in order to develop and use even better, even more economical, even more innovative technologies? Reason more to focus on the success criteria for sustainability and the "healthy" human-technic interaction within all technological development and its application in daily practice.

The Governing Smart Cities Report (World Economic Forum, 2021) provides policy benchmarks for the ethical and responsible development of smart cities. Such ethically responsible action offers opportunities to strengthen the trust of everyone: citizens as well as planners, manufacturers, users and operators of buildings and infrastructure.

Sustainability also means to fully integrate the Cradle-to-Cradle Principle (Braungart et al., 2014) into project planning and implementation: using the environment instead of just reducing harmful effects.

The concept is based on three principles: understanding waste as food, using renewable energy and promoting diversity. A practical example: digital twins – of both buildings and infrastructure – automatically generate real-time information and quantities of materials, the data is stored and can be called up at any time if the building is to be dismantled and the recycled material returned for use in the new building should be used. Cradle-to-Cradle means full resource and energy efficiency. It means building in the material cycle in compliance with sustainability goals.

Setting up a digital urban mining cadaster is one option - with an evaluation scheme standardized by building and goods passports. In this way, stocks of durable goods in the construction sector can be recorded, future material flows can be forecast at an early stage and the best possible recycling routes can be derived. This also applies to landfill mining, the extraction of valuable materials from old landfills.

Urban development and building planning could take the material and energy consumption in the life cycle much more into account and drastically reduce the consumption of mineral raw materials during construction and the ongoing energy consumption of buildings by combining resource-saving lightweight construction with the use of reusable and recyclable building materials. Especially through concrete recycling, 60 million tons of recycled building rubble could be used as aggregate for concrete production; but only about 0.6 million tons are used. A nationwide mandatory reuse as part of the Circular Economy standardization roadmap and the guarantee of maintaining the necessary quality represents one approach (Jehle, 2022). Through the increased use of existing technologies, resources could suddenly be saved immensely worldwide, and emissions reduced to a fraction, and CO_2 neutrality could be further accommodated (Fajga, 2022).

Digital twins in combination with AI-based systems automatically learn the behavior of a building and a system and can create forecasts. They reduce CO_2 by reducing energy consumption by up to 30% and optimizing existing buildings to become carbon neutral. In view of the many challenges in dealing with climate change caused by natural disasters, it is clear that cities and buildings will have to be planned and built even more "with nature" in the future. When planning smart cities, particular attention must be paid to less surface sealing, intelligent wastewater management, warning systems, e.g., in the event of flooding (e.g., Ahrweiler Germany in 2021), flood protection, stability and safety of tunnels, bridges and dams. The adaptation goes far beyond the newly required building materials and standards that are adapted to the requirements of climate change (e.g., acid rain, periods of heat).

Innovative, sensibly used digital and AI technologies support the Construction Industry to work sustainably applying efficient methods. Clients and contractors create added value these key technologies are offering. In addition, they promote a more diverse, inclusive environment in the close interface work that human-machine interaction triggers. New innovative technologies are signposts for the design of a Construction Branch with sustainable growth that is fair for people, society and the environment (ElMassah and Mohieldin, 2020).

SOLUTIONS AND RECOMMENDATIONS

The sensible use of innovative technologies sets an important signal for the trustworthy planning, design and operation of safe buildings and safe cities. The digital transformation in the construction sector makes a significant contribution to the protection and safety of people and society. In this sense, digitization and AI could stand for a seal of quality in Construction and sustainably strengthen the reputation of this unique industry.

Fire Protection in the Digital Era by BOSCH (Bosch, 2021)

Bosch Energy and Building Solutions plays a leading role in the design of innovative fire protection concepts incorporating the latest digital methods and AI-based technologies. Their technical highlights consist in:

- Detecting fire in the early stage (early fire detection, reduction of cost-intensive false alarms) or even before they occur (sensors as "sensory organs of future buildings")
- Targeted alarm and evacuation (e.g. fire detection identifies the number and location of people in the building at the same time = voice alarm and evacuation that is as efficient as possible)
- Optimizing the support of emergency services in fighting fires
- Documenting cases of damage without loss of data
- More networked, smarter and safer: there is a lot of potential in the buildings of tomorrow
- Fire protection protects against existential crises and has many other advantages
- based on intelligent networking and sensors

Various digital innovations enable intelligent, smart fire protection in Smart Cities.

Example 1 Intelligent fire detectors:

They collect a wide range of data: smoke density, pollution, temperatures, CO (carbon monoxide), EMV pollution. Building management systems can control the flow of people during an evacuation or ensure that elevators go down automatically in the event of a fire.

Example 2 Networked buildings in Smart Cities:

- In the event of a fire, a building can "warn" the neighboring building and, for example, have windows closed automatically

- From Digitization to AIoT: there is high potential of these new technologies for fire protection by optimizing building operation solutions with help of digital twins
- New fire protection solutions offer significantly more safety
- Fire protection of the future changes people's perception
- Modern fire protection in practice: uncompromising security solutions with added value
- Intelligent fire protection solutions fit the most diverse building types and
- Requirements (e.g. needs-based planning and building use. Remote services using Machine Learning (ML) and Deep Learning (DL): The ability of fire alarm systems to evaluate data independently, analyse states and provide forecasts.)
- Changing standards: the new European framework improves the quality of service for remote access
- Efficient and non-visible integration in buildings
- Up to 50% lower energy costs and significantly lower maintenance costs
- Reduction of misunderstandings and panic (security becomes directly visible and audible)
- Modular, expandable systems with software updates as the basis of a modern security infrastructure
- Implementation of a holistic and comprehensive building protection solution

Severe challenges in realizing the technical innovations raise and, thus, require save processes:

- Networked fire alarm technology must offer security against two dangers apart from fire: the misuse of personal data and Cyberattacks. Less important in fire protection because personal data is rarely collected.
- Legal requirements for maximizing security and safety.
- New fire risks from electrification
- Increasing e-mobility means that more and more battery-powered vehicles are parked in garages of commercial and residential buildings. If batteries catch fire, significantly more extinguishing water is required AND pollutants cause extreme heat. Both can damage the building`s fabric severely
- Without Sensors, no data. No data, no detection.
- Missing compatible technical interfaces = No networking

When it comes to Cyber Security, especially companies that offer digital fire protection building solutions in particular need competent advice when it comes to cyber security. It is therefore advisable to place value on data processing in Europe

(Information and Assistance from the Federal Office for Information Security (BSI) on the Cyber Security of Servers). Intelligent video systems allow personal data to be pixelated. In order to meet legal requirements and ensure maximum protection against Cyberattacks, networked systems require a carefully drawn up information security concept, regular checks for vulnerabilities and immediate action to be taken in the event of security gaps. Since threats from Cyberattacks are constantly changing, companies that offer digital fire alarm technology need a competent partner for IT security and data protection at this point, who constantly ensures maximum security.

Digital twins represent realization of a technological dream to ensure even more safety of live in smart cities and avoid potential safety traumata.

CONCLUSION

This review of state of the art reveals fundamental digitally methodological gap. The gap has been filled by defining the digital twin as most efficient and inclusive method for simulating fire scenarios and developing customized fire protection measures in Smart Cities environments. The method validated considering pros and cons of intelligent, smart fire engineering solutions for both buildings and Smart Cities. Scientific basis for further approaches of responsible use of digital twins and AI-based fire engineering concepts to ensure highest safety standards in Smart Cities.

Intelligent smart building operation with customized fire protection concepts is not about new technologies only. Recognizing the high potential of upgrading existing buildings to Smart Cities standards it is a question about financing such costly expenses. However, the advancements in IoT mean this is no longer the case. Modern sensors can be connected wirelessly, without the need for expensive installation costs. The reduced energy and maintenance costs allow for a quick return on investment, with figures showing that intelligent buildings use 20 - 40% less energy and result in 8 - 9% lower operating expenses than those with older systems (Iltisberger, 2021). All of these aspects show, that recognizing and shaping the holistic ecosystem is key of both success and sustainability.

Emerging technologies relieve people in a meaningful way, support and make work processes safer. Education and awareness-raising are important for orientation in order to consider opportunities and risks in a balanced manner. Overcoming this includes not only the technical and legal aspects, but in particular the necessary corporate digital responsibility (CDR) in the design of safe working and living environments. Anyone who has their hat on here and wants to be one step ahead is not only on their own path to success, but also fulfills sustainability goals (SDGs).

In addition, it is evident that the digital transformation is changing qualification profiles, creating new professional fields and requiring the adaptation of the educational landscape and teaching qualifications.

Predictive technologies are increasingly providing the basis for business decisions. Virtualized promises of performance and digital tours ensure more transparency, access and trust in public relations. Tailor-made fire protection concepts help to meet the new technical and social challenges in smart cities and new technological innovations that require special fire protection handling.

Although the pros outweigh the cons, the limitations stretch across the lack of consistency of digital processes, the lack of and legally uncovered data security and the human underestimation of the risk of fire.

There is a lot of potential and even more opportunities in digital change, and in how fire protection in smart cities creates "ignite" innovations.

However, despite the evaluation and consideration of advantages and disadvantages in technical decision-making processes, preventing damage and securing the common good comes first.

It is therefore guidance to reflect the moral and ethical questions when weighing up whether or not such outstandingly suitable methods as BIM should (have) be used for safer, more structured work that essentially supports people.

Innovation means taking responsibility to use technical innovations in a targeted manner to protect people, life and in the sense of the sustainable, efficient operation of buildings and entire cities.

The discourse on the design of corporate responsibility in the digital society requires a new quality of debate. The safe design of Smart Cities fire protection plays a prominent role in this, it not only enhances, it drives the discourse.

ACKNOWLEDGMENT

I would like to thank all of the experts interviewed in this research. They not only shared valuable insights but share the vision of the "Excellence Initiative for Sustainable, human-led AI in Construction" and addressing the diverse ethics debate for moving digital innovations forward.

In particular I would like to thank my supervisor, Prof. Marzia Traverso (PhD), Head of Chair of the Institute of Sustainability in Civil Engineering, Faculty of Civil Engineering at RWTH Aachen University, Germany.

Statistics and reports were of great support with data and facts to underlay the statements and recommendations made in this paper. Some public statements, which come from internet, literature and archive research, also underline the quality and statistical values of the expertise and survey values obtained, as well as limitations and urgently necessary measures. The author of this study conducts external research, is company-independent and is not financially supported by third-party funds, companies or other institutions. She is free in her research and shares her findings at the interface of "application practice - applied technical research - economic and social transfer". In this way, the author is researching the responsible use of digitization and AI -neutrally, critically and inclusive - and promoting the ethical debate about the AI technologies.

This research received no specific grant from any funding agency in the public, commercial, or not-for-profit sectors.

REFERENCES

Abdul, K., Hangjung, Z., & Ananth, C. (2021). Digital transformation and environmental sustainability. *MDPI Sustainability*, *13*(1530), 2071–1050.

Agnusdei, G. P., Elia, V., & Gnoni, M. G. (2021). A classification proposal of digital twin applications in the safety domain. *Computers & Industrial Engineering*, *154*, 107137. doi:10.1016/j.cie.2021.107137

Agnusdei, G. P., Elia, V., & Gnoni, M. G. (2021). Is digital twin technology supporting safety management? A bibliometric and systematic review. *Applied Sciences (Basel, Switzerland)*, *11*(6), 2767. doi:10.3390/app11062767

Bosch. (2021). *Digital twins for buildings. More transparency and efficiency for building operation*. https://www.boschbuildingsolutions.com/de/de/trends-und-themen/smart-buildings/digitaler-gebaeudezwilling/?gclid=EAIaIQobChMI9oL1zvy49QIVxeJ3Ch3rZgCtEAAYASABEgIqxvD_BwE

Bosch Energy and Building Solutions. (2021). *Fire Safety of the Future*. White Paper. Bosch Publishing. https://media.boschsecurity.com/fs/media/ib/media_1/solutions/Bosch-Whitepaper-Brandschutz_1.pdf

Braungart, M., & McDonough, W. (2014). *Cradle to Cradle: Einfach intelligent produzieren*. Piper Publishing.

buildingSMART. (2021). *White Paper - Fire Protection with BIM*. buildingSMART Publishing. https://buildingsmart-verlag.de/produkt/whitepaper-fg-brandschutz/

Cheng, M., Chiu, K.-C., Hsieh, Y.-M., Yang, I.-T., Chou, J.-S., & Wu, Y.-W. (2017). BIM integrated smart monitoring technique for building fire prevention and disaster relief. *Journal for Automation in Construction.*, *84*, 14–30. doi:10.1016/j.autcon.2017.08.027

Deutsche Rohstoffagentur, D. E. R. A. (2016). *Rohstoffe für Zukunftstechnologien 2016. Bundesanstalt für Geowissenschaften und Rohstoffe (BGR)* [Raw materials for future technologies 2016. Federal Institute for Geosciences and Natural Resources (BGR)]. https://www.deutsche-rohstoffagentur.de/DERA/DE/Downloads/Studie_Zukunftstechnologien-2016.html

ElMassah, S., & Mohieldin, M. (2020). Digital transformation and localizing the sustainable development goals (SDGs). *Ecological Economics*, *169*, 106490. doi:10.1016/j.ecolecon.2019.106490

Elshkaki, A., Graedel, T. E., Ciacci, L., & Reck, B. K. (2016). Copper demand, supply, and associated energy use to 2050. *Global Environmental Change*, *39*, 305–315. doi:10.1016/j.gloenvcha.2016.06.006

Ernstsen, S. N., Whyte, J., Thuesen, C., & Maier, A. (2021). How innovation champions frame the future: Three visions for digital transformation of construction. *Journal of Construction Engineering and Management*, *147*(1), 05020022. doi:10.1061/(ASCE)CO.1943-7862.0001928

European Commission. (2010). *Grundlagendokument Brandschutz Nr. 2 der Europäischen Kommission* [Basic Document Fire Protection No. 2 of the EuropeanCommission]. https://eur-lex.europa.eu/legal-content/DE/TXT/PDF/?uri=CELEX:32010D0738&rid=2

European Commission. (2020). *Im Blickpunkt – Energieeffizienz von Gebäuden*. European Commission. https://ec.europa.eu/info/news/focus-energy-efficiency-buildings-2020-lut-17_de#:~:text=Insgesamt%20entfallen%20auf%20Geb%C3%A4ude%20in,%2C%20Nutzung%2C%20Renovierung%20und%20Abriss

European Commission. (2021). *Status Quo and updating the reform recommendations for the Professional regulation of 2017*. https://eur-lex.europa.eu/legal-content/DE/TXT/PDF/?uri=CELEX:52021DC0385&from=ES

Fajga, K. W. (2022). Ehrgeizige Ziele gesetzt [Ambitious goals set]. *Allgemeine Bauzeitung ABZ*, (18), 3.

Fan, C., Zhang, C., Yahja, A., & Mostafavi, A. (2021). Disaster City Digital Twin: A vision for integrating artificial and human intelligence for disaster management. *International Journal of Information Management*, *56*, 102049. doi:10.1016/j.ijinfomgt.2019.102049

Gartner, Inc. (2018). *Four best practices to avoid digital twin failures.* Gartner Survey Report. https://www.gartner.com/en/newsroom/press-releases/2018-03-13-gartner-survey-reveals-nearly-half-of-organizations-implementing-iot-are-using-or-plan-to-use-digital-twin-initiatives-in-2018

Group, D. M. T. (2021). *Brandschutz – der stille Erfolgsfaktor* [Fire Protection – the smart key for success]. https://www.dmt-group.com/de/engineering-performance/engineering-performance-storie/archive/brandschutz-der-stille-erfolgsfaktor.html

Höfer, T., Bierwirth, S., & Madlener, R. (2020). Energie Mehrverbrauch in Rechenzentren bei Einführung des 5G Standards [Energy consumption in data centers with the introduction of the 5G standard]. Nachhaltige Rechenzentren–Chancen und Entwicklungsmöglichkeiten am. Stuttgart.

Iltisberger, O. (2021). *The role of smart buildings in the development of smart cities.* https://www.smart-energy.com/industry-sectors/iot/the-role-of-smart-buildings-in-the-development-of-smart-cities/

Jehle, P. (2022). Inkonsequenz bremst Betonrecycling [Inconsistency slows down concrete recycling]. *Der Bauingenieur, 97*(4). DOI: doi:10.37544/0005-6650-2022-04

Kirchschläger, P. G. (2021). Ethische Verantwortung in der digitalen Transformation [Ethical Responsibility in the digital transformation]. *Journal of Risk, Fraud and Compliance.*

Li, N., Becerik-Gerber, B., Krishnamachari, B., & Soibelman, L. (2014). A BIM-centered indoor localization algorithm to support building fire emergency response operation. *Journal for Automation in Construction., 42*, 78–89. doi:10.1016/j.autcon.2014.02.019

Liu, Z., Zhang, A., & Wang, W. (2020). A framework for an indoor safety management system based on digital twin. *Sensors (Basel), 20*(20), 5771. doi:10.339020205771 PMID:33053719

Lo, C. K., Chen, C. H., & Zhong, R. Y. (2021). A review of digital twin in product design and development. *Advanced Engineering Informatics, 48*, 101297. doi:10.1016/j.aei.2021.101297

Ludwig, K. (2021). *Rechtliche Beschränkungen bei der Digitalisierung der Bauwirtschaft* [Legal limitations within digitization in Construction]. Academic Press.

Max, D. (2021). *Building Information Modeling im Brandschutz. Bauphysik Kalender 2021 –Brandschutz*. Wilhelm Ernst & Sohn Publishing Berlin.

Metallidou, C. K., Psannis, K. E., & Egyptiadou, E. A. (2020). Energy efficiency in smart buildings: IoT approaches. *IEEE Access: Practical Innovations, Open Solutions*, *8*, 63679–63699. doi:10.1109/ACCESS.2020.2984461

Minarčík, P., Procházka, H., & Gulan, M. (2020). Advanced supervision of smart buildings using a novel open-source control platform. *Sensors (Basel)*, *21*(1), 160. doi:10.339021010160 PMID:33383746

Minoli, D., Sohraby, K., & Occhiogrosso, B. (2017). IoT Considerations, Requirements, and Architectures for Smart Buildings—Energy Optimization and Next-Generation Building Management Systems. *IEEE Internet of Things Journal*, *4*(1), 269–283. doi:10.1109/JIOT.2017.2647881

Müller, R. (2019). *Branchenbarometer Brandschutz* [Industry Barometer Fire Protection]. https://www.baufachmedien.de/branchenbarometer-brandschutz.html

Pilgrim, H. (2017). *The Dark Side of Digitalization: Will Industry 4.0 Create New Raw Materials Demands? PowerShift e.* V. Publishing.

Ponnewitz, J. (2019). Die BIM-basierte Baugenehmigungsprüfung-eine Grundlagenbetrachtung [BIM-based Building Permit Check – A Basic Consideration]. *Conference Proceedings of the 30th BBB-Assistance Meeting*.

Schulthess, P., Neuenschwander, M., Mosalam, K. M., & Knobloch, M. (2020). A computationally rigorous approach to hybrid fire testing. *Computers & Structures*, *238*, 106301. doi:10.1016/j.compstruc.2020.106301

Sheikhnejad, Y., Gonçalves, D., Oliveira, M., & Martins, N. (2020). Can buildings be more intelligent than users?-The role of intelligent supervision concept integrated into building predictive control. *Energy Reports*, *6*, 409–416. doi:10.1016/j.egyr.2019.08.081

Sovacool, B. K., Del Rio, D. F., & Griffiths, S. (2021). Policy mixes for more sustainable smart home technologies. *Environmental Research Letters*, *16*(5), 054073. doi:10.1088/1748-9326/abe90a

Van der Auweraer, H. (2021). *Digital Twins – Simulations by Siemens*. Siemens Publishing. https://new.siemens.com/de/de/unternehmen/stories/forschung-technologien/digitaler-zwilling/digital-twin.html

Weber-Lewerenz, B. (2021e). Unternehmerische Compliance im digitalen Transformationsprozess Teil 1 [Corporate Compliance in Digital Transformation Process - Part 1]. *Journal of Risk, Fraud and Compliance.* DOI: doi:10.37307 /j.1867-8394.2021.06

Weber-Lewerenz, B. (2021a). Corporate Digital Responsibility in Construction Engineering. *International Journal of Responsible Leadership and Ethical Decision-Making, 2*(1), 32–49. Advance online publication. doi:10.4018/ IJRLEDM.2020010103

Weber-Lewerenz, B. (2021b). Corporate digital responsibility (CDR) in construction engineering - Ethical guidelines for the application of digital transformation and artificial intelligence (AI) in user practice. *Springer Nature SN Applied Sciences., 3*(10), 801. Advance online publication. doi:10.100742452-021-04776-1

Weber-Lewerenz, B. (2021c). Ethical Aspects in AI in Construction. In *Conference Presentation and Proceedings of Reser Conference*. Fraunhofer Institute Publishing. https://publica.fraunhofer.de/dokumente/N-642928.html

Weber-Lewerenz, B. (2021d). Die unternehmerisch verantwortungsvolle Digitalisierung im Bauwesen. *Journal Der Bauingenieur., 96*(1-2), 19–25. doi:10.37544/0005-6650-2021-01-02-45

Weber-Lewerenz, B. (2021f). *Ethik in der KI im Bauwesen in Deutschland und KI Methoden und Erfahrungswerte* [Study by Fraunhofer Institute Stuttgart IAO on AI in Construction in Germany]. Fraunhofer Institute Publishing. https://publica.fraunhofer.de/eprints/urn_nbn_de_0011-n-6306 697.pdf

Weber-Lewerenz, B. (2021g). *KI in Lehre und Ausbildung, BIM und KI in Wissenschaft und Unternehmenspraxis.* buildingSMART Deutschland bSD Publishing

Weber-Lewerenz, B. (2022a). *Accents of Added Value in Construction 4.0 – Ethical Observations in Digitization and AI.* Springer Nature Publishing.

Weber-Lewerenz, B. (2022b). Thinking otherwise: Integrating Existing Buildings and Monument Protection in Smart Cities -Experience shared from user practice. In I. Vasiliu-Feltes (Ed.), *Impact of Digital Twins in Smart Cities Development.* IGI Global Publishing. doi:10.4018/978-1-6684-3833-6

Weber-Lewerenz, B. (2022c). Impact of education and training for diversity, inclusion and sustainable digital innovations. In I. Vasiliu-Feltes (Ed.), *Impact of Women Empowerment on SDGs in the Digital Era.* IGI Global Publishing. doi:10.4018/978-1-6684-3637-0.ch012

Wiggers, K. (2022). *PassiveLogic, which creates digital twins of buildings, lands $34M*. https://venturebeat.com/2022/01/11/passivelogic-which-creates-digital-twins-of-buildings-lands-34m/

World Economic Forum. (2021). *Governing Smart Cities: Policy Benchmarks for Ethical and Responsible Smart City Development*. White Paper. https://www3.weforum.org/docs/WEF_Governing_Smart_Cities_2021.pdf

Wrick, T. (2020). *Wenn die Digitalisierung den Brandschutz fordert und fördert – 5 Beispiele* [When digitization demands and promotes fire protection - 5 examples]. https://industrie-wegweiser.de/wenn-die-digitalisierung-den-brandschutz-fordert-und-foerdert-5-beispiele/

Ye, Z., Yang, J., Zhong, N., Tu, X., Jia, J., & Wang, J. (2020). Tackling environmental challenges in pollution controls using artificial intelligence: A review. *The Science of the Total Environment*, *699*, 134279. doi:10.1016/j.scitotenv.2019.134279 PMID:33736193

Zohdi, T. I. (2021). A digital twin framework for machine learning optimization of aerial fire fighting and pilot safety. *Journal of Computer Methods in Applied Mechanics and Engineering.*, *373*, 113446. Advance online publication. doi:10.1016/j.cma.2020.113446

KEY TERMS AND DEFINITIONS

AI: Artificial Intelligence. AI is the so-called umbrella term for applications in which machines provide human-like intelligence such as learning, judgment and problem solving. The development, application and maintenance of such data-driven technologies requires orientation in order to be able to fulfill the associated digital responsibilities. AI is understood as the ability of machines to emulate human intelligence and to learn independently, without being dependent on instructions from a human developer. For this purpose, software uses machine learning to evaluate large amounts of data for matches and draws conclusions from them.

BIM: Building information modeling.

CDR: Corporate digital responsibility.

Digitization: There is no clear definition for the term. In the original sense, digitization refers to converting analog data values to digital formats. This data can be processed by information technology. Another meaning of digitization is the digital revolution, also known as digital change or digital transformation.

Ethics: Ethics deals with the prerequisites and the evaluation of human action and is the methodical reflection on morality. At the center of ethics is specifically moral action, especially with regard to its justification and reflection (ethics describes and judges morality critically). The focus of ethics is on the three questions of the "highest good", the right action in certain situations and freedom of will.

IoT: Internet of Things. The term describes that the computer in the digital world is increasingly being supplemented by "intelligent objects" to "AI", artificial intelligence. Instead of being the object of human attention itself, as is currently the case, the "Internet of Things" is intended to support people in their activities.

Performance-Based Fire Engineering: From an engineering perspective, the performance-based fire protection design framework is valuable as it integrates scientific knowledge, performance (design) criteria, fire scenarios, the building operation and building user`s performance into the design process so that fire protection engineers can do more than rely on code prescriptions that are usually implemented as a political response, after major accidents.

Section 3
Regional Perspectives

Chapter 8
How Digital Twin Technology Promotes the Development of Smart Cities:
Case Studies in China

Poshan Yu
https://orcid.org/0000-0003-1069-3675
Soochow University, China & Australian Studies Centre, Shanghai University, China

Zixuan Zhao
Independent Researcher, China

Michael Sampat
Independent Researcher, Canada

ABSTRACT

This chapter aims to study whether and how current practices based on digital twin technology can help the development of smart cities in China. This chapter will explore the roles of the government in promoting the application of digital twin technology in the development agenda of industrial infrastructure and smart cities through policies (i.e., institutions). How do these institutions promote the development of digital twin technology in various industries and how will these developments in turn affect stakeholders in various sectors? This chapter will investigate the impact of China's digital twin technology research and institutional changes on the sustainable development of smart cities. This chapter will use bibliometric analysis to perform statistical and correlation analysis on the digital twin/blockchain/AI/IoTs literature included in the Web of Science (WOS) database. Cases from China will be used for comparison. This chapter will provide suggestions for the development of digital twin technology-based ecosystems in emerging economies.

DOI: 10.4018/978-1-6684-3833-6.ch008

INTRODUCTION

After nearly 5 years of conceptual development in the construction of a technological framework, digital twin cities have become the best solution for the construction of new smart cities in China in recent years. China takes policy orientation as the standard, and several provinces and cities have set up pilot projects for digital twin cities. Smart city projects in Beijing and Guangzhou emphasize the construction of a unified perception management service platform for the whole city. In Shijiazhuang, the first national spatiotemporal data platform for smart cities comes into operation.

In times of increasing urbanization, local decision-makers must be prepared to maintain and increase the quality of life of a growing urban population. For instance, there are major challenges related to minimizing pollution, managing traffic as well as making efficient use of scarce energy resources. To address these challenges urban decision-makers, as well as citizens will need the capacity to make the right assessment of urban situations based on correct data, and, more importantly, they will need the key information contained in the data to assist them in their decision processes. As put by Neelie Kroes, EU commissioner for the Digital Agenda, data is the new gold, meaning that data is a valuable resource that can be mined for creating new values. In the context of smart cities, there is an abundance of data sources that can be mined by applying data analytics techniques and can generate value by offering innovative services that increase citizens' quality of life. Data may be provided by all stakeholders of a smart city (Strohbach, 2015), i.e., the society represented by citizens and businesses and governments represented by policymakers and administrations.

With the rapid development of information technology, thc construction and development of cities are increasingly inseparable from digital empowerment, and the construction of new smart cities is an important initiative for the development of major cities around the world, as well as an inevitable requirement for improving the effectiveness of urban governance. It has been more than ten years since the construction of smart cities in China started in 2008. With the development of information technology, the digital twin is gradually applied to manufacturing, transportation, medicine, and other fields. The development of cutting-edge technologies, such as cloud computing, blockchain, the Internet of Things (IoT), and big data, is rapidly transferring data from the physical world to the digital twin world. Digital twin technology can reflect the running state of a physical entity quickly and efficiently through virtual twin technology, and manage and detect its running state. Therefore, the more complex the system is, the more suitable it is to use digital twin technology for decision rehearsal, analog planning, and so on. As the most complex and huge system, the management efficiency of the city after the establishment of the digital twin will be the highest.

BACKGROUND

Analysis

The evolution of a new subject needs to build on the knowledge accumulation of relevant subjects. Similarly, no single research paper can be developed solely based on its content. The paper needs to draw on the past literature and research ðndings from its discipline or other relevant disciplines. Generally, research papers published in journals represent the frontiers of certain subjects, and the references cited in these papers provide the knowledge base of these papers. With the assistance of certain computer software, we could cluster references that are commonly cited in the smart city discipline and identify co-citation clusters. This step is important in identifying the foundation knowledge of smart city research and it could be achieved through visualizing the knowledge of smart city research using journal articles. To focus the subsequent research, we then used CiteSpace to search the existing literature in the WOS database to identify the core topics that are most closely related to smart cities today.

CiteSpace is one of the most popular tools for knowledge mapping (e.g., Chen, 2006; Chen et al., 2010). It is particularly designed to support the analytic process of visualizing and it can produce co-citation networks based on article citations to reveal the structure of a particular research field (Chen et al., 2010). Using CiteSpace to analyze existing smart city literature would help provide a clear picture of how knowledge and theories advanced and how to research topics evolved, therefore, this analysis is highly desirable.

Research Method

Building Database for CiteSpace

There are three steps in building a database for CiteSpace analysis. The first step is to build a database that is suitable for analysis by selecting reputable journals. Referred journal articles are reliable sources of knowledge and the quality of journals usually implies the quality of articles. Therefore, this study chose the top journals included in the Web of Science (WOS) database.

The second step is to use appropriate keywords to select articles from journals. The researcher should pay special attention to the validity and representativeness of keywords and check the relevance of each study returned. The types of articles should be checked, and book reviews and editorial letters should be excluded to eliminate 'noise' in the database. By setting keywords closely related to smart cities such as AI, Blockchain, IoT, and Digital twin, a total of 4450 articles (published from 2000 to 2022) were retained.

The third step is to extend the database. The database should contain the most complete literature coverage possible and, as suggested by Chen et al. (2010), co-citation networks as indicated by selection articles' references should also be included in the database. References cited by a given article provide valuable information regarding intellectual connections between various concepts and theories.

Mapping Knowledge Using CiteSpace

Using CiteSpace, major paths of smart city knowledge evolution were mapped first. The analytical process consists of categorizing knowledge clusters in the smart city discipline, identifying major clusters in the smart city knowledge map, identifying key studies that made signiðcant contributions to knowledge development, and performing a timeline analysis on these studies in terms of origin, development, and current status.

Figure 1. Frequency, Burst, Degree, and keywords of selected articles
Source: WOS

Freq	Burst	Degree	Keyword
1077	4.7	55	smart city
586		47	internet
436	5.32	53	internet of thing
260		47	system
228		42	challenge
227		35	big data
223		36	management
185		46	artificial intelligence
181		35	framework
170		34	iot
169		36	security
159		30	technology
157	9.53	45	internet of things (iot)
144		29	model
134		40	architecture
126		22	smart contract
119		18	network
108		18	machine learning
106		24	digital twin
105		38	design
97		17	blockchain
93		36	privacy
84		27	deep learning
82		30	cloud computing
75		43	future
73		9	scheme
72		38	wireless sensor network
70		28	edge computing
68		40	optimization
63		29	algorithm
63		30	smart grid
60		35	cloud
53		17	smart city
50		41	service
45		16	neural network
45		34	performance
44		27	artificial intelligence (ai)
41		27	information
39		22	supply chain
39		21	energy
37		24	authentication
37		31	efficient
36		23	protocol
36		11	internet of thing
36		26	health care
35		29	big data analytics
34		36	energy management
33		32	fog computing
33		22	vehicle
32		34	sustainable city
31		12	blockchain technology
31		28	energy efficiency
29		26	integration
28		26	computer architecture
28		35	opportunity
28		10	access control
28		39	smart home
28		32	prediction
27		21	communication
27		35	things iot
26		9	secure
26		17	classification
26		27	cyber security
26		26	data analytics
25		35	platform
25		11	intrusion detection
24		20	electric vehicle
24		16	5g mobile communication
23		16	task analysis
23		26	smart building
23		17	edge
23		13	data model
22		19	trust
22		23	impact
21		24	real-time system
21		20	resource allocation
21		11	trend
20	4.43	28	sensor network
20		29	simulation
20		15	data privacy
20		24	quality
20		26	attack
20		21	industry 4
19		41	sensor
19		8	smart contract
19		19	distributed ledger
19		10	waste management
19		24	resource management
19		16	smart
18		26	renewable energy
18		28	intelligence
18		16	computational modeling
17		15	artificial intelligence

Second, hot research topics and frontline research in the smart city ðeld were identiðed based on the frequency of popular keywords used in journal articles. The burst of certain keywords can be used to analyze the evolution of smart city research and identify the most recent research trends. The burst refers to a signiðcant change in a variable's values in a relatively short time (Figure 1). CiteSpace considers this type of change a way to identify the front line of research.

Knowledge Mapping Results

The following parameters in CiteSpace were used: (1) Time slice from 2000 to 2022; (2) Term source =keywords; (3) Node type = keywords; (4) Pruning = pathðnder/pruning the merged network; (5) Select 25 most cited articles per slice. After running CiteSpace, we got the map shown in Figure 2. The clustering function was performed by choosing 'T' as the labeling source and log-likelihood ratio as the method. The results returned thirteen knowledge clusters and two of them are major clusters based on co-citation cluster information.

By comparing Figure 1 and Figure 2, we can see that the frontier technologies used in the development of smart cities from 2010 to 2022 are IoT, Digital twin, Edge computing, Machine learning, AI, Cloud computing, 3D city model, etc. Hot topics related to smart cities in the last three years include the digital economy, urban planning, etc. The above findings help to determine the focus and main directions of this paper's research on smart cities.

Figure 2. The timeline of co-citation clusters from 2010 to 2022.
Source: WOS

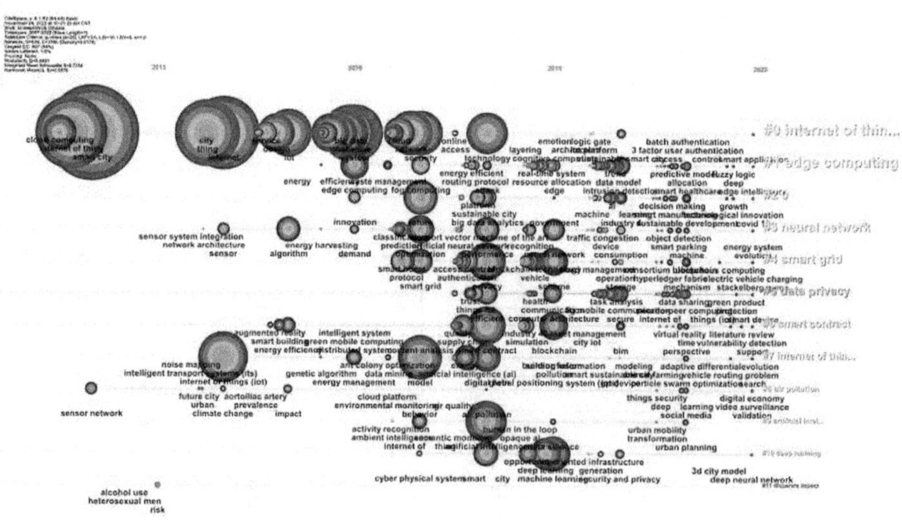

MAIN FOCUS OF THE CHAPTER

Smart City

Being unprecedentedly dense venues for the interactions — economic, social, and of other kinds – between people, goods, and services, megacities also entail significant challenges. These relate to the efficient use of resources across multiple domains, e.g., energy supply and demand, building and site management, public and private transportation, healthcare, safety, and security, etc. To address these challenges, a more intelligent approach to managing assets and coordinating the use of resources is envisioned, based on the pervasive embodiment of sensing and actuating technologies throughout the city fabric and supported by ubiquitous communication networks and the ample processing capacity of data centers. The umbrella term Smart City (European Smart Cities project) refers to the application of this approach in any of six dimensions (Benamrou et al., 2016):

- economic solidity;
- sustainable mobility;
- environmental protection;
- social quality;
- governance;
- digital transformation.

As already mentioned before for other indicators, devising a smart city should inevitably improve the life quality of citizens, maybe through the use of IoT technologies, thus increasing the efficiency of services (Yu et al., 2022g; 2022j) and including efforts to make the best use of existing resources, e.g. managing food production, congestion, and pollution (Crisostomi et al., 2016). A key point of the digital transformation process of a smart city is further related to the need to be able to collect data produced by heterogeneous sources, e.g., data streams from bicycle sharing schemes, car parks, air quality sensor data, etc. (Okrepilov et al., 2019), and analyze them in a close-to-real-time way (Sarker et al., 2018), so that city planners can employ up-to-date predictive models and use them in planning activities. It is noteworthy that citizens should actively exploit opportunities for growth and development, too, through appropriate policy implementation, human capital development, and a developed culture of innovation (Sharma et al., 2016).

Digital Transformation

From a general point of view, a digital transformation process can be described as "a change in the internal mechanisms of an organization, intended to provide new (or significantly better) products or services supply, by using information technology as a key enabler." According to this definition, a smart space, involving the interaction among humans and technology-enabled systems in a physical or digital environment, can be enhanced and become intelligent following different key dimensions, identifying the directions that a smart city should follow to shift from being an isolated environment to an intelligent municipality (Tomičić Pupek et al., 2019).

Industry 4.0

Industry 4.0 is the umbrella concept for a new industrial paradigm that includes several future industry achievements related to Cyber-Physical Systems (CPS), IoT, iPhone operating system (IoS), Robotics, Big Data, Cloud Manufacturing, and Augmented Reality. Adopting these technologies is key to the development of more intelligent manufacturing processes, including devices, machines, production modules, and products that can independently exchange information, stimulate actions and control each other, thus enabling an intelligent manufacturing environment (Vaidya et al., 2018; Pereira & Romero, 2017). One of the more important terms coming along with Industry 4.0 is IoT. The position of industrial production is related to the integration of Information and Communication Technology (ICT) systems with production processes, internal and external facilities, a network of suppliers and customers, machine-to-machine communication (M2M), and communication with employees (Smit et al., 2016). Integration of IoT and IoS into the production environment can promote industrial development.

Internet of Things (IoT)

IoT allows for a digitally connected real world, whereby connected devices can exchange collected data, interact with each other, and remotely control objects across the Internet, possibly without human intervention. IoT is where the Internet meets the physical world (Ammar et al., 2018) such that societies and industries can benefit from IoT to achieve a quantum shift towards a smart digitally controlled world. Due to the implementation of IoT technology, the way people interact with each other and the environment has been improved and reshaped. Consequently, one can say that people have reached a better understanding of the world while the IoT enables more efficient interaction with it (Sadawi, 2021).

With the continuous progress of technology, the development and application of IoT technology have achieved remarkable results in recent years. Through the introduction of IoT innovation, smart sustainable cities can improve different aspects of their urban management, for instance, urban mobility, public transportation, e-governance, safety, security, public lighting, and environmental monitoring (Yu et al., 2022a). The adoption of IoT technologies is expected to allow to monitor, control, and manage all the available resources, such as electric power, soil, water, people, and so on (Yu et al., 2022n).

The advent of IoT is enabling not only the connection and integration of devices that monitor physical world phenomena but also data-driven and AI-augmented intelligence. At all levels, synergies from advances in IoT, data analytics, and artificial intelligence (AI) are firmly recognized as strategic priorities for digital transformation (Cerf et al., 2012). Thanks to its advanced features of innovation, such as very low latency (less than 1 ms) and very high bandwidth (10 Gb/s), IoT is emerging as one of the primary use cases for 5G. Therefore, 5G can be considered as an enabling communication technology for smart cities, allowing more and more IoT devices to be connected to the Internet, regardless of their location and time, supporting applications such as smart traffic systems, public safety, security, and surveillance in the context of smart cities (Rao & Prasad, 2018).

In general, a complete infrastructure willing to integrate the sustainable mobility concept with IoT has to be able to perform several tasks, as shown in Figure 1, with a particular focus on the data resulting from these mobility-related scenarios, as follows:

- Support both local and isolated scenarios (not connected to the Internet), and fully on-Cloud architectures, to continuously collect a large amount of heterogeneous data from the city infrastructure (Tomaras, 2018), supporting several paradigms for data injection, e.g. on-demand data, proactive forward, scheduled forward, etc. (Sowe et al., 2014 and Fazio et al., 2012).
- Handle safety-critical scenarios, where involved entities (both IoT-related and not) should interact with the municipality's control centers (e.g., traffic congestion area, pedestrian crossing paths, event mass moving) safely and securely, even providing mechanisms to cope with the lack of connection/communication (Rathore, 2016).
- Provide distributed storage mechanisms, to decentralize the mobility network (and avoid bottlenecks often affecting centralized solutions), thus supporting efficient strategies for data retrieval and disaster recovery.

- Provide high-layer mechanisms for the execution of AI and other data analysis algorithms, to extract features and implement predictive mechanisms based on the collected environmental data, e.g. smart parking, predictions on traffic flow, and traffic flow reconstruction (Bucchiarone, 2019 and Poncela et al., 2014).
- Provide several easy-to-use visual consultation mechanisms for mobility data, through various representation technologies, based on the roles of the different users accessing them (Faber, 2018).

AI and Blockchain

AI and blockchain are also key drivers behind today's innovation. Both are expected to revolutionize every aspect of our lives and contribute trillions of dollars to the global economy. On the one hand, there are some flaws in the blockchain like scalability, security, and efficiency; on the other hand, AI has its fair share of issues with privacy and trustworthiness. The business integration of these two technologies will complement each other to revolutionize each other. The blockchain can provide privacy and trustworthiness and AI can build machine learning algorithms on the blockchain to achieve security, scalability, and effectiveness. Thus, the integration of AI and the blockchain is the blockchain for AI and AI for the blockchain. The key issues to be addressed in the convergence of the blockchain and AI are security and privacy, threats and attacks, intelligent infrastructure, technical and business challenges, lack of standards, interoperable regulations, smart contract vulnerabilities, and deterministic executions, as well as good governance (Yu et al., 2022c).

GOVERNMENT ROLES IN SHAPING SMART CITIES IN CHINA

China's smart city construction has been developing for decades. From system construction to brain construction to urban operation and management, a brand new vision of smart China has opened up. The policy support of the Chinese government is of great significance to the promotion of smart city construction. The factor of "government-led" urbanization in China is greater than that of "market evolution". Institution, as rules of the game, helps to interpret the performance of the player in the game (North, 1991). According to Table 1, policies (i.e. institutional changes) were introduced from 2009 to 2022, and the behaviors or decisions of smart city stakeholders changed accordingly. For instance, since 2016, the national and local 13th Five-Year Development plans have been issued successively, and many cities have taken the construction of smart cities as the focus of their future development.

Table 1. Important documents and conferences related to the development history of smart cities in China since 2008

Characteristics	Policy/ Meeting	Authority	Date	Main Contents and Significance
Exploration period: Scattered and disorganized	Notice On The Issuance Of The Guidelines For The Construction Of Digital City Management Model (For Trial Implementation)	Ministry of Housing and Urban-Rural Development	2009.7.25	1. The construction of the digital city management system should focus on applicability, expandability and compatibility to lay the foundation for the sustainable development of digital city management. 2. Stipulate the core of digital urban management construction.
	Notice On The Implementation Of National Smart City Pilot Work	General Office of Ministry of Housing and Urban-Rural Development	2012.12.5	1. Encourage local governments to actively develop smart cities, improve urban management capabilities and services, and promote industrial transformation and development. 2. Formulate feasible national smart city goals and implementation plans according to local conditions, and establish corresponding policies, organizations and funding support systems.
Adjustment period: National coordination and collaborative guidance	National New Urbanisation Plan (2014-2020)	State Council of the Central Committee of the Communist Party of China	2014.3.16	1. Clearly define the objectives of China's urbanization over the next six years, which will also influence China's urban and rural development for a longer period. 2. Established a system of 18 urbanization in four areas: "urbanization level", "basic public services", "infrastructure" and "resources and environment". 3. Promote the construction of the transportation network to become a hot investment. 4. Focus on exploring the establishment of an urban cluster management coordination model. 5. Promote coordination and linkage of the industrial division of labor,, infrastructure and environmental management among cities across the region.
	Guidance On Promoting The Healthy Development Of Smart Cities	National Development and Reform Commission	2014.8.27	1. Point out the general direction and construction priorities of China's smart city development. 2. Propose some basic and common requirements for the construction of smart cities in China and establish some common codes of conduct. 3. Build consensus in the community and reflect normative and guidance at the national level.
	Guidance On The Construction And Application Implementation Of The Smart City Standards System And Evaluation System	The Standardization Administration of China, Central Internet Information Office, National Development and Reform Commission	2015.11.11	1. A total of about 50 standards in the field of smart cities are required to be completed by 2020. 2. The standardization of smart cities has been put on the national agenda.

continued on following page

Table 1. Continued

Characteristics	Policy/ Meeting	Authority	Date	Main Contents and Significance
Breakthrough period: Strategic rise and system integration	Outline Of The 13th Five-Year Plan For National Economic And Social Development Of The People's Republic Of China	General Office of the CPC Central Committee, General Office of the State Council	2016.3.17	1. Propose to "build several new demonstrative smart cities" with full use of modern information technology and big data. 2. Actively build green, forested cities. Promote green buildings, and popularise green transport and new energy supply systems such as distributed energy and shallow geothermal energy. 3. Accelerate the electrification of public transport and launch a new green living initiative. The construction of smart cities is included as an important project of information urbanization.
	Outline Of The National Strategy For The Development Of Information Technology	General Office of the CPC Central Committee, General Office of the State Council	2016.7.27	1. Strengthen top-level design to improve information technology in urban infrastructure, operation and management, public services, and industrial development. 2. Promote the construction of a new type of smart city by grade and classification. A programmatic document to regulate and guide the development of national informatization in the next 10 years.
	Smart City IT Operations Guide	National Technical Committee for Information Technology Standardization	2018.10.10	1. Provide a general framework for smart city operations and related advice on ICT infrastructure operations, data operations, information system operations, and security operations. 2. Apply to the establishment and management, operational monitoring, and evaluation of IT operation systems in smart cities. A document to ensure the safety and efficiency of smart city operations.
Integration period: Fine management and comprehensive integration	Guidelines On Supporting State-Level New Areas In Deepening Reform And Innovation And Accelerating High-Quality Development	General Office of the State Council	2020.01.17	1. Push forward the construction of a smart city and improved the level of meticulous urban management. 2. Optimise the functional layout of the main urban area and the new area, and promote the new area to take over some of the functions of the main urban area in an orderly manner. 3. Improve the level of construction of infrastructure and public service facilities in the new zone, and enhance the supporting functions of education, healthcare, and culture. Re-emphasis on the in-depth promotion of the construction of intelligent cities, to enhance the level of urban refinement management.

continued on following page

Table 1. Continued

Characteristics	Policy/ Meeting	Authority	Date	Main Contents and Significance
Integration period: Fine management and comprehensive integration	Implement Urban Renewal Initiatives	Ministry of Housing and Urban-Rural Development	2020.12.21	1. Accelerate the construction and upgrading of new types of urban infrastructure based on information technology, digitalization and intelligence, and comprehensively raise the level and efficiency of urban construction. 2. Significantly improve urban governance in a scientific, refined and intelligent manner Emphasizing the role of accelerating information, digital, intelligent and other infrastructure in promoting the construction of new cities.
	Guidelines On Accelerating The Development Of Digital Homes To Improve The Quality Of Residents	The Ministry of Housing and Urban-Rural Development, the Cyberspace Administration of the CPC Central Committee, the Ministry of Education, the Ministry of Science and Technology and other 16 departments	2021.4.17	Promote the connection between the basic platform of the digital home system and the new smart city platforms of "one network for all", "one network for management", smart property management, smart community information system, and socialized professional services. An important policy that responds to the national digital economy strategy and achieves economic transformation and upgrading and digital economy goals.
	Notice on Promoting regional experience in the Third Batch of National Comprehensive Pilot Projects for New Urbanization	The General Office of the National Development and Reform Commission	2021.8.26	Strive to build livable, humane, innovative, smart, green and humanistic cities and continue to modernize the urban governance system and capacity. Based on the first and second batches of pilot projects in 2019 and 2020, the authority will promote the experience of the third batch of pilot projects to provide a replicable experience and models for the high-quality development of new urbanization nationwide.
	"The 14th Five-year Plan" Digital economy development plan	The State Council	2022.1.12	The added value of core industries in the digital economy is required to account for 10 percent of GDP by 2025. Under the guidance of this policy, the traditional industrial chain will realize a comprehensive digital transformation, and the development of China's digital economy industry will enter an accelerated period.
	The Implementation Plan For New Urbanization During The 14th Five-Year Plan Period	The State Council	2022.7.28	1. Enrich the application scenarios of digital technologies, and develop telecommuting, telemedicine, smart travel, smart communities, and smart business circles. 2. Promote smart transformation. It plays a leading role in guiding China's urbanization and may be a landmark document in the process of China's urbanization.

Source: www.gov.cn

BENEFITS OF USING BLOCKCHAIN TO ENHANCE DIGITAL TWIN TECHNOLOGIES

How blockchain technology can help enhance digital twins and steadily promote the development of smart cities in China can be summarized as follows:

Decentralized and Immutable Data Transactions of Digital Twins

Protection of digital twin platform and data is crucial. Blockchain strengthens security through cryptographic hashing algorithms, and by storing digital twin data on distributed ledgers which can neither be changed easily nor controlled by any central authority. Immutability is one of the distinctive features of blockchain which ensures that the blockchain ledger remains indelible and keeps the unalterable history of digital twin transactions (Huang et al., 2020). This feature not only brings an unprecedented level of trust and provides integrity of digital twin data but also transforms the digital twin audit process into an efficient and cost-effective procedure.

With no third-party credit backing, through blockchain technology we can make secure payments over long distances on an open platform. And keep a record of all historical transactions. During the entire transaction, all participants in the network keep an identical ledger, and if a change is made to a piece of data in the ledger, all copies of the data are quickly modified in parallel (Putz., 2021). Each transaction in this distributed ledger is uniquely timestamped, eliminating the need for duplicate payments (figure 3). The blockchain's value transfer process can be summarized

Figure 3. Blockchain value transfer methods
Source: China Academy of Information and Communication Research

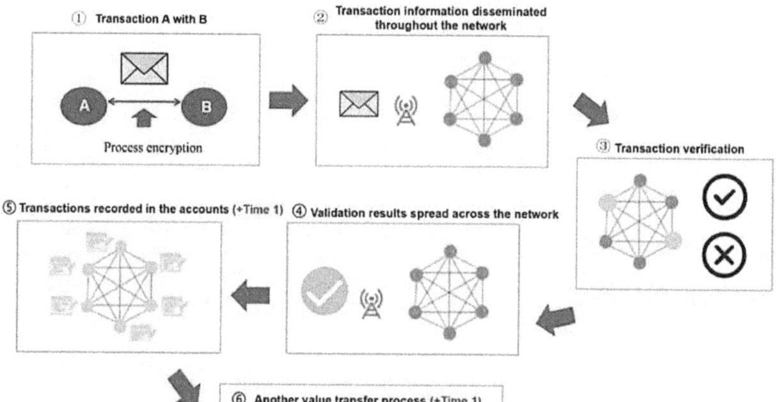

as a decentralized technology and means of replacing centralization to achieve the two most important aspects of corroboration and value transfer: bookkeeping and trust guarantees (Putz., 2021).

Digital Twin Technology's Identity and Legitimacy

Blockchain-enabled digital twin to store all information regarding products immutably. Furthermore, their transaction records are saved which allows buyers to know where the specific products were made, the ownership history, etc. This helps to create proof of the legitimacy and identity of the digital twin. Consider a scenario, where multiple fake replicas are available of digital twin along with fake documents which can prove their authenticity. Such type of problem can easily be addressed through digital certificates. Storing and issuing digital certificates through blockchain ensure that these certificates cannot be copied, modified, or corrupted (Suhail et al., 2022). Each physical product can have a transferable and updatable digital certificate. Besides ensuring identity and legitimacy, blockchain technology can act as an anti-theft mechanism because unverified digital twin technology cannot be used anywhere.

·Tracking digital twin technology globally with high accuracy

Creating digital twin technology combined with blockchain technology can help to identify and track them globally with high accuracy. Blockchain enables digital twin technology to access the distributed databases without any third party from multiple locations in a secure manner. In blockchain-based digital twin, record keeping, provenance tracking, and auditability become easy as the digital twin information can easily be accessed/tracked. The history of digital twin right from their origin to the present can be traced through blockchain (Yaqoob et al., 2020). Such type of accurate provenance tracking can be employed to analyze and find fraud in any part of digital twin (Suhail et al., 2022).

Access Privileges and Trusted Digital Twin Data Coordination

Most of the traditional access control mechanisms are vulnerable to cybersecurity attacks. However, blockchain eliminates risks of unauthorized access that often leads to data tampering. Blockchain-based access control manages all the digital twin information, that is, policies and rules in a distributed manner, where all the digital twins act autonomously. Moreover, blockchain-based access control enhances security through secure sockets layer certificates which are located on each digital twin, where the certificate data is managed on blockchains. On the other hand, digital twin usually coordinate/communicate with their physical objects through IoT-based technologies. Such coordination and communication of data is vulnerable to security threats which can be compensated through blockchain (Yaqoob et al., 2020).

THE ADDED VALUE OF INTEGRATING DIGITAL TWIN, IoT, BLOCKCHAIN FROM STAKEHOLDERS' PERSPECTIVE

Investors' Perspective

Data is the lifeblood of any business. Today, big data has applications in just about every industry – retail, healthcare, financial services, and government (Yu et al., 2022b; 2022m). Any organization that can assimilate data to answer nagging questions about its operations can benefit from big data (Yu et al., 2022d). Overall, the demand for big data transcends all sectors and businesses. Those who work to understand their customer's business and their problems will be able to proactively identify big data solutions appropriate to their needs and thus gain a competitive advantage over their competitors (Rabah, 2018). The IoT does not only enable a huge range of applications but covers a wide span of societies and industrial needs, as well. Specifically, IoT is expected to play a major role in transforming ordinary cities into smart ones, houses into smart homes, electrical grids into smart grids, and so on. Additionally, IoT has diverse applications including healthcare, sports, entertainment, as well as environmental applications and many more. On another front, IoT can be thought of as the backbone of digitizing the industrial sector by enabling optimized production and manufacturing processes in addition to cost reduction.

Regulators' Perspective

As an emerging technology, blockchain would provide an essential solution to the problems facing IoT, especially when utilizing smart contracts, which shall play an important role in managing and securing IoT devices, e.g. Yu et al. (2022h). Blockchain solves IoT issues as explained in what follows:

- Elimination of central authority: Blockchain provides trusted, unique, and distributed authentication of IoT devices where participants can identify every single device. As per data analysis, it could be executed with the aid of the smart contract facility provided by the blockchain. Those advantages are extremely important, especially for large-scale IoT systems.

- Peer to peer accelerated direct messaging: Thanks to the distributed ledger characteristics of blockchains, which not only eliminates the need for a central authority but enables the coordination and processing of transmitted data between devices and stores devices' interaction, state, and exchanged data immutably in the blockchain's ledger. Also, data flow in the centralized IoT system differs from that in the decentralized IoT-blockchain integrated system, especially since the integration takes different forms and designs.
- Automation and resource utilization: Blockchain enables direct and automated interaction between IoT devices using smart contracts. Also, blockchain's smart contract facilitates resource usage by running an on-demand code or smart algorithm to manage resource utilization and automate payments when the requested service is completed.

Management Perspective

Identity and access management (IAM) of IoT devices is facing multiple challenges such as the change of ownership during the lifetime of IoT devices from manufacturer to supplier then to retailer, until they end up in the hands of their consumers (Mahalle et al., 2013; Friese et al., 2014). Also, consumer ownership may change in case the IoT device is compromised, decommissioned, or resold. Another issue facing IAM is managing the attributes of the IoT devices such as serial number, manufacturer, make type, location, and deployment GPS coordinates. Another challenge related to IoT identity and access management is the IoT relationships, which may take the form of device-to-device, device-to-human, or device-to-service. Also, the types of IoT relationships could vary from deployed by to use by or sold by, shipped by, upgraded by, repaired by, and so on. Blockchain is capable of addressing the above challenges securely and effectively since it has been utilized to provide authorized and trusted identity registration and management, ownership tracking, and asset monitoring. Blockchain can register and provide identities to IoT devices with different attributes that are connected in a complex relationship and store all this information securely and immutably in a distributed manner. Therefore, blockchain supports a trusted and decentralized IoT identity governance and tracking throughout the life-cycle of the device (Khan & Salah, 2018).

Citizen Perspective

From the citizens' point of view, the design of a smart city should take into account people's quality of life, with particular attention to the privacy problem when personally identifiable information and household-level data about citizens are involved. This is notably important as citizens might fear the introduction of innovative technologies (Farooqi et al., 2019) or might perceive it as intrusive.

Figure 4. Digital twin interaction model
Source: Science Foundation Ireland (SFI)

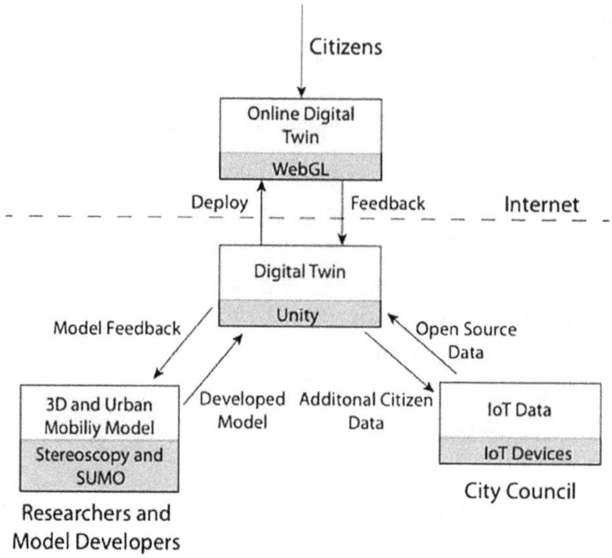

Citizens can browse through the simulated scenarios in the online digital twin and submit their comments and suggestions. For example, the diagram above shows the digital twin online interaction model, showing the technology and stakeholders involved in each session (Figure 4). In this green space planning simulation, the citizen suggested that they would like to include children's play facilities in the new park, an option that was not initially included in the planning scenario. Through the digital twin, information on urban governance can flow between citizens, experts, and city managers, and it becomes easy for citizens to participate in urban governance.

Issues, Controversies, Problems

There are insufficient case studies.

SOLUTIONS AND RECOMMENDATIONS

We showcase a few case studies from China.

After more than a decade of exploration and practice, China has explored a road of smart city construction in line with Chinese characteristics. In the new era, the demand for urban development is constantly changing (Guo et al., 2016). The construction of smart cities in China highlights the key points and strengthens the smart foundation. At the same time, the construction of smart cities should fully take into account the basic and supporting general infrastructure of the city, such as 5G, IoT, blockchain, industrialization degree, smart devices, intelligent manufacturing, digitalization, and business integration. Therefore, the construction of a smart city needs to take measures according to local conditions and construction characteristics of each province and city, highlighting their advantages (Guo et al., 2016).

Beijing Smart City: Focus on Four Regions to Create a Model City

Based on the original prominent industrial focus, Beijing focuses on the four key areas of the city sub-center administrative office area, Canal Business District, Zhangjiawan Design Town, and cultural tourism (Wang et al., 2016).

1. Smart Canal Business District focuses on FinTech and green finance (Yu et al., 2022f; 2022i), and cultivates the FinTech industry featuring "finance and technology" (Yu et al., 2022j). At the same time, Beijing has built an integrated financial service platform around the sub-service financial industry in the sub-center of the city.
2. Focusing on metaverse and digital culture and tourism, the Smart Culture and Tourism Zone takes "culture and consumption" as its core and creates the first metaverse digital consumption industry cluster. Relying on the influence of Universal Studios, this place will be built as an industrial ecological cluster for the aggregation of metaverse virtual space, content production, and animation games (Feng & Cai, 2021). By opening up the digital chain of the cultural tourism industry (Yu et al., 2022k), it creates a total digital synergy system of cultural tourism that is jointly built, jointly negotiated and jointly created.
3. Smart Zhangjiawan Design Town focuses on digital design and smart city life laboratory and takes the advantage of "innovation and low carbon" (Yu et al., 2022l) to create a demonstration model of digital design innovation in the design industry. At the same time, Zhangjiawan uses the AI design platform and digital resources to promote the development of the design industry and expands the scene of Zhangjiawan's "Smart City Life Laboratory" into a construction-level, park-level and block-level laboratory.
4. The smart administrative office area focuses on the application of autonomous driving, with "government affairs and service" as the core, and carries out pilot intelligent applications such as autonomous driving, three major buildings, and a smart office. By carrying out the demonstration and application of the intelligent IoT through autonomous driving, the Chinese government has built smart venues in urban sub-center theaters, libraries and museums to strengthen intelligent administrative offices (Feng & Cai, 2021).

Figure 5. Beijing Smart City Digital Transformation
Source: China Academy of Information and Communication Research

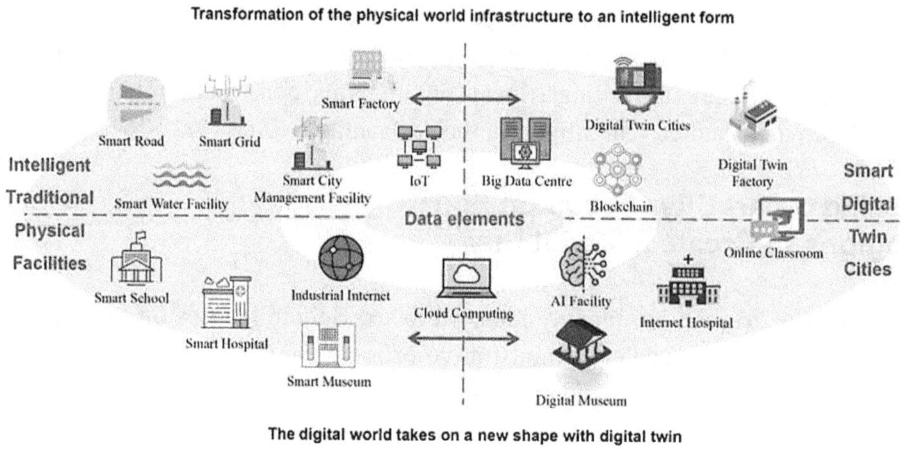

Beijing has been driven by three rounds of scene opening, data opening and technology polishing, giving full play to the role of "big incubators" for smart cities, and stimulating new business forms and models of smart economy. Through AI, blockchain, big data, privacy computing, and other public technology platforms, Beijing has set up a common technology service platform for the whole society to make full use of Beijing University's talent advantage, central national team and ministries and unicorn technology enterprises to provide basic support for the smart city industrial ecology (Qi & Ji, 2021). On the one hand, the smart city in Beijing actively attracts innovative elements and forms a government and enterprise alliance with government agencies, operators, scientific research institutions, developers, and investment institutions. On the other hand, it strengthens normative constraints and guarantees in industry management related to open scenarios, management and use of big data, and alliance operation (Qi & Ji, 2021). Through the innovation of relevant mechanisms and the implementation of a series of major applications of smart cities (Yu et al., 2022e), Beijing's smart city has formed several landing innovation achievements (Figure 5).

Shijiazhuang: The First National Spatiotemporal Data Platform for Smart Cities Comes Into Operation

The spatiotemporal big data platform is the core of the industrialization of spatiotemporal big data. It is the study and application of big data in space and time, based on geographic information, cloud computing, the IoT, big data, and other technologies, building an important spatiotemporal information substrate for the construction of smart cities. Spatiotemporal frame data and spatiotemporal change data, onto a specific platform and make them undergo a continuous aggregation effect (Zhang et al., 2022). This aggregation effect is to reveal the essential laws of things through multi-dimensional data fusion and correlation analysis and data mining and to make faster, more comprehensive, more accurate, and more effective judgments and predictions on things. In this sense, the spatiotemporal big data platform is the core value of big data, is the advanced form of big data development, and is the solution for the big data era (Zhang et al., 2022).

In September 2022, the Shijiazhuang smart city spatiotemporal big data platform through the completion of acceptance has been officially put into operation, becoming the first national smart city spatiotemporal big data platform built in Hebei Province. The platform integrates more than 700 small categories of data involving the city's basic space-time, public domain, natural resources, industry sectors, IoT real-time perception, Internet online capture, and other 6 major categories, covering the period since 2009, fusing the city's unified space-time big data "a map" (Li et al., 2022).

Shijiazhuang smart city spatiotemporal big data platform access to a variety of real-time sensing data and more than 12 million population data, more than 2 million enterprise legal person data, convergence management of billion stream data, terabytes of spatiotemporal data, but also docked Shijiazhuang city smart government, smart public security, smart city management, smart environmental protection, smart emergency, and other typical application systems to provide data services (Zhao et al., 2015). According to statistics, during the trial period, the platform accumulated 1,800 hours of use and 2.3 million times of service calls (Gov.cn, 2022).

In terms of operation and management, the platform relies on coordination and technical methods, and innovatively proposes a dynamic update mechanism for basic geographic data. Coordinating satellite image data, the Shijiazhuang Satellite Remote Sensing Application Centre guarantees semi-annual updates of sub-metre-level high-resolution remote sensing image data and quarterly updates of 2-metre resolution optical satellite remote sensing images, so that remote sensing image data can be dynamically updated (Gov.cn, 2022). Using the geographical names, business corporations and standard address data regularly provided by Shijiazhuang Civil Affairs Bureau, Market Supervision Bureau and Public Security Bureau, the platform can dynamically update the geographical names address database.

Shijiazhuang intelligent city spatiotemporal big data platform in the construction process, the use of 120 square kilometres of historical tilt photography data and historical three-dimensional fine model, as well as the convergence of the city of each district and county three-dimensional, realistic three-dimensional data, urban white film data, in the realistic three-dimensional construction path to explore, which also laid a certain foundation for the future construction of realistic three-dimensional Shijiazhuang (Hebei News).

Guangzhou: First City in China to Be Approved to Create an Early Demonstration Zone for Blockchain Development

In the past two years, Guangzhou has taken Huangpu District and Guangzhou Development Zone as the testing ground for reform and vigorously promoted the clustering and development of blockchain enterprises (Gzdaily, 2022). The "Implementation Opinions on Promoting the Innovative Development of Blockchain Industry in Guangzhou (2020-2022)", "Notice on Establishing Guangzhou Public Law Chain to Fully Implement Electronic Forensic Opinions", "Measures to Promote the Development of Blockchain Industry in Huangpu District and Guangzhou Development Zone of Guangzhou (10 Articles on Blockchain)", "Notice on Several Measures to Accelerate the Blockchain Industry in Huangpu District and Guangzhou Development Zone of Guangzhou to Lead Change" have been issued. As of May 2020, Guangzhou has gathered about 400 blockchain enterprises, covering various fields such as finance, government, people's livelihood and manufacturing (Gz.gov.cn). Guangzhou was officially approved by the Ministry of Industry and Information Technology to create a Blockchain Development Advance Demonstration Zone, becoming the first city in China to be approved to create a Blockchain Development Advance Demonstration Zone (Gzdaily, 2022).

There are already Tianhe District, Huangpu District, Panyu District, Baiyun District, Conghua District, Zengcheng District, the Municipal Financial Supervision Bureau and the Public Resources Transaction Center. Guangzhou has implemented more than thirty innovative applications of blockchain, such as blockchain electronic invoices, the platform of sharing chain for credible mutual recognition of educational talents in Guangdong, Hong Kong and Macao, blockchain + enterprise credit, blockchain + credit big data, etc. The agglomeration effect of the blockchain industry has been highlighted (Huanan et al., 2015).

As of May 2020, Guangzhou has gathered about 400 blockchain enterprises, covering various fields such as finance, government affairs, people's livelihood and manufacturing (Wang et al., 2019). Guangzhou was officially approved by the Ministry of Industry and Information Technology to create an early demonstration zone for blockchain development, becoming the first city in China to be approved to create an early demonstration zone for blockchain development (Wang et al., 2019).

FUTURE RESEARCH DIRECTIONS

Guangzhou is exploring blockchain + electronic license, through blockchain-empowered convergence of the four links of the license, license storage, license request, and license acquisition, to avoid the transmission process being susceptible to tampering. If the storage environment is tampered with or leaked, and the caller denies responsibility, the holder is not authorized to dispute (Guo, 2022). The people apply at the front desk with their identity documents or authorized use permits (QR codes written on the chain), and government staff check and extract data online according to their functions, reducing form entry and supporting one-window acceptance, online application, back-end review, automatic generation, privacy protection and online verification in five parts (Guo, 2022). The above work is very meaningful: the process of business start-up previously took two days for business registration, one day for seal engraving, five days for bank account opening, and two days for tax reporting, requiring the submission of seventeen sub-materials and at least six errands. With the blockchain, the industrial and commercial, public security, banking and taxation systems will be chained, and it will be possible to start a business with only three submissions, one errand and one day, supplemented by online filing by the individual applicant (Guo, 2022).

It can be seen that blockchain applications in Guangzhou focus on small and medium-sized enterprises and individuals, optimizing the business environment and realizing cross-sector and cross-industry enterprise and industry synergy (Choo, 2021). Guangzhou is exploring a new "Guangzhou model" of blockchain + enterprise chain code to solve the problems of loan financing for SMEs, risk control for banks and supervision for departments.

CONCLUSION AND RECOMMENDATIONS

Based upon the previous discussion, this chapter examines whether and how current practices based on digital twin technologies can help the development of smart cities in China. We also explore the role of government in promoting the application of digital twin technology in industrial infrastructure and smart city development agendas. How can these institutions facilitate the development of digital twin technologies in various industries? How will these developments in turn affect stakeholders in different sectors? Can blockchain technology secure the digital twin to steadily drive the development of China's smart cities? These issues have all been fully discussed in this chapter. We also use CiteSpace to analyze the literature related to digital twin technologies and smart cities in the WOS database to determine the focus of the research in this chapter. There are many cases of smart city development in China. By selecting three representative cities, Beijing, Shijiazhuang, and Guangzhou, we have demonstrated the construction and development model of China's smart cities and highlighted their unique characteristics.

ACKNOWLEDGMENT

The authors extend sincere gratitude to:

• Our colleagues from Soochow University, the Australian Studies Centre of Shanghai University and Krirk University as well as the independent research colleagues who provided insight and expertise that greatly assisted the research, although they may not agree with all of the interpretations/conclusions of this chapter.

• China Knowledge for supporting our research.

• The Editor and the International Editorial Advisory Board (IEAB) of this book who initially desk reviewed, arranged a rigorous double/triple blind review process and conducted a thorough, minute and critical final review before accepting the chapter for publication.

• All anonymous reviewers who provided very constructive feedbacks for thorough revision, improvement and fine tuning of the chapter.

REFERENCES

Al Sadawi, A., Hassan, M. S., & Ndiaye, M. (2021). A Survey On The Integration Of Blockchain With Iot To Enhance Performance And Eliminate Challenges. *IEEE Access: Practical Innovations, Open Solutions*, *9*, 54478–54497. doi:10.1109/ACCESS.2021.3070555

Ammar, M., Russello, G., & Crispo, B. (2018). Internet Of Things: A Survey On The Security Of IoT Frameworks. *Journal of Information Security and Applications*, *38*, 8–27. doi:10.1016/j.jisa.2017.11.002

Benamrou, B., Mohamed, B., Bernoussi, A. S., & Mustapha, O. (2016, October). Ranking models of smart cities. In *2016 4th IEEE International Colloquium on Information Science and Technology (CiSt)* (pp. 872-879). IEEE. 10.1109/CIST.2016.7805011

Bucchiarone, A. (2019). Collective Adaptation Through Multi-Agents Ensembles: The Case Of Smart Urban Mobility. *ACM Transactions on Autonomous and Adaptive Systems*, *14*(2), 1–28. doi:10.1145/3355562

Chen, C. (2006). CiteSpace II: Detecting And Visualizing Emerging Trends And Transient Patterns In Scientific Literature. *Journal of the American Society for Information Science and Technology*, *57*(3), 359–377. doi:10.1002/asi.20317

Chen, C. (2014). The CiteSpace Manual. *College of Computing and Informatics*, *1*(1), 1–84. doi:10.100711192-015-1576-8

Chen, C., Ibekwe-SanJuan, F., & Hou, J. (2010). The Structure And Dynamics Of Cocitation Clusters: A Multiple-Perspective Cocitation Analysis. *Journal of the American Society for Information Science and Technology*, *61*(7), 1386–1409. doi:10.1002/asi.21309

Choo, K. K. R., Gai, K., & Chiaraviglio, L. (2021). Blockchain-Enabled Secure Communications In Smart Cities. *Journal of Parallel and Distributed Computing*, *152*, 125–127. doi:10.1016/j.jpdc.2021.02.021

Crisostomi, E., Shorten, R., & Wirth, F. (2016). Smart Cities: A Golden Age For Control Theory? *IEEE Technology and Society Magazine*, *35*(3), 23–24. doi:10.1109/MTS.2016.2592782

European Smart Cities project. (n.d.). https://www.smart-cities.eu/model.html

Faber, A., Rehm, S. V., Hernandez-Mendez, A., & Matthes, F. (2018). Modeling And Visualizing Smart City Mobility Business Ecosystems: Insights From A Case Study. *Information (Basel)*, *9*(11), 270. doi:10.3390/info9110270

Faria, R., Brito, L., Baras, K., & Silva, J. (2017). Smart Mobility: A Survey. *Proceedings of the 2017 International Conference on Internet of Things for the Global Community*, 1–8.

Farooqi, N., Gutub, A., & Khozium, M. O. (2019). Smart Community Challenges: Enabling IoT/M2M Technology Case Study. *Life Science Journal*, *16*(7), 11–17. doi:10.1109/IoTGC.2017.8008972

Fazio, M., Paone, M., Puliafito, A., & Villari, M. (2012, July). Heterogeneous Sensors Become Homogeneous Things In Smart Cities. In *2012 Sixth International Conference on Innovative Mobile and Internet Services in Ubiquitous Computing* (pp. 775-780). IEEE. 10.1109/IMIS.2012.136

Feng, Q., & Cai, R. (2021). "Data Hegemony": Reflections for the Application and Development Direction of Metaverse Technology in Urban Design based on Digital. *Journal of World Architecture*, *5*(6), 52–61. doi:10.26689/jwa.v5i6.2780

Finck, M., Lamping, M., Moscon, V., & Richter, H. (2020). *Smart Urban Mobility*. Springer Berlin Heidelberg. doi:10.1007/978-3-662-61920-9

Friese, I., Heuer, J., & Kong, N. (2014, March). Challenges From The Identities Of Things: Introduction Of The Identities Of Things Discussion Group Within Kantara Initiative. In *2014 IEEE World Forum on Internet of Things (WF-IoT)* (pp. 1-4). IEEE. 10.1109/WF-IoT.2014.6803106

Gov.cn. (2022). *Shijiazhuang intelligent city spatio-temporal data platform completed*. http://news.mnr.gov.cn/

Guo, M., Liu, Y., Yu, H., Hu, B., & Sang, Z. (2016). An Overview Of Smart City In China. *China Communications, 13*(5), 203–211. doi:10.1109/CC.2016.7489987

Guo, Q., You, W. W., Hu, K., Chen, Y., & He, Z. (2022). Study On The Path And Countermeasures Of Blockchain Technology To Promote High-Quality Development Of Digital Economy In Guangzhou. *Open Journal of Business and Management, 10*(6), 3305–3326. doi:10.4236/ojbm.2022.106163

Gzdaily. (2022). *Give Full Play To The Development Of The Blockchain Industry To Make Digital More "Warm"*. https://www.gzdaily.cn/

Gz.gov.cn. (2022). *Guangzhou Creates A Pioneering Demonstration Zone For Blockchain Development.* https://www.gz.gov.cn/

Hebei News. (2022, September). Spatio-Temporal Big Data To Create A Smart City. *Brain*, 22.

Huanan, Z., Shijun, L., & Hong, J. (2015). Guangzhou smart city construction and big data research. In *2015 International Conference on Behavioral, Economic and Socio-cultural Computing (BESC)* (pp. 143-149). IEEE. 10.1109/BESC.2015.7365972

Huang, S., Wang, G., Yan, Y., & Fang, X. (2020). Blockchain-based data management for digital twin of product. *Journal of Manufacturing Systems, 54*, 361–371. doi:10.1016/j.jmsy.2020.01.009

Khan, M. A., & Salah, K. (2018). IoT Security: Review, Blockchain Solutions, And Open Challenges. *Future Generation Computer Systems, 82*, 395–411. doi:10.1016/j.future.2017.11.022

Li, Q., Du, Y., Luo, J., & Fang, Y. (2022, April). Research on Collaborative Development Level of Beijing-Tianjin-Hebei Smart City Based on Coupling Correlation Degree Analysis. In *2022 7th International Conference on Social Sciences and Economic Development (ICSSED 2022)* (pp. 1053-1064). Atlantis Press. 10.2991/aebmr.k.220405.177

Mahalle, P. N., Anggorojati, B., Prasad, N. R., & Prasad, R. (2013). Identity Authentication And Capability Based Access Control (IACAC) For The Internet Of Things. *Journal of Cyber Security and Mobility, 1*(4), 309–348. doi:10.13052/jcsm2245-1439.142

North, D. C. (1991). Institutions. *The Journal of Economic Perspectives, 5*(1), 97–112. doi:10.1257/jep.5.1.97

Okrepilov, V., Kuzmina, S., & Kuznetsov, S. (2019, March). Tools Of Quality Economics: Sustainable Development Of A 'Smart City' Under Conditions Of Digital Transformation Of The Economy. In IOP Conference Series: Materials Science and Engineering (vol. 497, no. 1, pp. 012134). IOP Publishing. doi:10.1088/1757-899X/497/1/012134

Oldenburg, R. (1999). *The Great Good Place: Cafes, Coffee Shops, Bookstores, Bars, Hair Salons, And Other Hangouts At The Heart Of A Community*. Da Capo Press.

Pereira, A. C., & Romero, F. (2017). A Review Of The Meanings And The Implications Of The Industry 4.0 Concept. *Procedia Manufacturing*, *13*, 1206–1214. doi:10.1016/j.promfg.2017.09.032

Poncela, J., Vlacheas, P., Giaffreda, R., De, S., Vecchio, M., Nechifor, S., Barco, R., Aguayo-Torres, M. C., Stavroulaki, V., Moessner, K., & Demestichas, P. (2014). Smart Cities Via Data Aggregation. *Wireless Personal Communications*, *76*(2), 149–168. doi:10.100711277-014-1683-5

Putz, B., Dietz, M., Empl, P., & Pernul, G. (2021). Ethertwin: Blockchain-based secure digital twin information management. *Information Processing & Management*, *58*(1), 102425. doi:10.1016/j.ipm.2020.102425

Qi, J., & Ji, J. (2021). Beijing as Smart City: Promotion of International Communication Center. *Philosophical Studies*, *11*(2), 143–148.

Rabah, K. (2018). Convergence of AI, IoT, Big Data And Blockchain: A Review. *Law Institute Journal*, *1*(1), 1–18.

Rao, S. K., & Prasad, R. (2018). Impact Of 5G Technologies On Smart City Implementation. *Wireless Personal Communications*, *100*(1), 161–176. doi:10.100711277-018-5618-4

Rathore, M. M., Ahmad, A., Paul, A., & Rho, S. (2016). Urban Planning And Building Smart Cities Based On The Internet Of Things Using Big Data Analytics. *Computer Networks*, *101*, 63–80. doi:10.1016/j.comnet.2015.12.023

Sarker, M. N. I., Wu, M., & Hossin, M. A. (2018, May). Smart Governance Through Bigdata: Digital Transformation Of Public Agencies. In 2018 international conference on artificial intelligence and big data (ICAIBD) (pp. 62-70). IEEE. doi:10.1109/ICAIBD.2018.8396168

Sharma, R., Fantin, A. R., Prabhu, N., Guan, C., & Dattakumar, A. (2016). Digital Literacy And Knowledge Societies: A Grounded Theory Investigation Of Sustainable Development. *Telecommunications Policy*, *40*(7), 628–643. doi:10.1016/j.telpol.2016.05.003

Smit, J., Kreutzer, S., Moeller, C., & Carlberg, M. (2016). Policy Department A: Economic and Scientific Policy–Industry 4.0. European Parliament.

Sowe, S. K., Kimata, T., Dong, M., & Zettsu, K. (2014, July). Managing Heterogeneous Sensor Data On A Big Data Platform: IoT Services For Data-Intensive Science. In *2014 IEEE 38th international computer software and applications conference workshops* (pp. 295-300). IEEE. 10.1109/COMPSACW.2014.52

Strohbach, M., Ziekow, H., Gazis, V., & Akiva, N. (2015). Towards A Big Data Analytics Framework For IoT And Smart City Applications. In *Modeling And Processing For Next-Generation Big-Data Technologies* (pp. 257–282). Springer. doi:10.1007/978-3-319-09177-8_11

Suhail, S., Hussain, R., Jurdak, R., Oracevic, A., Salah, K., Hong, C. S., & Matulevičius, R. (2022). Blockchain-based digital twins: Research trends, issues, and future challenges. [CSUR]. *ACM Computing Surveys*, *54*(11s), 1–34. doi:10.1145/3517189

Theodoridis, E., Mylonas, G., & Chatzigiannakis, I. (2013, July). Developing An IoT Smart City Framework. In *IISA 2013* (pp. 1–6). IEEE. doi:10.1109/IISA.2013.6623710

Tomaras, D. (2018, March). Crowdsourcing Techniques For Smart Urban Mobility. In *2018 IEEE International Conference on Pervasive Computing and Communications Workshops (PerCom Workshops)* (pp. 460-461). IEEE. 10.1109/PERCOMW.2018.8480244

Tomičić Pupek, K., Pihir, I., & Tomičić Furjan, M. (2019). Smart City Initiatives In The Context Of Digital Transformation–Scope, Services And Technologies. *Management*, *24*(1), 39–54. doi:10.30924/mjcmi.24.1.3

Vaidya, S., Ambad, P., & Bhosle, S. (2018). Industry 4.0–A Glimpse. *Procedia Manufacturing*, *20*, 233–238. doi:10.1016/j.promfg.2018.02.034

Wang, G., El Saddik, A., Lai, X., Perez, G. M., & Choo, K. K. R. (Eds.). (2019). Smart City and Informatization. In *Proceedings of the 7th International Conference, iSCI 2019, Guangzhou, China, November 12–15, 2019* (*Vol. 1122*). Springer Nature.

Wang, L. J., Zhang, Y., Li, Q., & Ruan, P. N. (2016). Under The Background Of Smart City Development Of E-Government Performance Evaluation: Study In Every Districts Of Beijing. *Proceedings of the 6th International Conference on Information Technology for Manufacturing Systems.*

Yaqoob, I., Salah, K., Uddin, M., Jayaraman, R., Omar, M., & Imran, M. (2020). Blockchain for digital twins: Recent advances and future research challenges. *IEEE Network*, *34*(5), 290–298. doi:10.1109/MNET.001.1900661

Yu, P., Chen, D., & Ahuja, A. (2022a). Smart and Sustainable Economy: How COVID-19 Has Acted as a Catalyst for China's Digital Transformation. In S. Kautish & G. Kaur (Eds.), *AI-Enabled Agile Internet of Things for Sustainable FinTech Ecosystems* (pp. 106–146). IGI Global. doi:10.4018/978-1-6684-4176-3.ch006

Yu, P., Chen, J., Sampat, M., & Misuko, N. (2022b). The Digital Transformation of Rural Agricultural Business Management: A Case Study of China. In S. Bilgaiyan, J. Singh, & H. Das (Eds.), *Empirical Research for Futuristic E-Commerce Systems: Foundations and Applications* (pp. 23–52). IGI Global. doi:10.4018/978-1-6684-4969-1.ch002

Yu, P., Gong, R., & Sampat, M. (2022c). Blockchain Technology in China's Digital Economy: Balancing Regulation and Innovation. In P. Tehrani (Ed.), *Regulatory Aspects of Artificial Intelligence on Blockchain* (pp. 132–157). IGI Global. doi:10.4018/978-1-7998-7927-5.ch007

Yu, P., Gu, H., Zhao, Y., & Ahuja, A. (2022d). Digital Transformation Driven by Internet Data Center: Case Studies on China. In D. Piaggesi, H. Landazuri, & B. Jia (Eds.), *Cases on Applying Knowledge Economy Principles for Economic Growth in Developing Nations* (pp. 203–230). IGI Global. doi:10.4018/978-1-7998-8417-0.ch011

Yu, P., He, K., & Sampat, M. (2022e). COVID-19's Impact on the Chinese Tourism and Hospitality Sector. In V. Costa & C. Costa (Eds.), *Sustainability and Competitiveness in the Hospitality Industry* (pp. 23–66). IGI Global. doi:10.4018/978-1-7998-9285-4.ch002

Yu, P., Jiao, A., & Sampat, M. (2022f). The Effect of Chinese Green Transformation on Competitiveness and the Environment. In P. Ordóñez de Pablos, X. Zhang, & M. Almunawar (Eds.), *Handbook of Research on Green, Circular, and Digital Economies as Tools for Recovery and Sustainability* (pp. 257–279). IGI Global. doi:10.4018/978-1-7998-9664-7.ch014

Yu, P., Liu, Z., & Hanes, E. (2022g). Supply Chain Resiliency, Efficiency, and Visibility in the Post-Pandemic Era in China: Case Studies of MeiTuan Waimai, and Ele.me. In Y. Ramakrishna (Ed.), *Handbook of Research on Supply Chain Resiliency, Efficiency, and Visibility in the Post-Pandemic Era* (pp. 195–225). IGI Global. doi:10.4018/978-1-7998-9506-0.ch011

Yu, P., Liu, Z., Hanes, E., & Mumtaz, J. (2022h). Integration of IoT and Blockchain for Smart and Secured Supply Chain Management: Case Studies of China. In S. Goyal, N. Pradeep, P. Shukla, M. Ghonge, & R. Ravi (Eds.), *Utilizing Blockchain Technologies in Manufacturing and Logistics Management* (pp. 179–207). IGI Global. doi:10.4018/978-1-7998-8697-6.ch010

Yu, P., Lu, S., Hanes, E., & Chen, Y. (2022i). The Role of Blockchain Technology in Harnessing the Sustainability of Chinese Digital Finance. In P. Swarnalatha & S. Prabu (Eds.), *Blockchain Technologies for Sustainable Development in Smart Cities* (pp. 155–186). IGI Global. doi:10.4018/978-1-7998-9274-8.ch009

Yu, P., Lu, S., Sampat, M., Li, R., & Ahuja, A. (2022j). How AI-Enabled Agile Internet of Things Can Enhance the Business Efficiency of China's FinTech Ecosystem. In S. Kautish & G. Kaur (Eds.), *AI-Enabled Agile Internet of Things for Sustainable FinTech Ecosystems* (pp. 190–223). IGI Global. doi:10.4018/978-1-6684-4176-3.ch009

Yu, P., Tang, H., Tang, H., & Hanes, E. (2022k). Using "Digitalization + Intellectual Property" Management to Realize Cultural Economy: Case Study of Chinese Museums. In R. Pettinger, B. Gupta, A. Roja, & D. Cozmiuc (Eds.), *Handbook of Research on Digital Transformation Management and Tools* (pp. 236–264). IGI Global. doi:10.4018/978-1-7998-9764-4.ch011

Yu, P., Weng, Y., & Ahuja, A. (2022l). Carbon Financing and the Sustainable Development Mechanism: The Case of China. In A. Rafay (Ed.), *Handbook of Research on Energy and Environmental Finance 4.0* (pp. 301–332). IGI Global. doi:10.4018/978-1-7998-8210-7.ch012

Yu, P., Xue, W., & Mahendran, R. (2022m). The Development and Impact of China's Digital Transformation in the Medical Industry. In M. Rodrigues & J. Proença (Eds.), *Impact of Digital Transformation on the Development of New Business Models and Consumer Experience* (pp. 97–128). IGI Global. doi:10.4018/978-1-7998-9179-6.ch006

Yu, P., Yu, M., & Sampat, M. (2022n). Smart Management for Digital Transformation in China. In B. Barbosa, S. Filipe, & C. Santos (Ed.), Handbook of Research on Smart Management for Digital Transformation (pp. 411-438). IGI Global. doi:10.4018/978-1-7998-9008-9.ch019

Zhang, H., Li, S., & Jin, H. (2015). Guangzhou Smart City Construction and Big Data Research. In *2015 International Conference on Behavioral, Economic and Socio-cultural Computing (BESC)* (pp. 143-149). IEEE. 10.1109/BESC.2015.7365972

Zhang, J., Bates, J., & Abbott, P. (2022). State-Steered Smart Mentality In Chinese Smart Urbanism. *Urban Studies (Edinburgh, Scotland)*, (14), 2933–2950. Advance online publication. doi:10.1177/00420980211062888

Zhang, Y. (2021). Game And Coordination Of Diverse Benefits In Smart City PPP Projects Based On Evolutionary Game. *Annals of Operations Research*, 1–24. doi:10.100710479-021-04290-y PMID:34776572

Zhao, Y., Zhang, H., Wu, Y., & Liu, Y. (2015). Public Service Demand Of Residents For Smart City Construction: Shijiazhuang City In Hebei Province As An Example. *Progress in Geography*, *34*(4), 473–481.

Chapter 9
Developing a Smart Community Index for Sustainability:
A Case Study of Lundu, Malaysia

Normalini Md Kassim
Unversiti Sains Malaysia, Malaysia

Nor Hazlina Hashim
https://orcid.org/0000-0003-4265-0557
Universiti Malaya, Malaysia

Jasmine A. L. Yeap
Unversiti Sains Malaysia, Malaysia

Saravanan Nathan Lurudusamy
Telekom Malaysia Berhad, Malaysia

T. Ramayah
Universiti Sains Malaysia, Malaysia

DOI: 10.4018/978-1-6684-3833-6.ch009

ABSTRACT

Lundu is in Southwest Sarawak, a Malaysian state on Borneo. Lundu is declared as one of the smart community initiatives that were launched by the Malaysian Communications and Multimedia Commission (MCMC) in February 2016. This chapter intended to understand and map with development index of a smart community in Lundu, and the involved stakeholders can be informed about the necessary criteria which enabled citizens to be deemed as smart. A quantitative research method has been employed whereby a self-administrated questionnaire was developed to gather information from the respondents which consisted of different sections with both open and close-ended questions. A total of 150 forms were distributed in different locations within the smart community site. Four key components in International Growth Centre (IGC) development index, Humanware, Technoware, Infoware, and Valueware, were used as a reference point in developing the smart community index (SCI), and recommendations on infrastructure and monitoring improvement are suggested at the end of the chapter.

INTRODUCTION

In modern times, there is a great need to change and improve the community's lifestyle through the use of ICT applications. The astonishing diffusion of digital technologies due to the expansion of the Internet is the main basis of the rollout of smart community facilities in Malaysia; supported by the integration of planning, new media and technologies. In the earlier days access to Internet was limited with very complicated information systems. This it has restricted the opportunity of exploring the potential use of new technologies and new media in giving shape to the new urban space.

However, now with the inception of mobile phones and the transition to other "smart" devices it has opened up new opportunities for creating a direct relationship between individuals and communities and between communities and decision-makers. Nonetheless, the digital revolution enabled cities and policymakers to realize the link between ICT and place in enhancing the local knowledge economy, where technology enables organizations to create responsible and sustainable solutions to problems faced by the community.

The idea of a smart community has been receiving increased attention worldwide in recent years. A smart community is regarded as a community where various next-generation technologies and advanced social systems are effectively integrated and utilized to improve and transform the everyday lives of the citizens in the designated area. The success of a smart community lies in how the technologies are adopted and used by the residents as well as how the lives of the residents are impacted as a result of the technology adoption particularly in smaller, less urban districts.

For instance, disaster management which is one of the flagship applications in the MCMC's Smart Community initiative that made that the underserved and the unserved areas in Malaysia to have access to ICT. The challenge is to provide the rural population of Malaysia with ICT services and applications at an affordable rate in order to boost penetration levels and bridge the digital divide between the rural and urban areas. This is being done incrementally and Malaysians will reap the benefits as well as to enhance their lifestyle in terms of both social and economy.

For this reason, this study aimed to develop an index that captures the essence of a Smart Community so that the involved stakeholders can be informed about the necessary criteria which enables citizens to be deemed as smart. This can then assist the stakeholders to create synergetic collaboration and partnerships that can further enhance the adoption and level of satisfaction of ICT technologies and ultimately form more smart communities among Malaysian citizens.

The Smart Community project site which assessed in this study's investigation was Lundu, Sarawak. Lundu is located in Southwest Sarawak, about 2 hours drive from Kuching with approximate 35,000 population. Smart Community Initiatives was launched in February 2016 and it was the second district chosen for the Smart Community Initiatives project site after Kemaman in March 2015.

BACKGROUND

Implementation of Smart Cities in Asia

China

Smart city development is favourable by governments and developers as to its benefits that been proven with the help of ICT it could improve the quality of life for the people. For instance, many smart cities projects have been carried out around the world while different developing models have been implemented and various initiatives were introduced to achieve respective targets. In China, smart community planning took place in Yishanwan, Jiangxia District, Wuhan. The community was developed alongside with a comprehensive framework to achieve the targets of construction of infrastructure, establish a sharing platform, develop and research of an application system and develop service portal (Anrong, Li, Li, & Kong, 2016). In the development plan, the synchronous development of informatization, new industrialisation, agricultural modernisation and new urbanisation act as the core of developing the smart community. The authority especially emphasis on agricultural modernisation as it plays as an important role in rural area and act as the fundamental for the suburban area to develop through urbanisation. (Anrong et al., 2016) agreed that it is insufficient to improve only on infrastructure but also stress on the innovation ability of the society by integrating the concept of sustainable to city development (Chourabi et al., 2012).

Japan

While in Japan, Smart Community often define as taking full advantage of IT technology to effectively control power flow and provide new services for power supplies and demand side users (Gao, Fan, Ushifusa, Gu, & Ren, 2016). And it is different from the definition in other countries and the smart community model can be described in four parts which are new information network, new energy system, new transportation system and new urban development. Emphasis is putting on smart energy grid implementation as a way to change citizen towards smarted lifestyle in term of daily life, work or office and transportation. There are four large scale demonstration cities in Japan which implementing smart community concepts which are Kyoto Keihanna District, Yokohama City, Kitakyushu City and Toyota City. All the demonstration cities having the similar target which is reducing the emission of Greenhouse Gas (GHG) and with the assistant of IT technology smart-grid electricity management was introduced and most of the researches about smart community are about using of renewable resource. for instance, eco-town project was introduced for the aim to build a system by using all the waste products of an industry as raw materials for another industry and realising a zero-waste recycling-oriented society.

Malaysia

Medini Smart City in Iskandar Puteri, Malaysia was chosen to be the pilot for smart city projects which focuses on three areas, economy, environment and social and promotes six dimensions: smart economy, smart governance, smart environment, smart mobility, smart people and smart living (Ghazali et al., 2016). This is a project collaborating the university and regional development agency. The objective of this project was focus on infrastructure provision and change of quality of life and environment as long term target by developing smart living through engaging local citizen for sustainable development. However Gil and Navarro (2013) argued that in terms of governance and policy context which were one of the factors in analyzing smart city initiatives, the participation of local seems lacking from the technology partnership. While there was a noticeable effort from the aspect of education where there are few internationally well-known universities such as University of Newcastle plan to open up campus in Medini.

Singapore

Singapore was ranked 7th in the world's smartest cities, according to the 2019 Cities in Motion Index (CIMI), issued by the IESE Business School. The index ranks across nine dimension including human capital, social cohesion, the economy, governance, the environment, mobility and transportation, urban planning, international outreach, and technology (Esri Singapore, 2022).

A digital twin for planning in Singapore is Urban challenge, innovation process, solution and continuity. Singapore has long faced challenges whilst planning for the built environment. Innovations in housing, mobility, and land use have been key priorities of the government. In 2013, 3D modelling was an emerging technology, and the state recognized its potential in offering planners a more comprehensive and accurate medium to design and pilot urban solutions. With access to live real-time monitoring data from IoT sensors, decision-making and urban interventions are better informed by accurate spatial data and real-life human behaviour. Some of the applications of Virtual Singapore include flood risk analysis, opportunities for solar panels and green roofs, and examining the impact of wind load on vegetation (UNDP, 2022)

Characteristic of Smart Community

Smart Citizen

Evidently, it shows the needs to shift to develop a smart community rather than smart city where social inclusion, enabling citizen, supporting community should be included in the city development agenda (Gurstein, 2014). Wongbumru and Dewancker (2014)'s study focusing on how next-generation technology could be utilise and integrated in community particularly through improving citizen innovation and participation in reaching the goal of improving quality of life demonstrating example from Japan that not only focusing on smart energy management but going further to smart community by involving all the stakeholders and aims towards behavioral change through lifestyle innovation. Moreover, author pointed the challenges where there is situation although infrastructure on smart energy system is provided, citizen shows lack of participation and concluded that to bring up "smart citizen" should put as priority before succeeding a smart community (Wongbumru & Dewancker, 2014).

A smarter city should be treated as an organic system - as a network, as a linked system and smart citizen is part of it (Nam & Pardo, 2011). In order to enable this part of "smart citizen" to be fully integrated to the entire system, first the citizen should be educated in a way to "infuse" themselves in the smart system. As without the knowledge, it could not maintain the sustainable cycle of the system and it will become a huge hinder for long term sustainable development. If only with a smart and innovative citizen that act as motivator for the entire system to ensure the smooth network and linkage among the smart initiatives; they themselves could only be the beneficiary from the smart system.

Training

Stratigea (2012)'s work focusing on relating smart cities concept to community development adapting the concepts from Intelligent Community Forum (ICF) that demonstrated the critical success factors for cities "going smart" and presented in a pyramid structure. As Broadband infrastructure act as the first fundamental base, knowledge-based workforce is the second level from bottom proven that apart from infrastructure and facilities, it is important to educate skilled workforce to support the further development. As it is agreed by Eger (2005) as well that internet and broadband access is just the first step and an integrated intelligent community is a catalyst to economic growth (Hughes & Spray, 2002). Undoubtedly, the installation of internet and broadband could offer citizen the accessibility to be linked and connected with each other as well as with the government and businesses that stimulated the local growth in terms of both economy and socially (Nam & Pardo, 2011). Moreover, introducing innovation energy system, information and communication technology could encourage and support private companies, government and SMEs in providing various services to the community (Wongbumru & Dewancker, 2014).

Smart Governance

Apart from installing ICTs to the city, the citizens should introduce with a new decision-making mechanism which incorporate the community into the city planning (Eger, 2005). Hollands (2008) emphasize that the way citizen interacts and becoming a member to the society is the goal of succeeding a smart community and information technology is playing a role not just as a physical infrastructure that drive towards smart cities but offering opportunity to empower and educate "smart citizen".

Besides, argument exists as according to (Gurstein, 2014) that current situation was centralized and top down approach and the voices and opinion of the citizen were put as lowest priority when comes to city development. While in the process of development towards smarter community, all stakeholders from every segment in the society should be included and each party should assume the responsibility and attention as none of the segment should be neglected. Public private partnership to involve locals and realizing a connected society is an example (Hughes & Spray, 2002). Smart community that should focus on cooperation, emphasize on shared governance and participation of the citizen is essential to include the voice and opinions of the community to local development. The role of government and policy in providing governance to the city is a strong support for smart city initiative (Nam & Pardo, 2011). Besides, in their study of conceptualizing smart city with dimensions pointed the element in institutional factors which includes integrated and transparent governance, strategic and promotional activities, networking and partnership. Other than that, with the launch of e-government in many smart cities as one of the initiatives, it encourages the participation of citizen to more community development projects. In such way, government should be more transparent and accountable to share information to citizen, so they could voice their opinion and be part of decision-making process which affect their daily lives. Smart governance it set to be the cornerstone for smart city, therefore, partnership and collaboration could prove as an effective approach to connect public, private and individual towards structuring an effective bottom-up mechanism where it is the key steps in transforming citizen towards a smarter way.

MAIN FOCUS OF THE CHAPTER

Digital Twin and Smart City

Digital twins first appeared and played a role in industries of product and manufacturing design, and later emerged in industries such as aerospace (Liu et al. 2020), automation (Talkhestani et al. 2018), shipbuilding (Mondoro and Grisso, 2019), healthcare (Jimenez et al., 2020) and energy (Enders and Hoßbach 2019; O'Dwyer et al. 2020). In recent years, with the rapid development of technologies and industries such as the Internet of Things(IoT) (Kaur et al. 2020), big data (Tao et al. 2018), cloud computing (Khatib et al. 2019) and artificial intelligence(AI) (Kharchenko et al. 2020), the construction basis of smart city has gradually evolved from the original static 3D modeling level towards the digital twin level that combines dynamic digital technology and static 3D model (Yan et al. 2019), which forms a new concept that digital twin city assists smart city construction. Obviously, the digital twin city is a wide application of the digital twin concept at the city level. It aims at constructing a complex giant system between the physical world and the virtual space that can map each other and interact with each other in both directions. It can match the physical city to the corresponded "twin city", forming a pattern of coexistence of both and integration of physical cities in the physical dimension and digital cities in the information dimension (Deren, Wenbo and Zhenfeng, 2021).

A digital twin is a virtual representation of the real world, including physical objects, processes, relationships, and behaviors. Geographic Information System (GIS) creates digital twins of the natural and built environments and uniquely integrates many types of digital models. Our world is transforming and rapidly becoming digital. Digital twins are abstracting and modelling everything. To improve business processes, reduce risk, optimize operational efficiencies, and enhance decision-making with automation to predict outcomes (esri Singapore, 2022).

In February 2022, Bentley Systems and GPS Lands Singapore created the world's first digital twin of entire nation in Singapore. Bentley's digital twin can be constantly updated in response to new data. To develop Singapore Digital Twin, aircraft with laser-scanning capabilities captured digital terrain and surface data. Which was combined with aerial images onto a single platform. Its models can map any city in the world without using photogrammetry drones by combining realistic photos from 64 different global sources with AI and machine learning (Tomorrow's World Today, 2022).

ISSUES, CONTROVERSIES AND PROBLEMS

Definition of Smart City vs. Smart Community

The 'Smart Community' will provide comprehensive solutions encompassing energy, water, and medical systems in order to realize a synergetic balance between environmental considerations and comfortable living. ITU-T Focus Group on Smart Sustainable Cities established a concrete definition for smart sustainable cities, which can be used worldwide. They reviewed 116 definitions of smart sustainable city from different sources such as academic, government, and corporate. In their report, smart community and smart city are considered as a same term. Similarly, Albino, Berardi, and Dangelico (2015) did not differentiate the terms of smart city and smart community and they treat these two terms with the same meaning.

They clarified the meaning of the word "smart" in the context of cities through an approach based on an in-depth literature review of relevant studies as well as official documents of international institutions. However, a smart city initiative needs to create a community where all citizens can engage more easily and effectively (Paskaleva, 2009). Ishida and Isbister (2000) debated some similar terms such digital community, smart community, digital city, information city and e-city. Tanabe, van den Besselaar, and Ishida (2003) considered all of those alternative terms are used to refer to a connected community that combines broadband communications infrastructure. The geographical dimension (space) of smart communities is varied; it can be extended from a city district up to a multimillion metropolis (van den Besselaar & Koizumi, 2005). And Harrison et al. (2010) agreed that Instrumented, Interconnected and Intelligent were the three fundamental factors for a smarter city.

While as referring to Nam and Pardo (2011) in their study that attempts to conceptualize smart city with multi-dimensions stated that there are various definition of smart city and majority of them are focusing on hardware development by providing infrastructures and services such as education, health care, transportation and etc. which congruent with other existing literature. Whereas Eger (2005) pointed out that the key factors in succeeding a smart community is all about collaborations and connectivity where internet, knowledge creation and innovation and the core elements in the community. Apparently, these presented the distinctiveness among smart city and smart community through definition and gap exists between these concepts that require further clarifications which agreed by researchers as well (Stratigea, 2012). And it is notable to take note as how it is defined will determine the development path.

Issue: Need to Develop an Index

Index is defined as a sign or measurement of something as stated in Oxford Dictionaries (n.d.). While in the context of research, an index is a point accumulation of multiple data from variable items that said to be related to the subject that intended to measure. Though, before an index is able to be developed, the items need to be selected, the empirical relationship need to be assessed, develop index scoring and lastly validating of the index (Crossman, 2018).

Index has been used widely in measuring broad social science concepts such as City Development Index (CDI), Human Development Index (HDI), ICT Development Index (IDI) and etc. As to the reasons why index is popular among the international organizations such as United Nations is that these measures can aggregate complex or multidimensional questions directly, so policymakers can get clear and representative images and able to be utilizes as a benchmark when comparing to other countries (Foa & Tanner, 2012). For instance, in order to measure a city development, it involves many dimensions such as development level, sustainability, congestion and inclusiveness while these could not be comprehended in a single indicator but a composite index could present these ideas and issues in a single estimate where CDI is developed (United Nations, 2001).

SOLUTIONS AND RECOMMENDATIONS

In current study, to restate one of the objectives is to develop an index to capture the essence of a Smart Community so that the involved stakeholders can be informed about the necessary criteria which enables citizens to be deemed as smart, the concept of Informative Global Community Development Index (IGC) is used as the foundation. There are four key components in IGC development index which are Humanware, Technoware, Infoware and Valueware. The level of human capability and skills in utilizing and adapting ICT in their daily activities are the key measurement points in Humanware component. For instance, measuring the usage and how frequent they used a particular ICT tool. Secondly, Technoware is about hardware and ICTs facilities. In the context of developing a Smart Community, the provision of ICTs infrastructure is essential to encourage the usage of technology, variables such as number of devices used is measured. Inforware component is focusing on all activities of collection, management, processing and info dissemination. For example, the information regarding the subscription of mobile data or home broadband is collected. Lastly, Valueware is value achievement aspect. Through the usage of ICT, the benefits or impacts that happened to local community is measured, variable such as the intention in adopting new technology or the benefits to their daily lives (Malek, 2009).

RESEARCH METHODOLOGY

Quantitative research method has been employed whereby the local residents in the four project sites are the main targeted respondents for this study. This study spread out across 18 months involving few stages. The first phrase is involving design and development of research framework and questionnaire and delivered the survey instrument by the end of this phase. In Phase 2, the survey instrument gone through pre-test and pilot test before finalizing the. Data collection has been carried out in Phase 3 followed by data analysis, interpretation, and index calculation.

Research Instrument

A self-administrated questionnaire was developed to gather information from the respondents which consisted of different sections with both open and close-ended questions. A screening question was included too as the first question to check if the respondent is staying and working in the selected smart community.

Below were the questions asked in different sections:

1. Section A – Demographic Profile
    a. Gender
    b. Age

   c. Race
   d. Education level
   e. Occupation
   f. Income level
   g. Marital status
   h. Number of children
2. Section B – Technological Profile
   a. Do you own a smartphone? And what is the brand?
   b. What other technological devices do you own?
   c. How many years have you been using internet?
   d. Do you subscribe to any mobile data (internet plan)? And what is the brand?
   e. How is the quality of service provided by the service provider? (broadband)
   f. Do you subscribe to a home broadband (internet) plan? And what is the brand?
   g. How is the quality of service provided by the service provider? (mobile data)
3. Section C – MCMC's Initiatives in Lundu/Kota Belud/Kemaman/ Putrajaya
   a. Awareness on their area being assigned as Smart Community project site.
   b. Awareness on the MCMC's initiatives (Infrastructure, Content Training and Digital Services) and their usage level (7-point Likert Scale).
   c. Satisfaction on the MCMC's initiatives (7-point Likert Scale).
4. Section D – Perceived Characteristic of Initiatives
   a. 7 items of performance expectancy, effort efficiency, social influence, facilitating condition, hedonic motivation, price value and habit were measure with 2 questions for each item using 5- points Likert Scale to each group of initiatives.
5. Section E – Intention to use / Continue using MCMC's Initiatives
6. Section F – Impact of MCMC's Initiatives
   a. 6 items of informational, psychological, social, economic, political and cultural impacts were measured with 2 questions for each item 5- points Likert Scale to each group of initiatives.
7. Section G – Opinion on technology
   a. 4 items of optimism, innovativeness, discomfort and insecurity were measured with 2 questions for each item 7- points Likert Scale to each group of initiatives.
8. Section H – About yourself (this section is included as an additional, purely for analysis purposes).

Sampling

Purposive sampling method was used to distribute survey forms. Respondents were selected based on their living area which only within the project site of Smart Community where MCMC's initiatives are implemented.

Data Collection

At survey site, one or more enumerators were appointed to assist in distributing survey forms to the target respondents. They have been briefed and trained by the researchers regarding the content of the survey forms as well as guidance to be given to the respondents.

At Lundu, similarly 150 survey forms were distributed to three areas which were Lundu, Sematan and Sampadi. 90 survey forms were distributed in Lundu including Kg. Pugu, Kg. Pandan, Kg. Sekambal, Bajo, Kg. Sileng, Kg. Sungai Lundu, Kg. Seketi and Kg. Stunggang. 40 survey forms for Sematan covering Kg. Tanah Hitam, Kg. Terusan Jaya, Kg. Tamaga and Kg. Siru, Lastly, 20 forms for Sampadi covering Kg. Sampadi, Kg. Rambungan, Kg. Sungai Cina and Kg. Stoh.

Index Calculation

As mentioned earlier, four key components in IGC development index which are Humanware, Technoware, Infoware and Valueware were used as a reference point in developing the index. In contrast to most of the indices measuring digital divide

Table 1. The underlying foundations of the smart community index

Theme/Component	Explanation and Importance
Human	The extent to which individuals regard ICTs in a favourable or unfavourable way. This component is important because the attitude held by person can determine how ready he/she is in embracing innovative technology-enabled changes in their lives.
Technology	The extent to which individuals have access to hardware/gadgets and also ICT facilities. This component is important because for ownership and access to such facilities and gadgets are tools for technology usage.
Information	The extent to which individuals use and apply the initiatives introduced MCMC. This component is important because it shows how engage individuals are with using technology usage.
Value	The extent to which individuals have used/participated in MCMC's Smart Community initiatives and have witness how the initiatives have helped them and improved their quality of life. This component is important because it shows the impact that technology has on the individuals' lives.

or preparedness's method of developing indices using ratios and rates obtained from secondary data, SCI's approach is a more personal one, where a survey was conducted to obtain pertinent information about the community via the residents themselves in the Smart Community initiatives project sites. From the data gathered at these sites, certain portions were used in the development of the SCI. The portions chosen captures the four themes of Human, Technology, Information and Value.

While the four themes may be adopted, nevertheless, the description of those 4 themes vary slightly to fit the context of study for this research. Thus for this research, the descriptions of the four themes are listed in Table 1.

With the four components being the foundation for this study, four sub-indexes were constructed to capture the essence of a Smart Community namely 'access and ownership', 'use-diffusion', 'impact' and 'attitude'.

Data Analysis

Respondent Profile

There is a rather equal representation of both males (53.2%) and females (46.8%) among the respondents, with the males slightly outnumbering the females (See Table 2). Most of the respondents were either from the <20-year-old group (39.2%) or 20–29-year-old group (28%) given that those two age ranges had the majority number of persons. Third age group from 30–39-year-old presented 19.2% from total sample. There are two age group where get lowest number which are age 40-49 and >50 years old presented 8% and 5.6%. Many respondents were Malay (73.6%) followed by the Bidayuh (10.4%) and other races (6.4%) such as Melanau, Jawa and Bajau.

Majority of the respondents had fundamental education background being majority had SPM or equivalent level of educational qualification (60%) followed by Bachelor's degree (8.8%) and Certificate or Diploma (8%). There was still a band of respondents who had or currently still at secondary school level (12%) as to the majority group age of the sample was below <20 years old. In term of employment, most of the respondents were student (41.3%) or self-employed (24.6%). Therefore, majority of them do not work (44%) and for those working their monthly income ranged between RM 1001 to RM 3000 (25.6%).

Respondents who were single (69.6%) far outnumber the married (27.2%) and widowed (3.2%). For those who are married, having two or three children seemed to be the norm (6.3%). A high percentage of the respondents surveyed did not have any children (72.4%) mainly due to the fact that they are still single/unmarried.

Almost all the respondents surveyed owned a smartphone (98.4%) (See Table 3). The brand of smartphones owned were predominantly Samsung (36.8%). Besides smartphones, majority of the respondents also own laptop (48%) and a tablet (22%).

Table 2a. Respondents' demographic profile

Variable	Category	Frequency	Percentage*
Gender	Male	66	53.2
	Female	58	46.8
Age	<20	49	39.2
	20-29	35	28.0
	30-39	24	19.2
	40-49	10	8.0
	>50	7	5.6
Race	Malay	92	73.6
	Chines	4	3.2
	Indian	1	0.8
	Iban	7	5.6
	Bidayuh	13	10.4
	Others	8	6.4
Eduation	SPM Or Equivalent	75	60.0
	STPM O quivalent	5	4.0
	Certificte/Diploma	10	8.0
	Bachelrs Degree	11	8.8
	Mastero PhD Degree	9	7.2
	Others	15	12.0

The average number of years of internet usage experience among the respondents was 6.8 years (std. dev.=4.21). While 45% of the respondents were being very new to the internet (1 to 5 years) and there were also users that had > 15 years of experience (5%).

98.4% of the respondents subscribed to a mobile data plan with Digi (50%) and Celcom (28.3%) being the major service providers preferred by respondents. 47.9% of the mobile data plan subscribers were happy and rated the mobile data service quality as good. Those who were not satisfied with the quality of mobile data service (rated bad and terrible = 6.7%) mentioned reasons such as 'limited coverage' and 'connections are slow'.

For the home broadband plan, a large percentage of the respondents (68.3%) did not subscribe to it. Among those who subscribed to the home broadband plan, TM (48.7%) and Maxis (35.9%) were the preferred choices. Majority were happy with quality of the home broadband service, with 63.4% respondents rating the service as good. For the few respondents who were not satisfied with the broadband service (rated bad = 7.3%), majority cited reasons such as the 'speed not living up to expectations' and 'slow connection'.

Table 2b. Respondents' demographic profile (cont.)

Variable	Category	Frequency	Percentage*
Occupation	Professional/Management	7	5.6
	Teacher	13	10.3
	Housewife/Husband	4	3.2
	Technical Staff	4	3.2
	Self-Employed	31	24.6
	Admin/Clerical Staff	7	5.6
	Student	52	41.3
	Others	8	6.3
Income	Not Working	55	44.0
	RM1000 or below	18	14.4
	RM1001-RM3000	32	25.6
	RM3001-RM5000	10	8.0
	RM5001-RM7000	7	5.6
	RM7001 and above	3	2.4
Marital Status	Single	87	69.6
	Married	34	27.2
	Widowed	4	3.2

**Note: n=127; *valid percentages are used which were calculated by excluding missing cases*

Descriptives for MCMC's Initiatives in Lundu

As many as 81.4% of respondents surveyed were aware of Lundu being a Smart Community project area. The various initiatives carried out by MCMC under the Smart Community project were grouped under three broad overarching themes namely Infrastructure, Content Training and Digital Services (See Table 4).

Awareness of two initiatives namely telecommunication (telco) data services such as Maxis/DiGi/Celcom/UMobile and Pusat Internet (PI) were the highest at 97.9% and 86.7% respectively; followed by Entrepreneurship training workshops and ICT training at 81.3% and 77.9% respectively. Usage of the initiatives were the highest for telco data services (94.4%), followed by PI (85.2%), ICT training workshops (70.5%), entrepreneurship training workshops (66.3%), MyCrib booking (38.5%) and e-magazine (37.8%).

In terms of the frequency of usage for infrastructure initiatives, majority reported that they used data services to a large extent (mode=7; mean=5.00; std. dev.=1.93) (See Table 5). While there was varied responses among the respondents regarding PI whereby there were some who answered that they use to a lesser degree, to those

Table 3. Respondents' technological profile

Variable	Category	Frequency	Percentage
Own A Smartphone	Yes	125	98.4
	No	2	1.6
Smartphone Brand	Samsung	46	36.8
	Apple	21	16.8
	Huawei	13	10.4
	Xiaomi	3	2.4
	Others	42	33.6
Other Device Ownership	Laptop	61	48.0
(upon 100)	Desktop	22	17.3
	Tablet	28	22.0
	Others	0	0.0
	None	42	33.1
Subscribe mobile data plan	Yes	125	98.4
	No	2	1.6
Mobile data plan brand	Digi	60	50.0
	Celcom	24	28.3
	Maxis	33	19.2
	UMobile	2	1.7
	Others	1	0.8
Mobile data service quality	Excellent	34	28.7
	Good	58	47.9
	Mediocre	21	17.4
	Bad	6	5.0
	Terrible	2	1.7
Subscribe broadband plan	Yes	40	31.7
	No	86	68.3
Broadband plan brand	TM	19	48.7
	Times	1	2.6
	Maxis	14	35.9
	Others	5	12.8
Broadband service quality	Excellent	6	14.6
	Good	26	63.4
	Mediocre	6	14.6
	Bad	3	7.3
Years of using Internet	Mean = 6.85		
	Std. deviation = 4.21		

Table 4. Respondents' awareness and usage of the smart community initiatives

Initiatives	Aware (%)	Not Aware (%)	Use (%)	Don't Use (%)
Infrastructure				
1. Pusat Internet	**104(86.7)**	16(13.3)	**92(85.2)**	16(14.8)
2. Data Services	**92(97.9)**	2(2.1)	**102(94.4)**	6(5.6)
Content Training				
1. ICT Training	**81(77.9)**	23(22.1)	**79(70.5)**	33(29.5)
2. Entrepreneurship	**78(81.3)**	18(18.8)	**69(66.3)**	35(33.7)
Digital Services				
1. MyCrib Booking	**42(41.2)**	60(58.8)	**40(38.5)**	64(61.5)
2. E-Magazine	**40(41.7)**	56(58.3)	**37(37.8)**	61(62.2)

**Note: n=127*

who use moderately and a large degree (mean=3.82, std. dev = 2.13). Followed by ICT training, although majority did not used this initiative (mode=1.00), there were some who used it moderately (mean=3.41, std.dev.= 2.05) and similar to entrepreneurship training (mode=1; mean=3.15; std. dev.=2.05). On the other hand, the MyCrib booking (mode=1.00; mean=2.17, std. dev. = 1.72) and e-magazine (mode=1.00, mean=2.12, std. dev. = 1.71) were either not used much or not used at all by the respondents.

For the respondents who use the initiatives, generally they felt satisfied with the initiatives, as seen from the mean scores of all the initiatives which ranged from

Table 5. Descriptives for usage frequency of the smart community initiatives

Initiatives	Mode	Mean	Median	Std. Deviation
Infrastructure				
1. Pusat Internet	1.00	3.89	4.00	2.13
2. Data Service	7.00	5.00	5.00	1.93
Content Training				
1. ICT Training	1.00	3.41	4.00	2.05
2. Entrepreneurship	1.00	3.15	3.00	2.05
Digital Services				
1. MyCrib Booking	1.00	2.17	1.00	1.72
2. E-Magazine	1.00	2.12	1.00	1.71

*Note: N=127; Items were measured on a scale 1=not at all to 7=a large extent

Table 6. Descriptives for satisfaction with usage of the smart community initiatives

Initiatives	Mode	Mean	Median	Std. Deviation
Infrastructure				
1. Pusat Internet	7.00	5.65	6.00	1.41
2. Data Services	7.00	5.62	6.00	1.33
Content Training				
1. ICT Training	7.00	5.27	5.00	1.65
2. Entrepreneurship	7.00	5.24	5.00	1.57
Digital Services				
1. MyCrib Booking	4.00	4.44	4.00	1.59
2. E-Magazine	4.00	4.50	4.00	1.45

*Note: N=127; Items measured on a scale 1=very dissatisfied to 7=very satisfied. Scores were calculated for those who use the initiatives.

5.24 to 5.65 except for MyCrib booking and e-magazine (See Table 6). Upon closer inspection, majority of the respondents were very satisfied with PI, data services, ICT training and entrepreneurship workshop (mode for the stated initiatives = 7). However, there is probably some room for improvement for initiatives like MyCrib booking (mode=4) and e-magazine (mode=4) as most of the respondents were only moderately satisfied with the initiatives.

Scoring of the Smart Community Index

Based on the data gathered from the survey, the indicators of the sub-indexes were compiled and aggregated into a score ranging from 0 to 100. The higher the score for the sub-index, the higher the access and ownership; 2) usage of smart community initiatives; 3) impact on lives experienced and 4) attitude or readiness towards embracing ICTs. The overall aggregated scores can be split into three quartiles (based on cut-off points) and are result in three major ranges – Low (0 – 33.33), Average (33.34 – 66.67) and High (66.68 – 100) level of savviness. Table 7 shows the results for the SCI tabulated for Lundu.

Infrastructure Improvement

Based on smart community initiatives, broadband services by many telecommunication providers have involved and collaborate to make sure improvement can be done. Especially cellular services like capacity and coverage can be access at Lundu area

Table 7. Smart community index (SCI)

Index Item	Score
Access & Ownership	59.22
Use-Diffusion	46.47
Impact	78.78
Attitude	69.71
SMART COMMUNITY INDEX	63.65
Sample Size (N)	127
Level of Savviness	
Low	0 (0%)
Average	84 (66.1%)
High	43 (33.9%)

which is located at rural area. The area with 4G LTE and 3G coverage is mostly 100% coverage for all areas. However, at more remote areas away the coverage should be improved by telecommunications company namely Telekom Malaysia, Maxis, Celcom, Digi and other providers.

To be specific basic facilities such as telephony is not made available and some of this area need urgently electricity and water supplies improvements beside the telecommunications infrastructure. Thus, improvements need to be planned carried out for the infrastructure to ensure better coverage and capacity for internet access for the more underdeveloped areas in Lundu.

Pusat Internet centres serve as a very good enabler for development of a smart community especially in more rural areas because in addition to providing means of technology and internet access to the community, the centres function well in developing technological literacy among children and housewives. In this sense, continuation, and expansions of pusat internet centres should be relinquished. In the more urban smart community project locations, Pusat Internet centres need to be positioned at strategic locations to ensure public participation is appreciated.

Monitoring the Progress and Achievement of Smart Community Project

This can be done via the use of the index that measures 4 important components (access and ownership, use-diffusion, impact, attitude) that define a smart community. The Smart Community Index uses a modular approach whereby the further criteria can be added apart from the 4 components should the need arise. Currently the index tracks the use of smart community initiatives which can serve as important proxies for technology-related activities; and the list of use-diffusion activities may vary a little depending on the initiatives set up according to the project location. Thus, once the project has gained some momentum, the current list of activities can be revised to include use-diffusion activities that reflect a more holistic use of ICTs.

FURTHER RESEARCH

Overall, the community in Lundu scored the highest index of 63.65% where 66.1% of the residents fell under the average level of smartness group and 33.9% fell under high level.

Involvement and participation in Smart Communities initiatives very much concerns that use-diffusion of MCMC smart community initiatives. The question of whether the community intends to continue using or participating in the MCMC smart community initiatives depends on the satisfaction they have gained from using the said initiatives. The analysis has proven that satisfaction is an important determinant for the community's continuous usage of MCMC smart community initiatives. This is consistent across communities in Lundu and other respective areas. The more the communities are involved in using or participating in the MCMC smart community initiatives, the higher the satisfaction garnered from the usage. Based on previous IS theories, factors like performance expectancy, effort expectancy, social influence, facilitating conditions, hedonic motivation, price value and habit serve as important variables that can affect community's usage of smart community initiatives. Given that each location may differ due to some contextual differences, different prerequisites for involvement and participation in smart community initiatives could emerge. Thus, prerequisites for community involvement and participation in Smart Community initiatives (infrastructure, content training and digital services) are best discussed according to location. In addition, participation and involvement in the initiatives can at times be augmented by a user's positive personality traits or hampered by the more negative traits.

For the Lundu's community, two factors are necessary to ensure usage of infrastructure initiatives, namely effort expectancy and optimism. Greater usage of infrastructure initiatives (i.e. data services, pusat internet) depend on how easy it is for the all layers of community to understand and use. The more users perceive that the infrastructure initiatives are easy to use, the higher the tendency for them to continue using the infrastructure especially for those with high innovativeness traits because these are people who love to tinker with all thing's technology. Similarly, the participation in content training rests on whether the training delivered is easy enough for the community to understand and later on apply in their lives for the community. However, this can only occur provided the community do not develop distrust (low insecurity) towards technology.

To uncover the impact of MCMC's initiatives on the community itself, as for Lundu, the overall score demonstrated that only impact from psychological and cultural were higher whereas the other impacts were rated lower. Lundu's community found that these initiatives improved their ability in solving daily problems by utilizing ICT's initiatives such as the content training provided. From the economic aspect, the result was found same as Kota Belud where the economic impact was rated as the lowest among these items and required the attention of stakeholders to put more effort in promoting these benefits and advantages of using ICT in their business.

The idea of a smart community is one where individuals residing and working in a particular locality have access to technology; are able to use and apply ICTs in their everyday lives and witness a positive change in their lives as a result of the usage. Also, individuals would have favorable attitudes towards embracing ICTs and they would feel empowered as a result of that usage. In this manner, four important aspects serve as essential criteria for the development of a smart community and can determine the level of technology savviness among the community members.

Attitude to Embrace Technology

The extent to which individuals regard ICTs in a favourable or unfavourable way. This component is important because the attitude held by person can determine how ready he/she is in embracing innovative technology-enabled changes in his/her life. Attitude is characterized by a person's enabling traits such as optimism and innovativeness as well as inhibiting traits such as the discomfort and insecurity felt towards technology.

Access to Infrastructure and Ownership of Technological Devices

The extent to which individuals have access to hardware/gadgets and also ICT facilities as measured by indicators such as the number of technology devices owned, subscription to mobile data and home broadband. This component is important because for ownership and access to such facilities and gadgets are basic requirements (tools) for technology usage.

ICT Use-diffusion

The extent to which individuals use and apply ICTs in their daily lives such as the Smart Community initiatives introduced MCMC. This component is important because it shows how engage individuals are with technology usage, for instance how they use technology to help them in their daily routine, work or leisure. The Smart Community initiatives are used as a proxy for the use-diffusion of ICTs (i.e., insfrastructure, content training and digital services).

Impact Derived From ICT Usage

The extent to which individuals have used ICTs in their daily routine, work or leisure (for instance, applied and participated in MCMC's Smart Community initiatives) and have witnessed how the initiatives have helped them and improved their quality of life. This component is important because it shows the impact that technology has on the individuals' lives. The change or improvement felt in their lives should encompass various types of impact namely informational, psychological, social, economic, political and cultural.

Based on the index which was calculated to determine the level of technology (tech) savviness among the residents of Lundu, it was found that the community in the generally exhibited an average level of tech savviness. Lundu seen to be having high score in its use-diffusion by virtue that its community are more engaged in the various ICT initiatives organized by MCMC to promote the usage of technology. Apart from the use-diffusion of ICTs, the community harbored favorable attitude towards technology and experienced a positive change in various aspects of their lives. In short, all four criteria are necessary criteria for the determination of a smart community.

CONCLUSION

Urbanization and the growth of human population are leading to increased complexities in the interactions of citizens with public spaces, creating cities that must be more responsive to community dynamics. Despite the critical need for community-based city management, current decision-making approaches are often uninformed by the collective behavior of communities across time and in different locations. It is pertinent to remember that the success of smart cities is still reliant on people as the communities inhabiting a geographical area will be users of the smart city technologies introduced to elevate modern living. Thus, communities need to be technologically savvy and smart in adopting, utilizing and integrating various technologies into their daily lives not just in metropolitan areas but also smaller, less urban districts such as Lundu.

This study has provided valuable insights on the state of Lundu community's access and ownership, use-diffusion, and attitude towards technologies as well as the impact derived from the technologies. Based on the SCI index, Lundu was found to be an average-level smart community. This is a community which has a fairly optimistic perspective on the use of technology, delighted at the change or improvement that technology has brought into their lives but still needs to catch up to the urban localities in terms of their access to, ownership of and usage of digital technologies. Hence, some of the key recommendations include strengthening the infrastructure availability at the area, extending IT resources to the rural ensuring homogeneous adoption and placement of the resources at the strategic location in order to benefit the public at large. On that note, MCMC's initiatives to educate and equip the Lundu community need to be continued and upgraded progressively.

With a smart community, prospects of developing the locality into a smart city will be brighter. The development of a smart city can then pave the way for digital twin technologies. As smart city and community initiatives continue to be deployed more widely across the nation and around the world, municipalities will be prone to seek out digital twins given digital twins technology enable the planning, management and optimization of cities across a range of applications, such as mobility and sustainability. In this sense, a smart community index provides great value because it shows where the population of a locality stands in (1) their level of human capability and skills in utilizing and adapting ICT in their daily activities; (2) the extent to which they have access to hardware/gadgets and also ICT facilities; (3) the extent to which they use and apply the ICT-related initiatives and; (4) the extent to which individuals have used/participated in ICT (Smart Community) initiatives and have witness how the initiatives have helped them and improved their quality of life. Ultimately, such insights gained from the smart community index can help to steer the community towards the right direction of becoming a smart city that employs digital twins.

ACKNOWLEDGMENT

The authors acknowledge all reviewers for their comments on this manuscript and are grateful to the committee members of the IGI Global for developing a book on The Impact of the Digital Twins in Smart Cities Development. This article is supported by the Research University Grant USM (No: 1001/PMGT/8016126).

REFERENCES

Albino, V., Berardi, U., & Dangelico, R. M. (2015). Smart cities: Definitions, dimensions, performance, and initiatives. *Journal of Urban Technology*, *22*(1), 3–21. doi:10.1080/10630732.2014.942092

Anrong, D., Li, G., Li, J., & Kong, X. (2016). Research on Smart Community Planning of Yishanwan, China towards New Urbanization. *International Review for Spatial Planning and Sustainable Development*, *4*(1), 78–90. doi:10.14246/irspsd.4.1_78

Chourabi, H., Nam, T., Walker, S., Gil-Garcia, J. R., Mellouli, S., Nahon, K., Pardo, T. A. & Scholl, H. J. (2012). *Understanding smart cities: An integrative framework.* Paper presented at the System Science (HICSS), 2012 45th Hawaii International Conference on System Sciences. doi: 10.1109/HICSS.2012.615

Crossman, A. (2018, July 3). *How To Construct an Index for Research.* Retrieved from https://www.thoughtco.com/index-for-research-3026543

Deren, L., Wenbo, Y., & Zhenfeng, S. (2021). Smart City based on digital twins. *Computers & Urban Society*, *1*(4), 1–11.

Eger, J. M. (2005). Smart communities, universities, and globalization: Educating the workforce for tomorrow's economy. *Metropolitan Universities*, *16*(4), 28–38.

Enders, M. R., & Hoßbach, N. (2019). Dimensions of digital twin applications-a literature review. *25th Americas Conference on Information Systems, AMCIS 2019.*

Foa, R., & Tanner, J. (2012). *Methodology of the indices of social development* (No. 2012-04). Academic Press.

Gao, W., Fan, L., Ushifusa, Y., Gu, Q., & Ren, J. (2016). Possibility and Challenge of Smart Community in Japan. *Procedia: Social and Behavioral Sciences*, *216*, 109–118. doi:10.1016/j.sbspro.2015.12.015

Ghazali, M., Okamura, T., Abdullah, T., Sunar, M. S., Mohamed, F., & Ismail, N. (2016). In the Quest of Defining Smart Digital City in Medini Iskandar Malaysia, Iskandar Puteri, Malaysia. *Proceedings of the SEACHI 2016 on Smart Cities for Better Living with HCI and UX.*

Gil, O., & Navarro, C. (2013). *Innovations of Governance in Cities and Urban regions: Smart Cities in China, Iskandar (Malaysia), Japan, New York and Tarragona (Spain).* Paper presented at the EURA Conference: Cities as Sheedbeds for Innovation.

Gurstein, M. (2014). Smart cities vs. smart communities: Empowering citizens not market economics. *The Journal of Community Informatics, 10*(3). Advance online publication. doi:10.15353/joci.v10i3.3438

Harrison, C., Eckman, B., Hamilton, R., Hartswick, P., Kalagnanam, J., Paraszczak, J., & Williams, P. (2010). Foundations for smarter cities. *IBM Journal of Research and Development, 54*(4), 1–16. doi:10.1147/JRD.2010.2048257

Hollands, R. G. (2008). Will the real smart city please stand up? Intelligent, progressive or entrepreneurial? *City, 12*(3), 303–320. doi:10.1080/13604810802479126

Hughes, C., & Spray, R. (2002). Smart communities and smart growth-Maximising benefits for the corporation. *Journal of Corporate Real Estate, 4*(3), 207–214. doi:10.1108/14630010210811831

Ishida, T., & Isbister, K. (2000). *Digital cities: technologies, experiences, and future perspectives*. Springer Science & Business Media. doi:10.1007/3-540-46422-0

Jimenez, J. I., Jahankhani, H., & Kendzierskyj, S. (2020). *Health care in the cyberspace: medical cyber-physical system and digital twin challenges*. Academic Press.

Kaur, M. J., Mishra, V. P., & Maheshwari, P. (2020). The convergence of digital twin, IoT, and machine learning: transforming data into action. Digital Twin Technologies and Smart Cities.

Kharchenko, V., Illiashenko, O., & Morozova, O. (2020). Combination of digital twin and artificial intelligence in manufacturing using industrial IoT. In *2020 IEEE 11th international conference on dependable systems, services and technologies (DESSERT)*. IEEE. 10.1109/DESSERT50317.2020.9125038

Khatib, M. M. E., Al-Nakeeb, A., & Ahmed, G. (2019). Integration of cloud computing with artificial intelligence and its impact on telecom sector—A case study. *iBusiness, 11*(1), 1–10.

Liu, S., Bao, J., & Lu, Y. (2020). Digital twin modeling method based on biomimicry for machining aerospace components. *Journal of Manufacturing Systems*.

Malek, J. A. (2009). Informative global community development index of informative smart city. *Proceedings of the 8th WSEAS International Conference on*.

Mondoro, A., & Grisso, B. (2019). On the integration of SHM and digital twin for the fatigue assessment of naval surface ships. Structural Health Monitoring 2019. doi:10.12783hm2019/32203

Nam, T., & Pardo, T. A. (2011). Conceptualizing smart city with dimensions of technology, people, and institutions. *Proceedings of the 12th annual international digital government research conference: digital government innovation in challenging times.* 10.1145/2037556.2037602

O'Dwyer, E., Pan, I., Charlesworth, R., Butler, S., & Shah, N. (2020). Integration of an energy management tool and digital twin for coordination and control of multi-vector smart energy systems. *Sustainable Cities and Society*, *62*, 102412. doi:10.1016/j.scs.2020.102412

Paskaleva, K. A. (2009). Enabling the smart city: The progress of city e-governance in Europe. *International Journal of Innovative Research and Development*, *1*(4), 405–422. doi:10.1504/IJIRD.2009.022730

Stratigea, A. (2012). The concept of 'smart cities'. Towards community development? Netcom. *Réseaux, communication et territoires, 26*(3/4), 375-388.

Talkhestani, B. A., Jazdi, N., & Schlögl, W. (2018). Consistency check to synchronize the digital twin of manufacturing automation based on anchor points. *Procedia CIRP*, *72*, 159–164. doi:10.1016/j.procir.2018.03.166

Tanabe, M., van den Besselaar, P., & Ishida, T. (2003). Digital Cities II. Computational and Sociological Approaches. In *Second Kyoto Workshop on Digital Cities, Kyoto, Japan, October 18-20, 2001. Revised Papers* (*Vol. 2362*). Springer.

Tao, F., Cheng, J., Qi, Q., Zhang, M., Zhang, H., & Sui, F. (2018). Digital twin-driven product design, manufacturing and service with big data. *International Journal of Advanced Manufacturing Technology*, *94*(9-12), 3563–3576. doi:10.100700170-017-0233-1

United Nations. (2001). *Epilogue The City Development Index (CDI).* Retrieved from https://www.un.org/ga/Istanbul+5/116.pdf

van den Besselaar, P., & Koizumi, S. (2005). Digital Cities III. Information Technologies for Social Capital: Cross-cultural Perspectives. In *Third International Digital Cities Workshop, Amsterdam, The Netherlands, September 18-19, 2003, Revised Selected Papers* (*Vol. 3081*). Springer.

Wongbumru, T., & Dewancker, B. (2014). *Smart Communities for Future Development: Lessons from Japan.* Academic Press.

Yan, J., Zlatanova, S., & Aleksandrov, M. (2019). Integration of 3D objects and terrain for 3D modelling supporting the digital twin. *14th 3D GeoInfo Conference.*

Chapter 10
Impact of Digital Twin on Smart Communities:
Insights from the Putrajaya, Malaysia

Normalini Md Kassim
Universiti Sains Malaysia, Malaysia

Nor Hazlina Hashim
https://orcid.org/0000-0003-4265-0557
Universiti Malaya, Malaysia

Jasmine A. L. Yeap
Universiti Sains Malaysia, Malaysia

Saravanan Nathan Lurudusamy
Telekom Malaysia Berhad, Malaysia

T. Ramayah
Universiti Sains Malaysia, Malaysia

ABSTRACT

Putrajaya is a Malaysian federal administrative center and is called also City in the Garden. Due to this factor, this chapter intended to access the insights of the Putrajaya smart community initiatives since a quantitative research approach has been employed where a self-administrated questionnaire was developed to gather information from the respondents which consisted of different sections with both open and close-ended questions. A total of 163 forms were distributed in different locations within the smart community sites. A descriptive analysis on the usage was carried out to access the initiatives, and recommendations are suggested at the end of the chapter.

DOI: 10.4018/978-1-6684-3833-6.ch010

INTRODUCTION

Smart community is a concept focusing on 'grooming' up or equipped the community with technology and innovation knowledge on information and communication technology (ICT), providing with 'smart' infrastructure, solving social problems, and improving community's quality of lives. The concept was first discussed in the year of 1993 in Silicon Valley, California, United Sates of America, where business leaders, community members, government officials and educators decided to help the region to overcome economic recession (Lindskog, 2014). At present, the smart community concept and definition evolves to meet the current need. This concept is famously being used worldwide, trending, and popular.

The concept and goals between 'smart' community' and 'smart city' is intertwined in many cases. Based on Yigitcanlar, Kamruzzaman, Buys, Ioppolo & Costa & Yun (2018), the smart city concept focuses on three main drivers that are smart cities-community, technology, policy in which, connects to productivity, sustainability, accessibility, wellbeing, liveability, and governance. This indicates that smart community focuses on micro-environment in which emphasis on the citizens with knowledge, training, and skill sets especially those who are living in the rural areas. On the other hand, smart city focal point is on the macro-environment, whereby the facilities, infrastructure of township and planning, policy makers and governance, multi-stakeholders from healthcare, educators, telecommunications, and the Internet broadband marketers are also vital. In this matter, both are inextricably related and equally important.

According to Australian Smart Community Association (ASCA, 2017), government initiatives depending on the uniqueness and dynamic of its own community. This depends on the history, composition, linked to and address community needs towards better living, health, and happiness. The goal for society is to synchronize the mission towards achieving the Industrial Revolution of 4.0 for better quality lifestyle. Most importantly, all the initiatives are meant for the community and the community must involve in determining needs, decisions making and assessing the benefits of smart projects (ASCA 2017: 6). Based on Kassim, Yeap, Nathan, Hashim, & Ramayah, (2019) the concept of smart city and smart community is centered upon the goals of governments, developers, social-culture, economic factors, technologies advancement, and values.

In addition, there was a comprehensive review was conducted by Chai, Cvetkovic, and Page (2020) on the initiatives and benefits of building a smart city. The application of technology within communities by providing information and communication technology (ICT) in all areas of the city will help to improve the resource utilization, urban management, and services. These initiatives will eventually bring together the various service functions of the city into diverse, complex, interconnected, and manageable system and the same time will improve the communities' living quality. Therefore, three main stakeholders should incorporate in achieving this mission are the community, government, and private sectors.

Furthermore, based on Nor Sallehi (2017) smart city is the application and territorialization of practices that brings ICTs and people together to enhance the innovation and knowledge. In Malaysia, there are four selected areas for the development of smart city that are Greater Kuala Lumpur (KL), Iskandar Malaysia (Johor), Smart Selangor, and Putrajaya Smart City. A national-level framework called Smart City Framework Malaysia (MSCF) designed to guide local authorities in planning and developing local smart cities. These framework covers government, economy, environment, people, living, mobility, and digital infrastructure as main mechanisms of Malaysian smart city (https://www.malaysia.gov.my/portal/content/30947).

Putrajaya is the federal government, administrative and judicial center of the Malaysian capital since 1999. Putrajaya is also called City in the Garden and contain around 89,000 residents in which majority (96%) are Bumiputera. For this reason, this study aimed to uncover the insights of smart community initiatives on the community of Putrajaya. The success of a smart community lies in how the technologies are adopted and used by the residents as well as how the lives of the residents are impacted. With the progression of Putrajaya becoming a Smart City, this article aims to quantitatively analyze the impact of Malaysian Communications and Multimedia Commission's (MCMC) smart city initiatives on communities' awareness, frequency of usage and satisfactions among Putrajaya community itself.

BACKGROUND

Literature Reviews

As stated in the introduction section, the smart community concept and definition evolves. This concept is popular and interlink with smart city in which has similar niche goal. According to Yigitcanlar et al. (2018) and Dameri (2013), there are no common definition to be agreed upon as there are too many definitions of smart city in the growing literature. In a nutshell, to have a smart community, the place needs to be in a smart city as it can be defined as a city which aims at a better quality of life and sustainability utilizing information and communication technology. The success of a smart community depends on how the technologies are adopted and used by the communities and how the lives of the communities are impacted because of the innovation and technology adoption. A smart city with a digitalized infrastructure, intelligent service network and data-driven decisions must be occupied with the participants of an ongoing reflection flow of citizens and communities (Feher, 2018).

Research by Sengboon, Jalaluddin, Mohd Yusof and Zurinah (2019) was on behaviors and job positions of citizens who participate in the development of smart cities indicated that citizens can be active and independent volunteer in public life, can be led by the local champions or co-producers of public values, and can be educated-intention human sensors who drive changes, instead of being passive users of data or beneficiaries of services. This is in line with Gurstein (2014) who stated that social inclusion, enabling citizen and supporting communities should be a part of city development agenda rather than focusing on the smart city itself. Aside from the community itself, research from an Indian context conducted by Rana, Luthra, Mangla, Islam, Roderick & Dwivedi (2018) indicated that one of the barriers that affect the development of smart cities is governance. The study also recommended policy makers to promote e-governance services to bring accountability and transparency in the development of smart cities. E-governance services can help to strengthen democracy, citizen participation and public welfare because all section of society can access the government information.

In 2014, the Prime Minister of Malaysia had endorsed Smart City Iskandar Malaysia as a pilot for smart city projects which focuses on three areas, economy, environment and social and promotes six dimensions: smart economy, smart governance, smart environment, smart mobility, smart people, and smart living (Ghazali, Okamura, Abdullah, Sunar, Mohamed & Ismail, 2016). The main objective of this project was developing smart living by engaging local citizen to take part for sustainable development. This project has focus on changing the quality of life and infrastructure provision. The project was a collaborating project between the university and regional development agency. There was an effort from the aspect of education where are few internationally well-known universities such as University of Newcastle plan to open a campus in Medini (Kassim, Nathan, Yeap, Ramayah, & Hashim, 2018).

Besides Iskandar, Johor, Kemaman, a district in Terengganu is one of the sites that have been selected for the deployment of smart community in Malaysia. Since then, Kemaman has successfully set up internet access centre, built an operations centre for flood response, and run a hackathon for local to develop mobile applications (Chin, 2016). Another smart community initiatives developed by MCMC involved places like Kota Belud (Sabah), Lundu (Sarawak), Putrajaya (Selangor), and Langkawi (Kedah). Overall, these initiatives mainly focusing on the local community quality of life and socio-economic status (Kassim et.al, 2018).

Putrajaya is designed by Malaysians and planned as a Malaysian city to represent the administrative capital for Malaysia. It is located next to Multimedia Super Corridor (MSC), Cyberjaya and reflecting a city-in-a-garden and "intelligent city". It is of proximity to the international airport, technology industry clusters and academic institutions, well-positioned to combine the industries, academic and government sectors. It is a fertile ground for new knowledge and innovation to build clusters of knowledge driven by high jobs opportunities that will be the foundation for new era economy (Chowdhury, 2020).

Based on Putrajaya Blueprint (2019), smart city Putrajaya objectives addressed four (4) urban dimensions that derived into seven (7) smart city domains highlighted as the catalysts in leading to the Putrajaya Smart City Blueprint. These 2025 long-term plans are key fundamental attributes of a smart city. The four urban dimensions highlighted; Society, Environment, Governance and Economy. It derived into seven domains including Smart Transportation and Mobility, Smart Home and Environment, Smart Government Services, Smart Infrastructure and Utilities, Smart Safety and Security, Smart Economy, and Smart Community.

Smart Transportation and Mobility are the effort to reduce carbon emission and air pollution using efficient modes of public transportation and to alleviate traffic congestion issues through effective planning and enforcement. Smart Home and Environment are to ensure that the environmental conditions of Putrajaya will be constantly monitored to allow for a high-quality environment. The third domain is Smart Governance Services. Some of the efforts are providing E-government services (e.g. online payment, online licence application, online registration system (forum/ workshop/training), and City Management Monitoring (e.g., Putrajaya Command Centre (PCC), an integrated centre that monitors the city 24-hours on traffic, safety, disaster, crowd control, and provides express monitoring and advisory systems such as vehicle recovery services to detect accidents or any incidences). Based on Lim & Yigitcanlar (2022), participatory governance by actively using e-participatory platforms of e-portals, website, mobile apps, and social media are the vital component in smart cities planning.

Smart Infrastructure and Utilities in the fourth domain is pertaining to cost-savings for the city in the long term through better management of streetlights and waste disposal. The fifth domain is Smart Safety and Security which provides emergency response system in times of emergencies or natural disasters (e.g., panic button, smart CCTV, and dam integrity assessment). Smart Economy in sixth domain provides an attractive city platform to encourage new ways of doing businesses (e.g., cashless payment through e-wallet and e-kiosk, business promotion and advertisement using digital billboard and smart application, tourism promotion through smart application for city attraction, and setting up incubator/lab as to encourage business community to adopt technology).

Lastly, Smart Community in domain seven is to spread information and awareness through digital technologies (e.g., smart education by providing digital library, online school admission and digital classroom; innovative community through innovative centre (POINT) to create tech community in Putrajaya; and healthcare monitoring program including tele-health services and information for dengue hotspot area [iDengue]. Hence, this research intended to uncover the initiatives and impact of smart community in the federal administrative center of Malaysia that is Putrajaya.

Underpinning Theories

Unified Theory of Acceptance and Use of Technology (UTAUT) was a unified model developed by Venkatesh, Morris, Davis, and Davis (2003). There were various types of models in information technology acceptance research and all proposing different sets of determinants hence in the study, Venkatesh et al. (2003) aims to review and integrate previous researchers and models related in user acceptance in new technologies and formulated a more comprehensive and unified model and empirically validated the model. The result of the test portrayed and supported the model with strong empirical evidence proving that there were three direct determinants of intention use namely performance expectancy, effort expectancy and social influence while there were two direct determinants of usage behaviour namely intention and facilitating conditions.

Meanwhile, the performance expectancy is defined as to which an individual considers that by using the system or technology will help to improve his or her job performance while effort expectancy is referring to the level of ease when using the system. The third construct, social influence is defined as the degree that an individual think others believe he or she should use the new system or technology and the other construct of facilitating conditions referring to the level of an individual considers that an organizational and technical facilities exist to support the system.

This model has widely used and accepted by many researchers and has been adapted in their information technology acceptance studies. In China, Min, Ji, and Qu (2008) has utilized the model to investigate mobile commerce (m-commerce) user acceptance which revised to include characteristic of m-commerce, user satisfaction and Chinese context while researchers recommended to include trust, privacy protection and cost to the model. Besides, Taiwo and Downe (2013) proposed to conduct an empirical study through a meta-analytic review. However, the results presented a different outcome compared to Venkatesh et al. (2003)'s studies demonstrated the relationship between effort expectation, social influence and behavioral intention are weak which Venkatesh et al. (2003)'s results suggested to be moderate and varies with gender and age. Despite the controversy, this model was still widely used which recognized its ability to account 70 percent of the variance in usage intention which outshine the other eight models in this area of studies (Venkatesh et al., 2003). Williams, Rana, and Dwivedi (2015) presented the results of a systematic review of the development of UTAUT since its inception in 2003 through reviewing 174 papers that adapting UTAUT model and suggested that UTAUT is mainly used for technology adoption and acceptance research in the areas of e-government, e-banking, e-learning, and e-commerce. And this found to be consisted with Al-Eryani and Rashed (2016)'s study in Kuwait that concluded UTAUT model was being the best model in measuring e-readiness for e-government.

Fridayani and Nurmandi (2019) in their study has analyzed the citizens' behaviour of using technology on Sleman smart citizen 2021 through the "Lapor Sleman" mobile application. Sleman Regency was one of the pilot projects of Ministry of Communication and Informatics of Republic Indonesia in terms of smart city application. By using UTAUT model, the results revealed that performance expectancy, social influence, and facilitating condition are three independent variables which affect the behavioral intention and users' behavior. Moreover, the behavioural intention has significant effect on users' behaviour. Results also indicated that the citizens' behaviour of using technology has affected the making of a smart citizen. In conclusion, using technology in terms of "Lapor Sleman" mobile application can make the citizens as active participants in public life, in terms of social cooperation, freedom of expression and flexibility. However, there are still some limitations in the use of "Lapor Sleman" mobile application including the limited access to only android smartphones, lack of technical know-how, and people's reluctance to use the "Lapor Sleman" mobile application.

The past study also clearly investigated the intention of university students to adopt and use e-government services in China using UTAUT as the theoretical framework for the study. Results have demonstrated that facilitating conditions, perceived service quality, trust in the government, and social influence were significant predictors of the intention of university students to adopt and use e-government services. However, performance expectancy, effort expectancy, and trust on the Internet were not significant determinants of the intention to use e-government services. Social influence was associated with trust on the Internet but not in predicting the trust in government (Mensah, 2019).

UTAUT 2, on the other hand, was an extended model of the original UTAUT model which focus on continuance usage intention (Venkatesh, Thong, Chan, Hu, and Brown, 2011). Authors incorporating UTAUT with two-stage expectation confirmation theory of information systems continuance in order to have more in depth understanding on acceptance theory not only on pre usage intention but to assess user's continuance behaviour. Authors asserted that attention should put on the long-term viability of the new information system which rely more on continuous usage intention rather than initial adoption decisions. Therefore, the results proven that the changes in the pre usage beliefs and attitudes could be understood through the constructs of disconfirmation and satisfaction which eventually influence information system continuance intention. This was agreed by many researchers which extending their research model by cooperating these constructs such as satisfaction. Maillet,

Mathieu, and Sicotte (2015)'s study on nurses to use electronic patient records by extending UTAUT model with actual use and satisfaction and results shown that three out of four UTAUT constructs proposed a strong relationship on satisfaction. While there was study proposing that satisfaction was being significantly mediated by behavioural intention in the research investigating the mediating effect of behavioural intention between UTAUT predictors and service satisfaction (Dhaha & Ali, 2014). However, Giannakos, Pateli, and Chorianopoulos (2013)'s study presented that behavioural intention and satisfaction both to have direct relationship rather than mediating with four factors influence users' acceptance on using Facebook where framework was developed by integrating UTAUT model and users and gratification theory. As evidence demonstrated in various studies, it was sufficient to prove that by extending the construct of satisfaction as suggested by Venkatesh et al. (2011) in UTAUT model, researchers can have better and more comprehensive understanding on users' acceptance and their continuous usage intention.

A study by Meizhura, Alamanda and Adi Prabowo (2017), have examined the affecting factors of interest and behavior in using application system called LAPOR. The factors were identified by the model of UTAUT 2. LAPOR is one of the smart government programmes implemented as an online public aspirations and complaints service in Bandung, Indonesia. Results indicated that the most influential factor of interest in using the LAPOR system is Price Value. Other influential factors of interest and behavior are Hedonic Motivation, Social Influence, Habit, and Facilitating Condition. There, this research used as the underpinning theories to uncover the initiatives and impact of smart community in Putrajaya.

MAIN FOCUS OF CHAPTER

Research Methodology

This research study applied a quantitative survey approach to achieve the objectives. The researcher opts for purposive sampling to ask topic in a specialised population. This study involved the residents of Putrajaya as the main targeted respondents. Respondents were selected based on their living area which only within the project site of Smart Community in Putrajaya, in which was initiated by MCMC. The residents in project sites of Putrajaya are the main targeted respondents for this study.

A survey method was done by using one or more enumerators that have been appointed to assist in distributing survey forms to the respondents. They have been briefed and trained by the researchers regarding the content and purpose of the survey forms as well as guidance to be given to the respondents. A total of 180 survey forms have been distributed with consent letter. After screening, only 163 questionnaires have been completed and used for data analysis.

A screening questions were distributed to check if the respondent is staying and working in the selected smart community. The questions were asked in 3 sections including Demographics Profile, Technological Profile and Evaluation which consists of questions regarding perceived characteristic of initiatives, intention to use / continue, the impact and opinion on technology. The perceived characteristic of initiatives including performance expectancy, effort efficiency, social influence and facilitating condition adapted from UTAUT 2 (Venkatesh, Thong, Chan, Hu, and Brown, 2011) research framework which consist of hedonic motivation, price value, habit, and impact of MCMC's initiatives that have been made in Putrajaya from the aspect of psychological, social, economic, political, and cultural.

Pre-test of the questionnaire then were conducted among respondents from the Malaysian household residents using convenience sampling method. A pilot study then was conducted to learn of any discrepancies within the questions, which included determining whether the format of the questionnaire and questions are suitable. Both validity and reliability were addressed to measure the error of the questionnaires. A content validity questionnaire that comprised definitions of the constructs and associated items were generated and sent to academic experts, to get comments if the items in the questionnaire are not understandable, required rewording or if new entries needed to be added. A self-administrated questionnaire was developed as a final questionnaire to gather information from the respondents which consisted of different sections with both open and close-ended questions.

Research Findings

Based on the results, there is a rather equal representation of both males (52.8%) and females (47.2%) among the respondents, with the males slightly outnumbering than the females (refer Table 1). Most of the respondents were either from the 30–39-year-old group (36.4%) or 20–29-year-old group (29%) given that those two age ranges had the majority number of persons. These two age groups were mainly working adults who were from the government administrative clerical and officers as Putrajaya is federal administrative center of Malaysia. The third largest group was the age group between 40-49-year-old (18.6%). Meanwhile, the minority groups were less than 20-year-old group and above 50-year-old group with 10.5% and 5.6% respectively. Majority of respondents were Malays (90.7%) followed by the Chinese (3.7%), Indians (2.5%) and others (3.1%) such as Bajau and Dusun.

Majority of the respondents had a high education background, having either a Certificate or Diploma (35%) or bachelor's degree qualification (30.7%). In term of employment, most of the respondents were either worked as Professional or at

Management level (23.9%), admin/clerical staff (22.1%) or student (20.2%). For most of them, their monthly income either ranged between RM 1001 to RM 3000 (30.9%) or RM3001 to RM5000 (29.6%).

Respondents who are married (64.4%) far outnumber the single (35.6%). For those who are married, having two (28.4%) or three (20.6%) children seemed to be the norm. However, there are 13.7% of the respondents stated in the survey without any children. This is mainly due to the fact that they are still single/unmarried.

Table 1a. Respondents' demographic profile in Putrajaya

Variable	Category	Frequency	Percentage*
Gender	Male	86	52.8
	Female	77	47.2
Age	<20	17	10.5
	20-29	47	29.0
	30-39	59	36.4
	40-49	30	18.5
	>50	9	5.6
Race	Malay	147	90.7
	Chinese	6	3.7
	Indian	4	2.5
	Others	5	3.1
Education	SPM Or Equivalent	36	22.1
	STPM Or Equivalent	7	4.3
	Certificate/Diploma	57	35.0
	Bachelor's Degree	50	30.7
	Master or PhD Degree	7	4.3
	Others	6	3.7
Occupation	Professional/Management	39	23.9
	Teacher	11	6.7
	Housewife/Husband	3	1.8
	Technical Staff	26	16.0
	Self-Employed	7	4.3
	Admin/Clerical Staff	36	22.1
	Retired/Pensioner	4	2.5
	Student	33	20.2
	Others	4	2.5

Table 1b. Respondents' Demographic Profile – Putrajaya (cont.)

Variable	Category	Frequency	Percentage*
Income	Not Working	35	21.6
	RM1000 or below	3	1.9
	RM1001-RM3000	50	30.9
	RM3001-RM5000	48	29.6
	RM5001-RM7000	20	12.3
	RM7001 and above	6	3.7
Marital Status	Single	58	35.6
	Married	105	64.4
Number of Children	0	14	13.7
	1	9	8.8
	2	29	28.4
	3	21	20.6
	4	12	11.8
	5	7	6.9
	6	8	7.8
	8	1	1.0
	10	1	1.0

*Note: n=163; *valid percentages are used which were calculated by excluding missing cases

Almost all the respondents surveyed owned a smartphone (98.8%) (See Table 2). The brand of smartphones owned were predominantly Samsung (36.6%). Besides smartphones, majority of the respondents also own a laptop (55.8%) and a desktop (42.3%). The average number of years of internet usage experience among the respondents was 10.74 years (std. dev.=5.59). Majority of the respondent had acquired some experiences in using internet with at least 2 years to 10 years of experience.

Total percentage of 99.4% are the respondents subscribed to a mobile data plan with Celcom (38.5%) and Maxis (24.2%) being the major service providers preferred by respondents. 58.4% of the mobile data plan subscribers were happy and rated the mobile data service quality as good. They were only few who were not satisfied with the quality of mobile data service (rated bad and terrible = 3.7%) mentioned reasons such as 'signal is not strong enough' and 'not being able to see much value in the data plan packages.

For the subscription of home broadband plan, there was equal amount of those who subscribed (54%) and who was not (45.4%). Among those who subscribed to the home broadband plan, TM was the preferred choice (83.9%). Majority were happy with quality of the home broadband service, with 51.1% respondents rating the service as good.

Table 2. Respondents' technological profile – Putrajaya

Variable	Category	Frequency	Percentage
Own A Smartphone	Yes	161	98.8
	No	2	1.2
Smartphone Brand	Samsung	58	36.6
	Apple	41	25.6
	Huawei	17	10.6
	Xiaomi	8	5.0
	Others	36	22.5
Other Device Ownership	Laptop	91	55.8
(upon 100)	Desktop	69	42.3
	Tablet	48	29.4
	None	26	16.0
Subscribe mobile data plan	Yes	162	99.4
	No	1	0.6
Mobile data plan brand	Digi	37	23.0
	Celcom	62	38.5
	Maxis	39	24.2
	UMobile	12	7.5
	Others	11	6.8
Mobile data service quality	Excellent	30	18.6
	Good	94	58.4
	Mediocre	31	19.3
	Bad	6	3.7
Subscribe broadband plan	Yes	88	54.0
	No	74	45.4
Broadband plan brand	TM	73	83.9
	Times	3	3.4
	Maxis	9	10.3
	Others	2	2.3
Broadband service quality	Excellent	25	28.4
	Good	45	51.1
	Mediocre	18	20.5
Years of using Internet	Mean = 10.74		
	Std. deviation = 5.59		

Descriptive Results of MCMC's Initiatives in Putrajaya

The result finding showed 69.6% of respondents were aware of Putrajaya being a Smart Community project area. The various initiatives carried out by MCMC under the Smart Community project were grouped under three broad overarching themes namely Infrastructure, Content Training and Digital Services (See Table 3). Awareness of two initiatives namely telecommunication (telco) data services such as Maxis/DiGi/Celcom/UMobile and Pusat Internet (PI) were the highest at 91.3% and 61.5% respectively; followed by cashless payment (42.6%), ICT training (39.8%), e-magazine (34.0%) and lastly, the Entrepreneurship training workshops at 30%. Meanwhile, the Usage of the initiatives were the highest for telco data services (91.9%), followed by PI (31%), cashless payment (30.6%), ICT training workshops (19.7%), e-magazine (17.5%), and the entrepreneurship training workshops (11%).

In terms of the frequency of usage for infrastructure initiatives, majority reported that they used data services to a large extent (mode=7), and it was the only initiative with the highest mode score of 7 outweighing the other initiatives (mode=1) and with high mean score of 5.57 (std. dev.= 1.88). However, these initiatives were still slight to moderately use by the community such as cashless payment (mean= 2.20, std. dev.= 2.07) and PI (mean=1.95, std. dev.= 1.75). Other than that, ICT training (mean=1.78, std. dev.= 1.77), e-magazine (mean=1.45, std. dev.= 1.16) and entrepreneurship workshop and training (mean=1.35, std. dev.= 1.16) were either not used much or not used at all by the respondents. For details, refer to Table 4 (below).

Table 3. Respondents' awareness and usage of the smart community initiatives – Putrajaya

Initiatives	Aware (%)	Not Aware (%)	Use (%)	Don't Use (%)
Infrastructure				
1. Pusat Internet	**99(61.5)**	62(38.5)	**48(31.0)**	107(69.0)
2. Data Services	**147(91.3)**	14(8.7)	**147(91.9)**	13(8.1)
Content Training				
1. ICT Training	**64(39.8)**	97(60.2)	**31(19.7)**	126(80.3)
2. Entrepreneurship	**48(30.0)**	112(70.0)	**17(11.0)**	138(89.0)
Digital Services				
1. E-Magazine	**55(34.0)**	107(66.0)	**27(17.5)**	127(82.5)
2. Cashless Payment	**69(42.6)**	93(57.4)	**48(30.6)**	109(69.4)

*Note: n=163

Table 4. Descriptive for usage frequency of the smart community initiatives – Putrajaya

Initiatives	Mode	Mean	Median	Std. Deviation
Infrastructure				
1. Pusat Internet	1.00	1.95	1.00	1.75
2. Data Service	7.00	5.57	6.00	1.88
Content Training				
1. ICT Training	1.00	1.78	1.00	1.77
2. Entrepreneurship	1.00	1.35	1.00	1.16
Digital Services				
1. E-Magazine	1.00	1.45	1.00	1.16
2. Cashless Payment	1.00	2.20	1.00	2.07

*Note: N=163; Items were measured on a scale 1=not at all to 7=a large extent

Table 5. Descriptive for satisfaction with usage of the smart community initiatives – Putrajaya

Initiatives	Mode	Mean	Median	Std. Deviation
Infrastructure				
1. Pusat Internet	7.00	5.34	5.00	1.71
2. Data Services	5.00	5.00	5.00	1.24
Content Training				
1. ICT Training	7.00	5.13	5.00	1.69
2. Entrepreneurship	4.00	4.89	5.00	1.53
Digital Services				
1. E-Magazine	7.00	5.00	5.00	1.67
2. Cashless Payment	4.00	5.33	5.00	1.30

*Note: N=163; Items measured on a scale 1=very dissatisfied to 7=very satisfied. Scores were calculated for those who use the initiatives.

For the respondents who use the initiatives, generally they felt satisfied with the initiatives, as seen from the mean scores of all the initiatives which ranged from 4.89 to 5.34 (See Table 5). Upon closer inspection, majority of the respondents were very satisfied with PI, ICT training and e-magazine (mode for the stated initiatives = 7). Satisfaction levels for data service (mode=5) was also on the high side. However, there is probably some room for improvement for initiatives like entrepreneurship training and cashless payment (mode for the stated initiatives = 4) as most of the respondents were only moderately satisfied with the initiatives.

SOLUTIONS AND RECOMMENDATIONS

The main purpose of this research study is to uncover, gather the viewpoints and insights of Putrajaya smart community initiatives. Results reveal on Putrajaya's smart community that performance expectancy and price value are the important factors for the use infrastructure initiatives. This supported by Venkatesh et al. (2003 and 2011) whom first tested performance expectancy followed by others such as Meizhura, Alamanda and Adi Prabowo (2017), and Fridayani and Nurmandi (2019) which utilised UTAUT in their studies. It is key that the communities have high usage of data services and internet facilities.

Meanwhile, the easy to use is more for those who have higher on innovativeness. The participation in content training initiatives would depend much on whether the training delivered is easy to follow and learning new ICT skills that is not too uncomplicated. For the community's usage of digital services such as e-magazines and cashless payment applications, it is imperative that they see it's ease-of-use and usefulness in helping to accomplish tasks more efficiently.

The overall mean score for impact were low compared to other sites. Informational was the only item that scored on the higher side. These demonstrated that the initiatives by MCMC were not directly benefiting the local community. The impact from social aspect was the lowest which it could be explained that the needs and wants of the community who live in urban were different compared to rural area such as the other three sites. The existence of the internet and other ICT technologies were not new to urban citizens therefore the impacts on their daily life were lesser.

Besides that, there must be a clear signage or direction to be built, more on promotion on the activities conducted, improvement on institutional collaborations across the flagships, and design informal communication tools or mediums such as leveraging mosques/surau/church/temples to disseminate information about smart community. In addition, there are still lack of active participation from the volunteers, and a "virtual platform" of complaint/ suggestion box related to smart community must be developed and made known to the community. Furthermore, not only associations, non-governmental organizations are needed but schools and institutions of higher learning must also be part of the smart community or other programs within the community areas. Programs and modules at internet centres should also include kids, women, and the elderly people. This will indeed reflect more holistic use of ICTs towards improving the communities' lifestyles and well beings.

In conclusion, to build smart communities in Malaysia, more smart cities effort should be in place. Despite of the geographic, infrastructure and technological situations as well as socio-cultural may dissimilar, the smart community project can be replicated to other locations within other states (Peninsular and East

Malaysia) especially in the more rural and suburban localities given that the project has impacted many areas in the lives of the community – informational (how they used ICTs to communicate); psychological (how they are able to analyse their own situation and solve problems as a result of technology usage); social (enhanced their ICT skills and program management skills); economic (provided a better livelihood for them); political (improved interaction with the local governments); cultural (helped them to increase their awareness of cultural identity). Furthermore, Malaysia smart cities and smart communities' effort should mirror initiatives that had or are being developed by other Asian countries, for example, Japan, Taiwan, China, Singapore, Thailand, Indonesia, Australia, or perhaps European countries.

FUTURE RESEARCH DIRECTIONS

This study has been conducted extensively within the scope of research but there are a few limitations to be addressed for the improvement of the future research. Future research should be focused on the cooperation of various stakeholders like authorities and local champions. Stakeholders support and funding is a very crucial component in ensuring the smart community projects be sustained in the long run. Apart from the funding allocation from the government, more corporate sponsors can be approached for their interest and participation in helping to improve the lives of the community via technology. There are some of challenges that should also be investigated to ensure sustainability and replicability of the smart community projects. Perhaps by looking at the comparative study between Southeast Asia or Asia Pacific countries on governance and legislation matters, stakeholders participatory as well as the communities' activities. Triangulate data would be best to evaluate as on qualitative perspectives.

On the other hand, there are also the micro level or internal perspective such as commitment from local champions and stakeholders, policies or guidelines, the need to design and develop a clearer and better management plan that must contain all the strategic action plans. Based on Yamashita (2022), smart city initiatives should encompass policy dimensions as well as stakeholders' initiatives. In addition, Su, Mioa, & Wang (2022) also stressed the important of coordinated development of intelligent environment which include transportation, energy utilization, resource recovery especially on renewable energy systems, and community management. Furthermore, in current and future research, other factors to be considered would be on community well-being (Irvine, Suwanarit, Likitswat, Srilertchaipanij, Ingegno, Kaewlai, Boonkam, Tontisirin, Sahavacharin, Wongwatcharapaiboon, & Janpathompong, 2022).

ACKNOWLEDGEMENT

The authors would like to extend appreciation to the reviewers for their feedback for this manuscript as well as respondents who have participated in this research. This article is supported by the Research University Grant USM (No: 1001/PMGT/8016126).

REFERENCES

Al-eryani, A., & Rashed, A. (2016). The Impact of Culture on E-Readiness for E-Government in Yemen. In Trends, Prospects, and Challenges in Asian E-Governance (pp. 95-115). IGI Global.

Chai, Z., Cvetkovic, V., & Page, J. (2020). How does ICT expansion drive "Smart" urban growth? A case study of Nanjing, China. *Urban Planning*, *5*(1), 129–139. doi:10.17645/up.v5i1.2561

Chin, C. (2016). *Malaysia to Build 'Smart Communities' Nationwide*. Retrieved 29 March, 2021, from https://govinsider.asia/innovation/malaysia-to-build-smart-communities-nationwide/

Chowdhury, S. (2020). *Green, Smart and Connected Putrajaya*. Retrieved November 1, 2020, from https://www.thesmartcityjournal.com/en/cities/238-green-smart-and-connected-putrajaya

Dameri, R. P. (2013). Searching for smart city definition: A comprehensive proposal. *International Journal of Computers and Technology*, *11*(5), 2544–2551. doi:10.24297/ijct.v11i5.1142

Dhaha, I. S. Y. A., & Ali, A. Y. S. (2014). Mediating effects of behavioral intention between 3g predictors and service satisfaction. *Malaysian Journal of Communication Jilid*, *30*, 107–128.

Feher, K. (2018) Contemporary Smart Cities: Key issues and best practices. In *SMART 2018: The Seventh International Conference on Smart Cities, Systems, Devices and Technologies*. IARIA.

Fridayani, H. D., & Nurmandi, A. (2019). Do smart citizens make a smart city? A case study on the factors influencing citizen behavior using *Lapor Sleman* online-based. *Cosmogov: Jurnal Ilmu Pemerintahan*, *5*(1), 71–100. doi:10.24198/cosmogov.v5i1.18455

Ghazali, M., Okamura, T., Abdullah, T., Sunar, M. S., Mohamed, F., & Ismail, N. (2016). In the Quest of Defining Smart Digital City in Medini Iskandar Malaysia, Iskandar Puteri, Malaysia. *Proceedings of the SEACHI 2016 on Smart Cities for Better Living with HCI and UX.*

Giannakos, M. N., Pateli, A. G., & Chorianopoulos, K. (2013). Investigating Facebook's acceptance and satisfaction: a study in the Greek university community. *International Journal of Social and Humanistic Computing, 14*(1-2), 104-117.

Gurstein, M. (2014). Smart cities vs. smart communities: Empowering citizens not market economics. *The Journal of Community Informatics, 10*(3). Advance online publication. doi:10.15353/joci.v10i3.3438

Irvine, K. N., Suwanarit, A., Likitswat, F., Srilertchaipanij, H., Ingegno, M., Kaewlai, P., Boonkam, P., Tontisirin, N., Sahavacharin, A., Wongwatcharapaiboon, J., & Janpathompong, S. (2022). Smart City Thailand: Visioning and Design to Enhance Sustainability, Resiliency, and Community Wellbeing. *Urban Science, 6*(1), 7. doi:10.3390/urbansci6010007

Kassim, N. M. D., Nathan, S., Yeap, J. A. L., Ramayah, T., & Hashim, N.H. (2018). Insights of smart community in Kemaman. *International Journal of Engineering & Technology, 7*(4.36), 1181-1188. . doi:10.14419/ijet.v7i4.36.25380

Kassim, N. M. D., Yeap, J. A. L., Nathan, S., Hashim, N. H., & Ramayah, T. (2019). A Conceptual Paper of the Smart City and Smart Community. In Eurasian Economic Perspectives. Eurasian Studies in Business and Economics. Springer. doi:10.1007/978-3-030-18565-7_4

Lim, S. B., & Yigitcanlar, T. (2022). Participatory Governance of Smart Cities: Insights from e-Participation of Putrajaya and Petaling Jaya, Malaysia. *Smart Cities, 5*(1), 71–89. . doi:10.3390/smartcities5010005

Lindskog, H. (2004). Smart communities initiatives. In *ISOne World Conference, 2004*. The Information Institute. Retrieved from http://urn.kb.se/resolve?urn=urn:nbn:se:liu:diva-22201

Malaysia Smart City Framework (MSCF). (n.d.). https://www.malaysia.gov.my/portal/content/30947

Meizhura, H., Alamanda, D. T., & Adi Prabowo, F. S. (2017). Technological adoption for the online system of public aspirations and complaints service (Lapor) in Bandung city. *AFEBI Management and Business Review, 2*(1), 62–79. doi:10.47312/ambr.v2i1.57

Mensah, I. K. (2019). *Factors influencing the intention of university students to adopt and use e-government services: An empirical evidence in China.* Retrieved December 8, 2020, Retrieved from https://journals.sagepub.com/doi/pdf/10.1177/2158244019855823

Min, Q., Ji, S., & Qu, G. (2008). Mobile commerce user acceptance study in China: A revised UTAUT model. *Tsinghua Science and Technology, 13*(3), 257–264. doi:10.1016/S1007-0214(08)70042-7

Nor Sallehi, K. (2017). *Smart City Initiative.* Retrieved November 11, 2021, from https://www.mcmc.gov.my/skmmgovmy/media/General/pdf/Sesi-5-Rangkakerja-Berteraskan-Aspek-Informasi-(PLANMalaysia).pdf

Putrajaya Blueprint. (2019). Retrieved from https://smart.putrajaya.my/blueprint

Rana, N. P., Luthra, S., Mangla, S. K., Islam, R., Roderick, S., & Dwivedi, Y. K. (2018). Barriers to the development of smart cities in indian context. *Information Systems Frontiers*, 1–23. doi:10.100710796-018-9873-4

Sengboon, L., Jalaluddin, A. M., Mohd Yusof, H., & Zurinah, T. (2019). The behaviours and job positions of citizens in smart cities' development. *Journal of the Malaysian Institute of Planners, 17*(2), 133–145.

Stratigea, A. (2012). The concept of 'smart cities'. Towards community development? Netcom. *Réseaux, Communication et Territories, 26*(3/4), 375-388.

Taiwo, A. A., & Downe, A. G. (2013). The theory of user acceptance and use of technology (UTAUT): A meta-analytic review of empirical findings. *Journal of Theoretical and Applied Information Technology, 49*(1).

Venkatesh, V., Morris, M. G., Davis, G. B., & Davis, F. D. (2003). User acceptance of information technology: Toward a unified view. *Management Information Systems Quarterly, 27*(3), 425–478. doi:10.2307/30036540

Venkatesh, V., Thong, J. Y. L., Chan, F. K. Y., Hu, P. J. H., & Brown, S. A. (2011). Extending the two-stage information systems continuance model: Incorporating UTAUT predictors and the role of context. *Information Systems Journal, 21*(6), 527–555. doi:10.1111/j.1365-2575.2011.00373.x

Yamashita, J. (2022). Smart City Initiatives in Japan: Achievements and Remaining Issues. In R. K. Mishra, C. L. Kumari, S. Chachra, P. S. J. Krishna, A. Dubey, & R. B. Singh (Eds.), *Smart Cities for Sustainable Development. Advances in Geographical and Environmental Sciences.* Springer. doi:10.1007/978-981-16-7410-5_6

Yigitcanlar, T., Kamruzzaman, M., Buys, L., Ioppolo, G., Marques, J., Costa, E., & Yun, J. (2018). *Understanding 'smart cities': Intertwining development drivers with desired outcomes in a multidimensional framework.* Cities. doi:10.1016/j.cities.2018.04.003

ADDITIONAL READING

Eger, J. M. (2005). Smart communities, universities, and globalization: Educating the workforce for tomorrow's economy. *Metropolitan Universities*, *16*(4), 28–38.

Fachinelli, A. C., Yigitcanlar, T., Sabatini Marques, J., Cortese, T. T. P., Sotto, D., & Libardi, B. (2022). *Smart cities of Brazil: Performance of Brazilian Capital Cities.* City Living Lab. doi:10.18226/9786500438604

Guo, M., Xia, M., & Chen, Q. (2022). A review of regional energy internet in smart city from the perspective of energy community. *Energy Reports*, *8*, 161–182. doi:10.1016/j.egyr.2021.11.286

Gurick, M., & Felger, S. (2022). Organisation and community intelligence in smart city leadership and beyond. *IET Smart Cities*, *4*(1), 47–55. doi:10.1049mc2.12022

Hollands, R. G. (2008). Will the real smart city please stand up? Intelligent, progressive, or entrepreneurial? *City*, *12*(3), 303–320. doi:10.1080/13604810802479126

Hughes, C., & Spray, R. (2002). Smart communities and smart growth-Maximising benefits for the corporation. *Journal of Corporate Real Estate*, *4*(3), 207–214. doi:10.1108/14630010210811831

Li, L., Taeihagh, A., & Tan, S. Y. (2022). What factors drive policy transfer in smart city development? Insights from a Delphi study. *Sustainable Cities and Society*, *84*, 104008. doi:10.1016/j.scs.2022.104008

Myeong, S., Park, J., & Lee, M. (2022). Research Models and Methodologies on the Smart City: A Systematic Literature Review. *Sustainability*, *14*(3), 1687. doi:10.3390u14031687

Nagarale, V., & Telang, P. (2022). Urban Rejuvenation and Social Sustainability in Smart City: An Empirical Study of Community Aspirations. In N. C. Jana, A. Singh, & R. B. Singh (Eds.), *Livelihood Enhancement Through Agriculture, Tourism and Health. Advances in Geographical and Environmental Sciences.* Springer. doi:10.1007/978-981-16-7310-8_15

Nam, T., & Pardo, T. A. (2011). Conceptualizing smart city with dimensions of technology, people, and institutions. *Proceedings of the 12th annual international digital government research conference: digital government innovation in challenging times*. 10.1145/2037556.2037602

Puron-Cid, G., & Gil-Garcia, J. R. (2022). Are Smart Cities Too Expensive in the Long Term? Analyzing the Effects of ICT Infrastructure on Municipal Financial Sustainability. *Sustainability*, *14*(10), 6055. doi:10.3390u14106055

Singh, D., Divan, M., & Singh, M. (2022). Internet of Things for Smart Community Solutions. *Sensors (Basel)*, *22*(2), 640. doi:10.339022020640 PMID:35062602

Singh, S. (2014). Smart Cities—A $1.5 Trillion Market Opportunity. *Forbes*. Retrieved May 5, 2020, Retrieved from http://www.forbes.com/sites/ sarwantsingh/2014/06/19/smart-cities-a-1-5-trillion-market-opportunity/#3b9495237ef9

Su, Y., Miao, Z., & Wang, C. (2022). The Experience and Enlightenment of Asian Smart City Development—A Comparative Study of China and Japan. *Sustainability*, *14*(6), 3543. . doi:10.3390/su14063543

Sung, M. S., Shih, S. G., & Perng, Y. H. (2022). Multi-Criteria Evaluation of Site Selection for Smart Community Demonstration Projects. *Smart Cities*, *5*(1), 22–33. doi:10.3390martcities5010002

Wongbumru, T., & Dewancker, B. (2014). Smart Communities for Future Development: Lessons from Japan. *International Journal of Building, Urban Interior and Landscape Technology*, *3*, 69–76.

KEY TERMS AND DEFINITIONS

Information and Communication Technology (ICT): The infrastructure, networking components, devices, systems, or applications which enable computing.

Malaysian Communication and Multimedia Commission (MCMC): A government authority which regulate issues pertaining communications and multimedia industries in Malaysia.

Putrajaya: A city, federal territory, and government administrative capital of Malaysia.

Smart City: A modern urban area with high, modern, and efficient ICT usage of devices, resources, and services.

Smart Community: A community that stay at a place, area or district that equipped with high technology and good infrastructure to help them to connect, and commute within the community for their social, economic, and environment sustainability benefits.

Compilation of References

Abdul, K., Hangjung, Z., & Ananth, C. (2021). Digital transformation and environmental sustainability. *MDPI Sustainability*, *13*(1530), 2071–1050.

About EBDVF. (2022, November 3). European Big Data Value Forum. Retrieved from https://european-big-data-value-forum.eu/2022-edition/about/

About Us. (2022, December 14). Green Data Hub. Retrieved from https://www.greendatahub.at/about-us/?lang=en

Accenture. (2003). eGovernment leadership: Engaging the customer. Accenture.

Agnusdei, G. P., Elia, V., & Gnoni, M. G. (2021). A classification proposal of digital twin applications in the safety domain. *Computers & Industrial Engineering*, *154*, 107137. doi:10.1016/j.cie.2021.107137

Agnusdei, G. P., Elia, V., & Gnoni, M. G. (2021). Is digital twin technology supporting safety management? A bibliometric and systematic review. *Applied Sciences (Basel, Switzerland)*, *11*(6), 2767. doi:10.3390/app11062767

Agustan, S. M., Yulianto, S., Sutrisno, D., Bintoro, O. B., & Alhasanah, F. (2018). *Participatory Scenario Planning for Indonesian Space Industry in 2025.* IEEE International Conference on Aerospace Electronics and Remote Sensing Technology (ICARES), Bali, Indonesia.

Al Sadawi, A., Hassan, M. S., & Ndiaye, M. (2021). A Survey On The Integration Of Blockchain With Iot To Enhance Performance And Eliminate Challenges. *IEEE Access: Practical Innovations, Open Solutions*, *9*, 54478–54497. doi:10.1109/ACCESS.2021.3070555

Alasem, A. (2009). An Overview of e-Government Metadata Standards and Initiatives based on Dublin Core. *Electronic Journal of E-Government*, *7*(1), 1–10.

Albino, V., Berardi, U., & Dangelico, R. M. (2015). Smart cities: Definitions, dimensions, performance, and initiatives. *Journal of Urban Technology*, *22*(1), 3–21. doi:10.1080/10630732.2014.942092

Al-eryani, A., & Rashed, A. (2016). The Impact of Culture on E-Readiness for E-Government in Yemen. In Trends, Prospects, and Challenges in Asian E-Governance (pp. 95-115). IGI Global.

Al-Khouri, A. M. (2011). *An innovative approach for e-government transformation.* arXiv preprint arXiv:1105.6358.

Allam, Z., & Jones, D. S. (2021). Future (post-COVID) digital, smart and sustainable cities in the wake of 6G: Digital twin technology, immersive realities, and new urban economies. *Land Use Policy*, *101*, 105201. doi:10.1016/j.landusepol.2020.105201

Alsmadi, I., Al-Kabi, M., & Wahbeh, A. (2012). A Conceptual Organization for Websites Metrics and E-Government Websites: A Case Study. *International Journal of Information Communication Technologies and Human Development*, *4*(1), 93–117. doi:10.4018/jicthd.2012010107

Ammara, Ul., Rasheed, K., Mansoor, A., Al-Fuqaha, A., & Qadir, J. (2022). *Smart Cities from the Perspective of Systems*. Systems.

Ammar, M., Russello, G., & Crispo, B. (2018). Internet Of Things: A Survey On The Security Of IoT Frameworks. *Journal of Information Security and Applications*, *38*, 8–27. doi:10.1016/j.jisa.2017.11.002

Amor, R., & Dimyadi, J. (2021). The promise of automated compliance checking. *Journal of Developments in the Built Environment*, *5*, 100039. Advance online publication. doi:10.1016/j.dibe.2020.100039

Andreasson, K., Millard, J., & Snaprud, M. (2012, October). Evolving e-government benchmarking to better cover technology development and emerging societal needs. In *Proceedings of the 6th International Conference on Theory and Practice of Electronic Governance* (pp. 430-439). 10.1145/2463728.2463811

Anrong, D., Li, G., Li, J., & Kong, X. (2016). Research on Smart Community Planning of Yishanwan, China towards New Urbanization. *International Review for Spatial Planning and Sustainable Development*, *4*(1), 78–90. doi:10.14246/irspsd.4.1_78

Anshari, M., Almunawar, M. N., & Masri, M. (2022). Digital Twin: Financial Technology's Next Frontier of Robo-Advisor. *Journal of Risk and Financial Management*, *15*(4), 163. doi:10.3390/jrfm15040163

Assar, S., Boughzala, I., & Isckia, T. (2011). eGovernment trends in the Web 2.0 era and the open innovation perspective: an exploratory field study. In *Electronic Government: 10th IFIP WG 8.5 International Conference, EGOV 2011, Delft, The Netherlands, August 28–September 2, 2011 Proceedings*, *10*, 210–222.

Baker, P. (2021). Upskilling and reskilling employees for a competitive edge. *SearchHRSoftware*. https://www.techtarget.com/searchhrsoftware/feature/Upskilling-and-reskilling-employees-for-a-competitive-edge

Bayer. (n.d.). *Making Science Make Sense*. https://www.bayer.com/en/us/makingsciencemakesense/home

Benamrou, B., Mohamed, B., Bernoussi, A. S., & Mustapha, O. (2016, October). Ranking models of smart cities. In *2016 4th IEEE International Colloquium on Information Science and Technology (CiSt)* (pp. 872-879). IEEE. 10.1109/CIST.2016.7805011

Benchmark Measurement of European eGovernment Services, 8th Benchmark Measurment. European Commission. (2009). Rep. Prep. Capgemini Rand Eur. IDC Sogeti DTI.

Benson, M. (2021). Digital twin technology will revolutionize healthcare. *Engineering & Technology*, *16*(2), 50–53. doi:10.1049/et.2021.0210

Berntzen, L., & Olsen, M. G. (2009, February). Benchmarking e-government-a comparative review of three international benchmarking studies. In *2009 Third International Conference on Digital Society* (pp. 77-82). IEEE. 10.1109/ICDS.2009.55

Bhattacharya, D., Gulla, U., & Gupta, M. P. (2012). E-service quality model for Indian government portals: Citizens' perspective. *Journal of Enterprise Information Management*, *25*(3), 246–271. doi:10.1108/17410391211224408

BMVI. (2021). *Smart Design, Smart Construction, Smart Operation*. BMWI Publication. https://www.digitale-technologien.de/DT/Redaktion/DE/Downloads/Publikation/SSW/2021/SSW_Baupublikation.pdf?__blob=publicationFile&v=3

Boată, A., Angelescu, R., & Dobrescu, R. (2021). Using digital twin technology in health care. *UPB Scientific Bulletin, Series C: Electrical Engineering and Computer Science*, 53–62.

Bodden, J., & Dittschar, F. (2020). *Umsetzung des Stufenplans Digitales Bauen* [Implementation of the step-by-step Plan for Digital Construction]. BMVI Publishing, Working Paper 3.1 – Klärung von Rechtsfragen.

Bolton, E., & Bauerlein, A. (2020). *Space Workforce 2023: Executive Leadership Pledges*. National Academies. Retrieved from https://www.nationalacademies.org/documents/embed/link/LF2255DA3DD1C41C0A42D3BEF0989ACAECE3053A6A9B/file/D3063C38C8FAB019DB855143D5B047EB77160FD3E974?noSaveAs=1

Bosch Energy and Building Solutions. (2021). *Fire Safety of the Future*. White Paper. Bosch Publishing. https://media.boschsecurity.com/fs/media/ib/media_1/solutions/Bosch-Whitepaper-Brandschutz_1.pdf

Bosch. (2021). *Digital twins for buildings. More transparency and efficiency for building operation*. https://www.boschbuildingsolutions.com/de/de/trends-und-themen/smart-buildings/digitaler-gebaeudezwilling/?gclid=EAIaIQobChMI9oL1zvy49QIVxeJ3Ch3rZgCtEAAYASABEgIqxvD_BwE

Bosch. (2021). *Fire Protection*. White Paper by Bosch Energy and Building Solutions. https://media.boschsecurity.com/fs/media/ib/media_1/solutions/Bosch-Whitepaper-Brandschutz_1.pdf

Braungart, M., & McDonough, W. (2014). *Cradle to Cradle: Einfach intelligent produzieren.* Piper Publishing.

Braun, M. (2021). Represent me: Please! towards an ethics of digital twin technology in medicine. *Journal of Medical Ethics*, *47*(6), 394–400. doi:10.1136/medethics-2020-106134

Bucchiarone, A. (2019). Collective Adaptation Through Multi-Agents Ensembles: The Case Of Smart Urban Mobility. *ACM Transactions on Autonomous and Adaptive Systems*, *14*(2), 1–28. doi:10.1145/3355562

buildingSMART. (2021). *BIM Professional Certification Program.* https://education.buildingsmart.org/

buildingSMART. (2021). *White Paper - Fire Protection with BIM.* buildingSMART Publishing. https://buildingsmart-verlag.de/produkt/whitepaper-fg-brandschutz/

Caldwell, B., Cooper, M., Reid, L. G., Vanderheiden, G., Chisholm, W., Slatin, J., & White, J. (2008). Web content accessibility guidelines (WCAG) 2.0. *WWW Consortium (W3C)*, *290*, 1-34.

Cao, L. (2022). Decentralized ai: Edge intelligence and smart blockchain, metaverse, web3, and desci. *IEEE Intelligent Systems*, *37*(3), 6–19. doi:10.1109/MIS.2022.3181504

Capgemini, I. (2010). *Digitizing public services in Europe: Putting ambition into action: 9th benchmark measurement.* European Commission.

Chai, Z., Cvetkovic, V., & Page, J. (2020). How does ICT expansion drive "Smart" urban growth? A case study of Nanjing, China. *Urban Planning*, *5*(1), 129–139. doi:10.17645/up.v5i1.2561

Chase, J. G., Zhou, C., Knopp, J. L., Shaw, G. M., Näswall, K., Wong, J. H. K., Malinen, S., Moeller, K., Benyo, B., Chiew, Y. S., & Desaive, T. (2021). Digital twin technology in critical care: What, when, how, where, why? *IFAC-PapersOnLine*, *54*(15), 310–315. doi:10.1016/j.ifacol.2021.10.274

Chen, C. (2006). CiteSpace II: Detecting And Visualizing Emerging Trends And Transient Patterns In Scientific Literature. *Journal of the American Society for Information Science and Technology*, *57*(3), 359–377. doi:10.1002/asi.20317

Chen, C. (2014). The CiteSpace Manual. *College of Computing and Informatics*, *1*(1), 1–84. doi:10.100711192-015-1576-8

Chen, C., Ibekwe-SanJuan, F., & Hou, J. (2010). The Structure And Dynamics Of Cocitation Clusters: A Multiple-Perspective Cocitation Analysis. *Journal of the American Society for Information Science and Technology*, *61*(7), 1386–1409. doi:10.1002/asi.21309

Cheng, M., Chiu, K.-C., Hsieh, Y.-M., Yang, I.-T., Chou, J.-S., & Wu, Y.-W. (2017). BIM integrated smart monitoring technique for building fire prevention and disaster relief. *Journal for Automation in Construction.*, *84*, 14–30. doi:10.1016/j.autcon.2017.08.027

Chin, C. (2016). *Malaysia to Build 'Smart Communities' Nationwide*. Retrieved 29 March, 2021, from https://govinsider.asia/innovation/malaysia-to-build-smart-communities-nationwide/

Choo, K. K. R., Gai, K., & Chiaraviglio, L. (2021). Blockchain-Enabled Secure Communications In Smart Cities. *Journal of Parallel and Distributed Computing*, *152*, 125–127. doi:10.1016/j.jpdc.2021.02.021

Choudrie, J., & Weerakkody, G. G. V. (2004). Evaluating Global e-Government Sites: A View using Web Diagnostic Tools. *Electronic Journal of E-Government*, *2*(2), 105–114.

Chourabi, H., Nam, T., Walker, S., Gil-Garcia, J. R., Mellouli, S., Nahon, K., Pardo, T. A. & Scholl, H. J. (2012). *Understanding smart cities: An integrative framework*. Paper presented at the System Science (HICSS), 2012 45th Hawaii International Conference on System Sciences. doi: 10.1109/HICSS.2012.615

Chowdhury, S. (2020). *Green, Smart and Connected Putrajaya*. Retrieved November 1, 2020, from https://www.thesmartcityjournal.com/en/cities/238-green-smart-and-connected-putrajaya

Committee on Science, Space, and Technology. (2019 May 17). *Chairwoman Johnson and Ranking Member Lucas Introduce STEM Opportunities Act*. https://science.house.gov/news/press-releases/chairwoman-johnson-and-ranking-member-lucas-introduce-stem-opportunities-act

Council of the European Union. (2016). *Urban Agenda for the EU, Pact of Amsterdam*. https://e3p.jrc.ec.europa.eu/articles/pact-amsterdam-urban-agenda-eu-european-cities-get-their-say-eu-policy-making

Crisostomi, E., Shorten, R., & Wirth, F. (2016). Smart Cities: A Golden Age For Control Theory? *IEEE Technology and Society Magazine*, *35*(3), 23–24. doi:10.1109/MTS.2016.2592782

Crossman, A. (2018, July 3). *How To Construct an Index for Research*. Retrieved from https://www.thoughtco.com/index-for-research-3026543

Dameri, R. P. (2013). Searching for smart city definition: A comprehensive proposal. *International Journal of Computers and Technology*, *11*(5), 2544–2551. doi:10.24297/ijct.v11i5.1142

Data Ethic Commission of German Government. (2019) Assessment on Data Ethic. Brandenburgische Universitätsdruckerei und Verlagsgesellschaft Potsdam mbH (bud).

DeLone, W. H., & McLean, E. R. (1992). Information systems success: The quest for the dependent variable. *Information Systems Research*, *3*(1), 60–95. doi:10.1287/isre.3.1.60

DeLone, W. H., & McLean, E. R. (2003). The DeLone and McLean model of information systems success: A ten-year update. *Journal of Management Information Systems*, *19*(4), 9–30. doi:10.1080/07421222.2003.11045748

Demir Altıntaş, A. (2021). Loose coupling of GIS and BIM data models for automated compliance checking against zoning codes. *Journal of Automation in Construction*. DOI: doi:10.1016/j.autcon.2021.103743

Deren, L., Wenbo, Y., & Zhenfeng, S. (2021). Smart City based on digital twins. *Computers & Urban Society*, *1*(4), 1–11.

Deutsche Rohstoffagentur, D. E. R. A. (2016). *Rohstoffe für Zukunftstechnologien 2016. Bundesanstalt für Geowissenschaften und Rohstoffe (BGR)* [Raw materials for future technologies 2016. Federal Institute for Geosciences and Natural Resources (BGR)]. https://www.deutsche-rohstoffagentur.de/DERA/DE/Downloads/Studie_Zukunftstechnologien-2016.html

Dhaha, I. S. Y. A., & Ali, A. Y. S. (2014). Mediating effects of behavioral intention between 3g predictors and service satisfaction. *Malaysian Journal of Communication Jilid*, *30*, 107–128.

Dhanapal, A. C. T. A., Subapriya, S., Subramaniam, K., & Appukutty, M. (2022). Virtual Digital twin technology: A Renaissance in Precision Nutrition Approach. In *Integrating AI in IoT Analytics on the Cloud for Healthcare Applications* (pp. 1–23). IGI Global. doi:10.4018/978-1-7998-9132-1.ch001

Digital Matters. (2022, January 7.). *Mobilithek - Germany's data plattform that gets you moving*. Federal Ministry for Digital and Transport. Retrieved from https://bmdv.bund.de/SharedDocs/EN/Articles/DG/mobilithek.html

Digital Twin Market Size, Share, Growth & Trends Report, 2030. (n.d.). Retrieved January 10, 2023, from https://www.grandviewresearch.com/industry-analysis/digital-twin-market#:~:text=Report%20Overview

Does Job Hopping Help Or Hurt Your Career ? (2018 April 5). Robert Half. Retrieved from https://press.roberthalf.com/2018-04-05-Does-Job-Hopping-Help-Or-Hurt-Your-Career

Eagan, K., Hurtado, S., Figueroa, T., & Hughes, B. E. (2014). *Examining STEM pathways among students who begin college at four-year institutions*. National Academy of Sciences.

Eger, J. M. (2005). Smart communities, universities, and globalization: Educating the workforce for tomorrow's economy. *Metropolitan Universities*, *16*(4), 28–38.

e-Government of Korea Best Practices. (2014). Ministry of Public Administration and Security and National Information Society Agency. Retrieved from https://www.mois.go.kr/html/site/eng/eng_mov_2.html

eGovernment: International Best Practices. (2008). Forfás.

ElMassah, S., & Mohieldin, M. (2020). Digital transformation and localizing the sustainable development goals (SDGs). *Ecological Economics*, *169*, 106490. doi:10.1016/j.ecolecon.2019.106490

Elshkaki, A., Graedel, T. E., Ciacci, L., & Reck, B. K. (2016). Copper demand, supply, and associated energy use to 2050. *Global Environmental Change*, *39*, 305–315. doi:10.1016/j.gloenvcha.2016.06.006

Employee Tenure Summary. (2022 September 22). U.S. Bureau of Labor Statistics. Retrieved from https://www.bls.gov/news.release/tenure.nr0.htm

Enders, M. R., & Hoßbach, N. (2019). Dimensions of digital twin applications-a literature review. *25th Americas Conference on Information Systems, AMCIS 2019*.

Ernstsen, S. N., Whyte, J., Thuesen, C., & Maier, A. (2021). How innovation champions frame the future: Three visions for digital transformation of construction. *Journal of Construction Engineering and Management*, *147*(1), 05020022. doi:10.1061/(ASCE)CO.1943-7862.0001928

European Commission. (2010). *Grundlagendokument Brandschutz Nr. 2 der Europäischen Kommission* [Basic Document Fire Protection No. 2 of the EuropeanCommission]. https://eur-lex.europa.eu/legal-content/DE/TXT/PDF/?uri=CELEX:32010D0738&rid=2

European Commission. (2020). *Im Blickpunkt – Energieeffizienz von Gebäuden*. European Commission. https://ec.europa.eu/info/news/focus-energy-efficiency-buildings-2020-lut-17_de#:~:text=Insgesamt%20entfallen%20auf%20Geb%C3%A4ude%20in,%2C%20Nutzung%2C%20Renovierung%20und%20Abriss

European Commission. (2020). *White Paper On Artificial Intelligence - A European approach to excellence and trust*. European Commission.

European Commission. (2021). *Status Quo and updating the reform recommendations for the Professional regulation of 2017*. https://eur-lex.europa.eu/legal-content/DE/TXT/PDF/?uri=CELEX:52021DC0385&from=ES

European Smart Cities project. (n.d.). https://www.smart-cities.eu/model.html

European Union. (2018). *General Data Protection Regulation*. https://www.datenschutz-grundverordnung.eu/

European Union. (2021). *The Circular Cities Declaration*. https://circularcitiesdeclaration.eu/

Expo 2020 Dubai. (2020). *The human-centric city of the future*. https://www.expo2020dubai.com/

Faber, A., Rehm, S. V., Hernandez-Mendez, A., & Matthes, F. (2018). Modeling And Visualizing Smart City Mobility Business Ecosystems: Insights From A Case Study. *Information (Basel)*, *9*(11), 270. doi:10.3390/info9110270

Fair principles. (2022, January 21). GO FAIR. Retrieved from https://www.go-fair.org/fair-principles/

Fajga, K. W. (2022). Ehrgeizige Ziele gesetzt [Ambitious goals set]. *Allgemeine Bauzeitung ABZ*, (18), 3.

Fan, C., Zhang, C., Yahja, A., & Mostafavi, A. (2021). Disaster City Digital Twin: A vision for integrating artificial and human intelligence for disaster management. *International Journal of Information Management*, *56*, 102049. doi:10.1016/j.ijinfomgt.2019.102049

Faria, R., Brito, L., Baras, K., & Silva, J. (2017). Smart Mobility: A Survey. *Proceedings of the 2017 International Conference on Internet of Things for the Global Community*, 1–8.

Farooqi, N., Gutub, A., & Khozium, M. O. (2019). Smart Community Challenges: Enabling IoT/M2M Technology Case Study. *Life Science Journal*, *16*(7), 11–17. doi:10.1109/IoTGC.2017.8008972

Fath-Allah, A., Cheikhi, L., Al-Qutaish, R. E., & Idri, A. (2014b). E-government portals best practices: a comprehensive survey. *Electronic Government, an International Journal, 11*(1/2), 101-132.

Fath-Allah, A., Cheikhi, L., Al-Qutaish, R. E., & Idri, A. (2014c, October). Towards a best practice based e-government portals maturity model. In *Proceedings of the International Conference on Computer Science and Information Systems (ICSIS'14)* (pp. 100-106). Academic Press.

Fath-Allah, A., Cheikhi, L., Al-Qutaish, R. E., & Idri, A. (2016, August). A Validation of a Measurement Based E-Government Portals' Maturity Model. In *2016 42th Euromicro Conference on Software Engineering and Advanced Applications (SEAA)* (pp. 100-107). IEEE.

Fath-Allah, A., Cheikhi, L., Al-Qutaish, R. E., & Idri, A. (2014a). E-government maturity models: A comparative study. *International Journal of Software Engineering and Its Applications*, *5*(3), 71–91.

Fath-Allah, A., Cheikhi, L., Al-Qutaish, R. E., & Idri, A. (2015). E-Government Portals Maturity Models: A Best Practices' Coverage Perspective. *Journal of Software*, *10*(7), 805–824. doi:10.17706//jsw.10.7.805-824

Fath-Allah, A., Cheikhi, L., Al-Qutaish, R. E., & Idri, A. (2016). A Measurement-Based E-Government Portals' Maturity Model. In *International Perspectives on Socio-Economic Development in the Era of Globalization* (pp. 129–149). IGI Global. doi:10.4018/978-1-4666-9908-3.ch008

Fazio, M., Paone, M., Puliafito, A., & Villari, M. (2012, July). Heterogeneous Sensors Become Homogeneous Things In Smart Cities. In *2012 Sixth International Conference on Innovative Mobile and Internet Services in Ubiquitous Computing* (pp. 775-780). IEEE. 10.1109/IMIS.2012.136

Federal Institute for Building, Urban and Spatial Research (BBSR) in the Federal Office for Building and Regional Planning (BBR), Bonn. (2017). *Smart City Charta - The sustainable design of Digital transformation.* https://www.bmi.bund.de/SharedDocs/downloads/DE/veroeffentlichungen/themen/bauen/wohnen/smart-city-charta-kurzfassung-de-und-en.pdf?__blob=publicationFile&v=4#:~:text=Die%20Smart%20City%20Charta%20richtet,Bund%20und%20die%20L%C3%A4nder%20angesprochen

Federal Ministry of the Interior, Building, and Homeland. (2021). *Model projects Smart Cities*. https://www.bmi.bund.de/SharedDocs/kurzmeldungen/DE/2021/07/smart-city.html

Feher, K. (2018) Contemporary Smart Cities: Key issues and best practices. In *SMART 2018: The Seventh International Conference on Smart Cities, Systems, Devices and Technologies*. IARIA.

Feng, Q., & Cai, R. (2021). "Data Hegemony": Reflections for the Application and Development Direction of Metaverse Technology in Urban Design based on Digital. *Journal of World Architecture*, *5*(6), 52–61. doi:10.26689/jwa.v5i6.2780

Ferdousi, R., Laamarti, F., Hossain, M. A., Yang, C., & El Saddik, A. (2022). Digital twin technology for well-being: An overview. *Digital Twin*, *1*(7), 7. doi:10.12688/digitaltwin.17475.2

Finck, M., Lamping, M., Moscon, V., & Richter, H. (2020). *Smart Urban Mobility*. Springer Berlin Heidelberg. doi:10.1007/978-3-662-61920-9

Foa, R., & Tanner, J. (2012). *Methodology of the indices of social development* (No. 2012-04). Academic Press.

Fridayani, H. D., & Nurmandi, A. (2019). Do smart citizens make a smart city? A case study on the factors influencing citizen behavior using *Lapor Sleman* online- based. *Cosmogov: Jurnal Ilmu Pemerintahan*, *5*(1), 71–100. doi:10.24198/cosmogov.v5i1.18455

Friese, I., Heuer, J., & Kong, N. (2014, March). Challenges From The Identities Of Things: Introduction Of The Identities Of Things Discussion Group Within Kantara Initiative. In *2014 IEEE World Forum on Internet of Things (WF-IoT)* (pp. 1-4). IEEE. 10.1109/WF-IoT.2014.6803106

Gao, W., Fan, L., Ushifusa, Y., Gu, Q., & Ren, J. (2016). Possibility and Challenge of Smart Community in Japan. *Procedia: Social and Behavioral Sciences*, *216*, 109–118. doi:10.1016/j.sbspro.2015.12.015

Gartner, Inc. (2018). *Four best practices to avoid digital twin failures*. Gartner Survey Report. https://www.gartner.com/en/newsroom/press-releases/2018-03-13-gartner-survey-reveals-nearly-half-of-organizations-implementing-iot-are-using-or-plan-to-use-digital-twin-initiatives-in-2018

Gauss Centre for Supercomputing. (2020, May 7). *Using Digital Twins to design more sustainable cities*. Retrieved January 26, 2023, from https://phys.org/news/2020-05-digital-twins-sustainable-cities.html

Georgescu, M. R., & Georgescu, I. (2008). Do we need a powerful e-government? *Communications of the IBIMA*, *5*, 242–249.

Ghazali, M., Okamura, T., Abdullah, T., Sunar, M. S., Mohamed, F., & Ismail, N. (2016). In the Quest of Defining Smart Digital City in Medini Iskandar Malaysia, Iskandar Puteri, Malaysia. *Proceedings of the SEACHI 2016 on Smart Cities for Better Living with HCI and UX*.

Giannakos, M. N., Pateli, A. G., & Chorianopoulos, K. (2013). Investigating Facebook's acceptance and satisfaction: a study in the Greek university community. *International Journal of Social and Humanistic Computing, 14*(1-2), 104-117.

Gil, O., & Navarro, C. (2013). *Innovations of Governance in Cities and Urban regions: Smart Cities in China, Iskandar (Malaysia), Japan, New York and Tarragona (Spain).* Paper presented at the EURA Conference: Cities as Sheedbeds for Innovation.

Gov.cn. (2022). *Shijiazhuang intelligent city spatio-temporal data platform completed.* http://news.mnr.gov.cn/

Government on the Web II. (2002). National Audit Office. Retrieved from https://www.nao.org.uk/wp-content/uploads/2002/04/0102764.pdf

Gras, N. (2022, September 8). *Catena-X: Building trust to ensure resilience.* International Data Spaces. https://internationaldataspaces.org/catena-x-building-trust-to-ensure-resilience/

Group, D. M. T. (2021). *Brandschutz – der stille Erfolgsfaktor* [Fire Protection – the smart key for success]. https://www.dmt-group.com/de/engineering-performance/engineering-performance-storie/archive/brandschutz-der-stille-erfolgsfaktor.html

Grove, J. (2020, October 15). *Second issue of data SPACES_NOW available.* International Data Spaces. https://internationaldataspaces.org/second-issue-of-data-spaces_now-available/

Guo, M., Liu, Y., Yu, H., Hu, B., & Sang, Z. (2016). An Overview Of Smart City In China. *China Communications*, *13*(5), 203–211. doi:10.1109/CC.2016.7489987

Guo, Q., You, W. W., Hu, K., Chen, Y., & He, Z. (2022). Study On The Path And Countermeasures Of Blockchain Technology To Promote High-Quality Development Of Digital Economy In Guangzhou. *Open Journal of Business and Management*, *10*(6), 3305–3326. doi:10.4236/ojbm.2022.106163

Gupta, D., Kayode, O., Bhatt, S., Gupta, M., & Tosun, A. S. (2021). Hierarchical federated learning-based anomaly detection using digital twin technology for smart healthcare. *2021 IEEE 7th International Conference on Collaboration and Internet Computing (CIC)*, 16–25.

Gurstein, M. (2014). Smart cities vs. smart communities: Empowering citizens not market economics. *The Journal of Community Informatics*, *10*(3). Advance online publication. doi:10.15353/joci.v10i3.3438

Gz.gov.cn. (2022). *Guangzhou Creates A Pioneering Demonstration Zone For Blockchain Development.* https://www.gz.gov.cn/

Gzdaily. (2022). *Give Full Play To The Development Of The Blockchain Industry To Make Digital More "Warm"*. https://www.gzdaily.cn/

Harrison, C., Eckman, B., Hamilton, R., Hartswick, P., Kalagnanam, J., Paraszczak, J., & Williams, P. (2010). Foundations for smarter cities. *IBM Journal of Research and Development*, *54*(4), 1–16. doi:10.1147/JRD.2010.2048257

Hebei News. (2022, September). Spatio-Temporal Big Data To Create A Smart City. *Brain*, 22.

Henriksson, A., Yi, Y., Frost, B., & Middleton, M. (2007). Evaluation instrument for e-government websites. *Electronic Government, an International Journal, 4*(2), 204-226.

Hernandez-Boussard, T., Macklin, P., Greenspan, E. J., Gryshuk, A. L., Stahlberg, E., Syeda-Mahmood, T., & Shmulevich, I. (2021). Digital twin technology for predictive oncology will be a paradigm shift for precision cancer care. *Nature Medicine*, *27*(12), 2065–2066. doi:10.103841591-021-01558-5 PMID:34824458

Hiller, J. S., & Bélanger, F. (2001). Privacy strategies for electronic government. *E-government*, *200*, 162–198.

Höfer, T., Bierwirth, S., & Madlener, R. (2020). Energie Mehrverbrauch in Rechenzentren bei Einführung des 5G Standards [Energy consumption in data centers with the introduction of the 5G standard]. Nachhaltige Rechenzentren–Chancen und Entwicklungsmöglichkeiten am. Stuttgart.

Hollands, R. G. (2008). Will the real smart city please stand up? Intelligent, progressive or entrepreneurial? *City*, *12*(3), 303–320. doi:10.1080/13604810802479126

Hóu, Z., Li, Q., Foo, E., Dong, J. S., & de Souza, P. (2022, March). A digital twin runtime verification framework for protecting satellites systems from cyber attacks. In *2022 26th International Conference on Engineering of Complex Computer Systems (ICECCS)* (pp. 117-122). IEEE. 10.1016/j.jmsy.2019.10.001

Huanan, Z., Shijun, L., & Hong, J. (2015). Guangzhou smart city construction and big data research. In *2015 International Conference on Behavioral, Economic and Socio-cultural Computing (BESC)* (pp. 143-149). IEEE. 10.1109/BESC.2015.7365972

Huang, S., Wang, G., Yan, Y., & Fang, X. (2020). Blockchain-based data management for digital twin of product. *Journal of Manufacturing Systems*, *54*, 361–371. doi:10.1016/j.jmsy.2020.01.009

Hughes, C., & Spray, R. (2002). Smart communities and smart growth-Maximising benefits for the corporation. *Journal of Corporate Real Estate*, *4*(3), 207–214. doi:10.1108/14630010210811831

Iltisberger, O. (2021). *The role of smart buildings in the development of smart cities*. https://www.smart-energy.com/industry-sectors/iot/the-role-of-smart-buildings-in-the-development-of-smart-cities/

International Data Spaces. (2023, January 9). Gaia-X. Retrieved from https://internationaldataspaces.org/we/gaia-x/

Introduction . (n.d.). Catena-X. Retrieved from https://catena-x.net/en/angebote/edc-die-zentrale-komponente-fuer-die

Iribarren, M., Concha, G., Valdes, G., Solar, M., Villarroel, M. T., Gutiérrez, P., & Vásquez, Á. (2008). Capability maturity framework for eGovernment: A multi-dimensional model and assessing tool. In *Electronic Government: 7th International Conference, EGOV 2008, Turin, Italy, August 31-September 5, 2008 Proceedings*, *7*, 136–147.

Irvine, K. N., Suwanarit, A., Likitswat, F., Srilertchaipanij, H., Ingegno, M., Kaewlai, P., Boonkam, P., Tontisirin, N., Sahavacharin, A., Wongwatcharapaiboon, J., & Janpathompong, S. (2022). Smart City Thailand: Visioning and Design to Enhance Sustainability, Resiliency, and Community Wellbeing. *Urban Science*, *6*(1), 7. doi:10.3390/urbansci6010007

Ishida, T., & Isbister, K. (2000). *Digital cities: technologies, experiences, and future perspectives*. Springer Science & Business Media. doi:10.1007/3-540-46422-0

James, L. (2021). Digital twin technology will revolutionize healthcare: Digital twin technology has the potential to transform healthcare in a variety of ways—improving the diagnosis and treatment of patients, streamlining preventative care, and facilitating new approaches for hospital planning. *Engineering & Technology*, *16*(2), 50–53.

Jarke, M., Otto, B., & Ram, S. (2019, August 19). *Data sovereignty and Data Space Ecosystems - Business & Information Systems Engineering*. SpringerLink. https://link.springer.com/article/10.1007/s12599-019-00614-2

Jehle, P. (2022). Inkonsequenz bremst Betonrecycling [Inconsistency slows down concrete recycling]. *Der Bauingenieur, 97*(4). DOI: doi:10.37544/0005-6650-2022-04

Jimenez, J. I., Jahankhani, H., & Kendzierskyj, S. (2020). *Health care in the cyberspace: medical cyber-physical system and digital twin challenges*. Academic Press.

Kaisara, G., & Pather, S. (2011). The e-Government evaluation challenge: A South African Batho Pele-aligned service quality approach. *Government Information Quarterly*, *28*(2), 211–221. doi:10.1016/j.giq.2010.07.008

Kamel Boulos, M. N., & Zhang, P. (2021). Digital twin technology: From personalized medicine to precision public health. *Journal of Personalized Medicine*, *11*(8), 745. doi:10.3390/jpm11080745 PMID:34442389

Kasim, T. (2013). Automated sustainability compliance checking process: proof of concept. In *Proceedings of the 13th International Conference on Construction Applications of Virtual Reality* (pp. 30-31). School of Engineering, Cardiff University Press.

Kassim, N. M. D., Nathan, S., Yeap, J. A. L., Ramayah, T., & Hashim, N.H. (2018). Insights of smart community in Kemaman. *International Journal of Engineering & Technology, 7*(4.36), 1181-1188. . doi:10.14419/ijet.v7i4.36.25380

Kassim, N. M. D., Yeap, J. A. L., Nathan, S., Hashim, N. H., & Ramayah, T. (2019). A Conceptual Paper of the Smart City and Smart Community. In Eurasian Economic Perspectives. Eurasian Studies in Business and Economics. Springer. doi:10.1007/978-3-030-18565-7_4

Kaur, M. J., Mishra, V. P., & Maheshwari, P. (2020). The convergence of digital twin, IoT, and machine learning: transforming data into action. Digital Twin Technologies and Smart Cities.

Khan, M. A., & Salah, K. (2018). IoT Security: Review, Blockchain Solutions, And Open Challenges. *Future Generation Computer Systems*, *82*, 395–411. doi:10.1016/j.future.2017.11.022

Kharchenko, V., Illiashenko, O., & Morozova, O. (2020). Combination of digital twin and artificial intelligence in manufacturing using industrial IoT. In *2020 IEEE 11th international conference on dependable systems, services and technologies (DESSERT)*. IEEE. 10.1109/DESSERT50317.2020.9125038

Khatib, M. M. E., Al-Nakeeb, A., & Ahmed, G. (2019). Integration of cloud computing with artificial intelligence and its impact on telecom sector—A case study. *iBusiness, 11*(1), 1–10.

Kim, D. Y., & Grant, G. (2010). E-government maturity model using the capability maturity model integration. *Journal of Systems and Information Technology*, *12*(3), 230–244. doi:10.1108/13287261011070858

Kirchschläger, P. G. (2021). Ethische Verantwortung in der digitalen Transformation [Ethical Responsibility in Digital Transformation]. *Journal ZRFC, 6*, 253.

Kirchschläger, P. G. (2021). Ethische Verantwortung in der digitalen Transformation [Ethical Responsibility in the digital transformation]. *Journal of Risk, Fraud and Compliance*.

Klischewski, R., & Askar, E. (2010, October). Success factors of developing G2G services: the case of Egypt. In *Proceedings of the 4th International Conference on Theory and Practice of Electronic Governance* (pp. 152-160). 10.1145/1930321.1930354

Kobyakova, O. S., Starodubov, V. I., Kurakova, N. G., & Tsvetkova, L. A. (2021). Digital twin technology in Healthcare: An Assessment of Technological and Practical Prospects. *Annals of the Russian Academy of Medical Sciences*, *76*(5), 476–487. doi:10.15690/vramn1717

Kumar, V., Mukerji, B., & Butt, I. (2007). Factors for Successful E-Government Adoption: A Conceptual Framework. *Electronic. Journal of E-Government*, *5*(1), 63–76.

Laubenbacher, R., Niarakis, A., Helikar, T., An, G., Shapiro, B., Malik-Sheriff, R. S., Sego, T. J., Knapp, A., Macklin, P., & Glazier, J. A. (2022). Building digital twin technology of the human immune system: Toward a roadmap. *NPJ Digital Medicine*, *5*(1), 1–5. doi:10.103841746-022-00610-z PMID:35013539

Lauer-Schmaltz, M. W., Cash, P., Hansen, J. P., & Maier, A. (2022). Designing human digital twin technology for behavior-changing therapy and rehabilitation: A systematic review. *Proceedings of the Design Society*, *2*, 1303–1312. doi:10.1017/pds.2022.132

Layne, K., & Lee, J. (2001). Developing fully functional E-government: A four stage model. *Government Information Quarterly*, *18*(2), 122–136. doi:10.1016/S0740-624X(01)00066-1

Lee, G., & Kwak, Y. H. (2012). An open government maturity model for social media-based public engagement. *Government Information Quarterly*, *29*(4), 492–503. doi:10.1016/j.giq.2012.06.001

Li, Q., Du, Y., Luo, J., & Fang, Y. (2022, April). Research on Collaborative Development Level of Beijing-Tianjin-Hebei Smart City Based on Coupling Correlation Degree Analysis. In *2022 7th International Conference on Social Sciences and Economic Development (ICSSED 2022)* (pp. 1053-1064). Atlantis Press. 10.2991/aebmr.k.220405.177

Li, D., Jiao, J., Wang, S., & Zhou, G. (2022). *Satellites in Emergency Management: A Bibliometric Analysis*. Research Square., doi:10.21203/rs.3.rs-1533351/v1

Lim, S. B., & Yigitcanlar, T. (2022). Participatory Governance of Smart Cities: Insights from e-Participation of Putrajaya and Petaling Jaya, Malaysia. *Smart Cities, 5*(1), 71–89. . doi:10.3390/smartcities5010005

Li, N., Becerik-Gerber, B., Krishnamachari, B., & Soibelman, L. (2014). A BIM-centered indoor localization algorithm to support building fire emergency response operation. *Journal for Automation in Construction.*, *42*, 78–89. doi:10.1016/j.autcon.2014.02.019

Lindskog, H. (2004). Smart communities initiatives. In *ISOne World Conference, 2004*. The Information Institute. Retrieved from http://urn.kb.se/resolve?urn=urn:nbn:se:liu:diva-22201

Liu, S., Bao, J., & Lu, Y. (2020). Digital twin modeling method based on biomimicry for machining aerospace components. *Journal of Manufacturing Systems.*

Liu, Z., Zhang, A., & Wang, W. (2020). A framework for an indoor safety management system based on digital twin. *Sensors (Basel), 20*(20), 5771. doi:10.339020205771 PMID:33053719

Lo, C. K., Chen, C. H., & Zhong, R. Y. (2021). A review of digital twin in product design and development. *Advanced Engineering Informatics*, *48*, 101297. doi:10.1016/j.aei.2021.101297

Ludwig, K. (2021). *Rechtliche Beschränkungen bei der Digitalisierung der Bauwirtschaft* [Legal limitations within digitization in Construction]. Academic Press.

Lund, S., Manyika, J., Hilton Segel, L., Dua, A., Hancock, B., Rutherford, S., & Macon, B. (2019 July 11). *The future of work in America: People and places, today and tomorrow.* McKinsey & Company. https://www.mckinsey.com/featured-insights/future-of-work/th e-future-of-work-in-america-people-and-places-today-and-tomo rrow

Mahalle, P. N., Anggorojati, B., Prasad, N. R., & Prasad, R. (2013). Identity Authentication And Capability Based Access Control (IACAC) For The Internet Of Things. *Journal of Cyber Security and Mobility, 1*(4), 309–348. doi:10.13052/jcsm2245-1439.142

Main Features Reviewed by the United Nations e-Government Survey. (2008). United Nations.

Makolm, J. (2002). Best practices in e-government. In Electronic Government. Springer. doi:10.1007/978-3-540-46138-8_60

Malaysia Smart City Framework (MSCF). (n.d.). https://www.malaysia.gov.my/portal/content/30947

Malek, J. A. (2009). Informative global community development index of informative smart city. *Proceedings of the 8th WSEAS International Conference on.*

Manufacturing Institute, NAM and Samsung Seek Creators at MFG Day Event in South Carolina . (2019, October 4). National Association of Manufacturers.

Manufacturing Institute. (2021). *The Aging of the Manufacturing Workforce.* Manufacturing Institute. https://www.themanufacturinginstitute.org/research/the-aging-of-the-manufacturing-workforce/

Manyika, J., Lund, S., Chui, M., Bughin, J., Woetzel, J., Batra, P., Ko, R., & Sanghvi, S. (2017 November 28). *Jobs lost, jobs gained: What the future of work will mean for jobs, skills, and wages.* McKinsey & Company. https://www.mckinsey.com/featured-insights/future-of-work/jobs-lost-jobs-gained-what-the-future-of-work-will-mean-for-jobs-skills-and-wages

Martinez, A., & Christnacht, C. (2021). *Women Are Nearly Half of U.S. Workforce but Only 27% of STEM Workers.* United States Census Bureau. https://www.census.gov/library/stories/2021/01/women-making-gains-in-stem-occupations-but-still-underrepresented.html

Max, D. (2021). *Building Information Modeling im Brandschutz. Bauphysik Kalender 2021 – Brandschutz.* Wilhelm Ernst & Sohn Publishing Berlin.

Meizhura, H., Alamanda, D. T., & Adi Prabowo, F. S. (2017). Technological adoption for the online system of public aspirations and complaints service (Lapor) in Bandung city. *AFEBI Management and Business Review*, *2*(1), 62–79. doi:10.47312/ambr.v2i1.57

Mensah, I. K. (2019). *Factors influencing the intention of university students to adopt and use e-government services: An empirical evidence in China.* Retrieved December 8, 2020, Retrieved from https://journals.sagepub.com/doi/pdf/10.1177/2158244019855823

Meraghni, S., Benaggoune, K., Masry, Z. A., Terrissa, L. S., Devalland, C., & Zerhouni, N. (2021). Towards digital twin technology driven breast cancer detection. In *Intelligent Computing* (pp. 87–99). Springer. doi:10.1007/978-3-030-80129-8_7

Metallidou, C. K., Psannis, K. E., & Egyptiadou, E. A. (2020). Energy efficiency in smart buildings: IoT approaches. *IEEE Access: Practical Innovations, Open Solutions*, *8*, 63679–63699. doi:10.1109/ACCESS.2020.2984461

Minarčík, P., Procházka, H., & Gulan, M. (2020). Advanced supervision of smart buildings using a novel open-source control platform. *Sensors (Basel)*, *21*(1), 160. doi:10.339021010160 PMID:33383746

Minoli, D., Sohraby, K., & Occhiogrosso, B. (2017). IoT Considerations, Requirements, and Architectures for Smart Buildings—Energy Optimization and Next-Generation Building Management Systems. *IEEE Internet of Things Journal*, *4*(1), 269–283. doi:10.1109/JIOT.2017.2647881

Min, Q., Ji, S., & Qu, G. (2008). Mobile commerce user acceptance study in China: A revised UTAUT model. *Tsinghua Science and Technology*, *13*(3), 257–264. doi:10.1016/S1007-0214(08)70042-7

Mission statement . (2021, July 6). DIO. Retrieved from https://www.dataintelligence.at/en/mission-statement/

Mobility Data Space - ITS World Congress . (2021, October 13). CARUSO. Retrieved from https://www.caruso-dataplace.com/mobility-data-space-its-world-congress/

Mondoro, A., & Grisso, B. (2019). On the integration of SHM and digital twin for the fatigue assessment of naval surface ships. Structural Health Monitoring 2019. doi:10.12783hm2019/32203

Moon, M. J. (2002). The evolution of e-government among municipalities: Rhetoric or reality? *Public Administration Review*, *62*(4), 424–433. doi:10.1111/0033-3352.00196

Müller, R. (2019). *Branchenbarometer Brandschutz* [Industry Barometer Fire Protection]. https://www.baufachmedien.de/branchenbarometer-brandschutz.html

Nam, T., & Pardo, T. A. (2011). Conceptualizing smart city with dimensions of technology, people, and institutions. *Proceedings of the 12th annual international digital government research conference: digital government innovation in challenging times.* 10.1145/2037556.2037602

NASA. (n.d.). *Patent Portfolio.* https://technology.nasa.gov/patents

National Association of Manufacturers. (2021). *Manufacturing Institute, NAM and Samsung Seek Creators at MFG Day Event in South Carolina.* https://www.nam.org/manufacturing-institute-nam-and-samsung-seek-creators-at-mfg-day-event-in-south-carolina-6047/

Nemitz, P. (2021). *Executive Interview.* Aitrends. https://www.aitrends.com/executive-interview/executive-interview-paul-nemitz-principal-adviser-on-justice-policy-for-the-european-commission-brussels/

Nor Sallehi, K. (2017). *Smart City Initiative.* Retrieved November 11, 2021, from https://www.mcmc.gov.my/skmmgovmy/media/General/pdf/Sesi-5-Rangkakerja-Berteraskan-Aspek-Informasi-(PLANMalaysia).pdf

North, D. C. (1991). Institutions. *The Journal of Economic Perspectives*, *5*(1), 97–112. doi:10.1257/jep.5.1.97

O'Dwyer, E., Pan, I., Charlesworth, R., Butler, S., & Shah, N. (2020). Integration of an energy management tool and digital twin for coordination and control of multi-vector smart energy systems. *Sustainable Cities and Society*, *62*, 102412. doi:10.1016/j.scs.2020.102412

Oakland, J. S. (1991). The European centre for total quality management. *Total Quality Management*, *2*(3), 305–308. doi:10.1080/09544129100000035

Okrepilov, V., Kuzmina, S., & Kuznetsov, S. (2019, March). Tools Of Quality Economics: Sustainable Development Of A 'Smart City' Under Conditions Of Digital Transformation Of The Economy. In IOP Conference Series: Materials Science and Engineering (vol. 497, no. 1, pp. 012134). IOP Publishing. doi:10.1088/1757-899X/497/1/012134

Oldenburg, R. (1999). *The Great Good Place: Cafes, Coffee Shops, Bookstores, Bars, Hair Salons, And Other Hangouts At The Heart Of A Community*. Da Capo Press.

Page Speed Insights. (2015). Google. Retrieved from http://developers.google.com/speed/pagespeed/insights/

Papadomichelaki, X., & Mentzas, G. (2012). e-GovQual: A multiple-item scale for assessing e-government service quality. *Government Information Quarterly*, *29*(1), 98–109. doi:10.1016/j.giq.2011.08.011

Paskaleva, K. A. (2009). Enabling the smart city: The progress of city e-governance in Europe. *International Journal of Innovative Research and Development*, *1*(4), 405–422. doi:10.1504/IJIRD.2009.022730

Pearson. (2020, August). *The Global Learner Survey*. https://news.prudential.com/presskits/pulse-american-worker-survey-post-pandemic-work-life.htm

Pearson. (2021). *The Global Learner Survey*. https://www.pearson.com/content/dam/one-dot-com/one-dot-com/global/Files/news/gls/Pearson_Global_Learner_Survey_2019.pdf

Pedersen, O. L. (2021). *A proof-of-concept Digital Twin implementation for monitoring patients through the Clinical Pathway for Prostate Cancer in the Norwegian Health and Care Service*. NTNU.

People's Republic of China. (2021). *Personal Information Protection PIP Law*. https://usercentrics.com/knowledge-hub/china-personal-information-protection-law/

Pereira, A. C., & Romero, F. (2017). A Review Of The Meanings And The Implications Of The Industry 4.0 Concept. *Procedia Manufacturing*, *13*, 1206–1214. doi:10.1016/j.promfg.2017.09.032

Peters, R. M., Janssen, M., & Van Engers, T. M. (2004, March). Measuring e-government impact: existing practices and shortcomings. In *Proceedings of the 6th international conference on Electronic commerce* (pp. 480-489). 10.1145/1052220.1052281

Pilgrim, H. (2017). *The Dark Side of Digitalization: Will Industry 4.0 Create New Raw Materials Demands? PowerShift e.* V. Publishing.

Pinto, F., da Silva, C. F., & Moro, S. (2022). People-Centered Distributed Ledger Technology-IoT Architectures: A Systematic Literature Review. *Telematics and Informatics*, *70*, 101812. doi:10.1016/j.tele.2022.101812

Poncela, J., Vlacheas, P., Giaffreda, R., De, S., Vecchio, M., Nechifor, S., Barco, R., Aguayo-Torres, M. C., Stavroulaki, V., Moessner, K., & Demestichas, P. (2014). Smart Cities Via Data Aggregation. *Wireless Personal Communications*, *76*(2), 149–168. doi:10.100711277-014-1683-5

Ponnewitz, J. (2019). Die BIM-basierte Baugenehmigungsprüfung-eine Grundlagenbetrachtung [BIM-based Building Permit Check – A Basic Consideration]. *Conference Proceedings of the 30th BBB-Assistance Meeting*.

Popa, E. O., van Hilten, M., Oosterkamp, E., & Bogaardt, M.-J. (2021). The use of digital twin technology in healthcare: Socio-ethical benefits and socio-ethical risks. *Life Sciences, Society and Policy*, *17*(1), 1–25. doi:10.118640504-021-00113-x PMID:33397487

Posch, K. C., Posch, R., Tauber, A., Zefferer, T., & Zwattendorfer, B. (2011). *Secure and privacy-preserving eGovernment—best practice Austria*. Springer Berlin Heidelberg. doi:10.1007/978-3-642-19391-0_19

PR Newswire. (2019 July 16). *LEGO Group Kicks Off Global Program To Inspire The Next Generation Of Space Explorers As NASA Celebrates 50 Years Of Moon Landing*. Author.

Prinz, W., Rose, T., & Urbach, N. (1970, January 1). *Blockchain technology and International Data Spaces*. SpringerLink. https://link.springer.com/chapter/10.1007/978-3-030-93975-5_10

Prudential. (n.d.). *Pulse of the American Worker Survey: Post-Pandemic Work & Life*. https://news.prudential.com/presskits/pulse-american-worker-survey-post-pandemic-work-life.htm

Putrajaya Blueprint. (2019). Retrieved from https://smart.putrajaya.my/blueprint

Putz, B., Dietz, M., Empl, P., & Pernul, G. (2021). Ethertwin: Blockchain-based secure digital twin information management. *Information Processing & Management*, *58*(1), 102425. doi:10.1016/j.ipm.2020.102425

Qi, J., & Ji, J. (2021). Beijing as Smart City: Promotion of International Communication Center. *Philosophical Studies*, *11*(2), 143–148.

Rabah, K. (2018). Convergence of AI, IoT, Big Data And Blockchain: A Review. *Law Institute Journal*, *1*(1), 1–18.

Ramu, S. P., Boopalan, P., Pham, Q.-V., Maddikunta, P. K. R., Huynh-The, T., Alazab, M., Nguyen, T. T., & Gadekallu, T. R. (2022). Federated learning enabled digital twin technology for smart cities: Concepts, recent advances, and future directions. *Sustainable Cities and Society*, *79*, 103663. doi:10.1016/j.scs.2021.103663

Rana, N. P., Luthra, S., Mangla, S. K., Islam, R., Roderick, S., & Dwivedi, Y. K. (2018). Barriers to the development of smart cities in indian context. *Information Systems Frontiers*, 1–23. doi:10.100710796-018-9873-4

Rao, S. K., & Prasad, R. (2018). Impact Of 5G Technologies On Smart City Implementation. *Wireless Personal Communications*, *100*(1), 161–176. doi:10.100711277-018-5618-4

Rathje, J. M., & Katila, R. (2020). Enabling Technologies and the Role of Private Firms: A Machine Learning Matching Analysis. *Strategy Science*.

Rathore, M. M., Ahmad, A., Paul, A., & Rho, S. (2016). Urban Planning And Building Smart Cities Based On The Internet Of Things Using Big Data Analytics. *Computer Networks*, *101*, 63–80. doi:10.1016/j.comnet.2015.12.023

Razmjoo, A. (2021). Investigating Smart City Development Based on Green Buildings, Electrical Vehicles and Feasible Indicators. *Sustainability Journal 2021, 1*(14), 7808. doi:10.3390/su13147808

Ricci, A., Croatti, A., Mariani, S., Montagna, S., & Picone, M. (2022). Web of digital twin technology. *ACM Transactions on Internet Technology*.

Ricci, A., Croatti, A., & Montagna, S. (2021). Pervasive and Connected Digital twin technology—A Vision for Digital Health. *IEEE Internet Computing*.

Robinson, D. K. R., & Mazzucato, M. (2018). The evolution of mission-oriented policies: Exploring changing market creating policies in the US and European space sector. *Research Policy*, 936–948.

Rodríguez, R. A., Welicki, L., Giulianelli, D. A., & Vera, P. M. (2008, December). Measurement framework for evaluating e-governance on municipalities websites. In *Proceedings of the 2nd international conference on Theory and practice of electronic governance* (pp. 381-387). 10.1145/1509096.1509176

Rohleder, S. J. (2004). *The Government Executive Series. eGovernment Leadership: High Performance*. Maximum Value.

Russo, A. (2020 October 20). *Recession and Automation Changes Our Future of Work, But There are Jobs Coming, Report Says*. World Economic Forum. Retrieved fromhttps://www.weforum.org/press/2020/10/recession-and-automation-changes-our-future-of-work-but-there-are-jobs-coming-report-says-52c5162fce/

Sarker, M. N. I., Wu, M., & Hossin, M. A. (2018, May). Smart Governance Through Bigdata: Digital Transformation Of Public Agencies. In 2018 international conference on artificial intelligence and big data (ICAIBD) (pp. 62-70). IEEE. doi:10.1109/ICAIBD.2018.8396168

Schulthess, P., Neuenschwander, M., Mosalam, K. M., & Knobloch, M. (2020). A computationally rigorous approach to hybrid fire testing. *Computers & Structures, 238*, 106301. doi:10.1016/j.compstruc.2020.106301

Seifert, J. W., & McLoughlin, G. J. (2007, July). *State e-government strategies: Identifying best practices and applications*. Library of Congress.

Sengboon, L., Jalaluddin, A. M., Mohd Yusof, H., & Zurinah, T. (2019). The behaviours and job positions of citizens in smart cities' development. *Journal of the Malaysian Institute of Planners, 17*(2), 133–145.

Shahkooh, K. A., Saghafi, F., & Abdollahi, A. (2008, April). A proposed model for e-Government maturity. In *2008 3rd International conference on information and communication technologies: From theory to applications* (pp. 1-5). IEEE. 10.1109/ICTTA.2008.4529948

Shangguan, D., Chen, L., Ding, J. (2020). A Digital Twin-Based Approach for the Fault Diagnosis and Health Monitoring of a Complex Satellite System. *Symmetry in Mechanical Engineering, 2.*

Sharma, R., Fantin, A. R., Prabhu, N., Guan, C., & Dattakumar, A. (2016). Digital Literacy And Knowledge Societies: A Grounded Theory Investigation Of Sustainable Development. *Telecommunications Policy, 40*(7), 628–643. doi:10.1016/j.telpol.2016.05.003

Sheikhnejad, Y., Gonçalves, D., Oliveira, M., & Martins, N. (2020). Can buildings be more intelligent than users?-The role of intelligent supervision concept integrated into building predictive control. *Energy Reports, 6*, 409–416. doi:10.1016/j.egyr.2019.08.081

Shi, Y., Gao, Y., Luo, Y., & Hu, J. (2022). Fusions of industrialization and digitalisation (FID) in the digital economy: Industrial system digitalization, digital technology industrialization, and beyond. *Journal of Digital Economy.*

Siau, K., & Long, Y. (2005). Synthesizing e-government stage models–a meta-synthesis based on meta-ethnography approach. *Industrial Management & Data Systems, 105*(4), 443–458. doi:10.1108/02635570510592352

Simple answers about life in Switzerland. (2014). Retrieved from https://www.ch.ch/fr/

Smart Cities Market Size, Share & Growth Report, 2030. (n.d.). Retrieved January 10, 2023, from https://grandviewresearch.com/industry-analysis/smart-cities-market#:~:text=The%20global%20smart%20cities%20market

SmartSat Cooperative Research Centre. (2021). *Space industry skills gap analysis*. Author.

Smit, J., Kreutzer, S., Moeller, C., & Carlberg, M. (2016). Policy Department A: Economic and Scientific Policy–Industry 4.0. European Parliament.

Sørum, H. (2011). An empirical investigation of user involvement, website quality and perceived user satisfaction in eGovernment environments. *Electronic Government and the Information Systems Perspective: Second International Conference, EGOVIS 2011, Toulouse, France, August 29–September 2, 2011 Proceedings*, 2, 122–134.

Sovacool, B. K., Del Rio, D. F., & Griffiths, S. (2021). Policy mixes for more sustainable smart home technologies. *Environmental Research Letters*, *16*(5), 054073. doi:10.1088/1748-9326/abe90a

Sowe, S. K., Kimata, T., Dong, M., & Zettsu, K. (2014, July). Managing Heterogeneous Sensor Data On A Big Data Platform: IoT Services For Data-Intensive Science. In *2014 IEEE 38th international computer software and applications conference workshops* (pp. 295-300). IEEE. 10.1109/COMPSACW.2014.52

Space Foundation Editorial Team. (2022, July 27). *Space Foundation Releases The Space Report 2022 Q2 Showing Growth Of Global Space Economy*. Space Foundation. Retrieved from https://www.spacefoundation.org/2022/07/27/the-space-report-2022-q2/

Standard, I. (2002). *Ergonomics of human-system interaction–usability methods supporting human-centred design. ISO standard tr 16982: 2002*. International Organization for Standardization.

Stanford University. (2016). *Urban Information Modelling*. https://cife.stanford.edu/Seed2016_UrbanInfoModeling

Stratigea, A. (2012). The concept of 'smart cities'. Towards community development? Netcom. *Réseaux, communication et territoires, 26*(3/4), 375-388.

Stratigea, A. (2012). The concept of 'smart cities'. Towards community development? Netcom. *Réseaux, Communication et Territories*, *26*(3/4), 375-388.

Strohbach, M., Ziekow, H., Gazis, V., & Akiva, N. (2015). Towards A Big Data Analytics Framework For IoT And Smart City Applications. In *Modeling And Processing For Next-Generation Big-Data Technologies* (pp. 257–282). Springer. doi:10.1007/978-3-319-09177-8_11

Suhail, S., Hussain, R., Jurdak, R., Oracevic, A., Salah, K., Hong, C. S., & Matulevičius, R. (2022). Blockchain-based digital twins: Research trends, issues, and future challenges. [CSUR]. *ACM Computing Surveys*, *54*(11s), 1–34. doi:10.1145/3517189

Taiwo, A. A., & Downe, A. G. (2013). The theory of user acceptance and use of technology (UTAUT): A meta-analytic review of empirical findings. *Journal of Theoretical and Applied Information Technology*, *49*(1).

Taj, I., & Zaman, N. (2022). Towards Industrial Revolution 5.0 and Explainable Artificial Intelligence: Challenges and Opportunities. *International Journal of Computing and Digital Systems*, *12*(1), 295–320. doi:10.12785/ijcds/120128

Talkhestani, B. A., Jazdi, N., & Schlögl, W. (2018). Consistency check to synchronize the digital twin of manufacturing automation based on anchor points. *Procedia CIRP*, *72*, 159–164. doi:10.1016/j.procir.2018.03.166

Tanabe, M., van den Besselaar, P., & Ishida, T. (2003). Digital Cities II. Computational and Sociological Approaches. In *Second Kyoto Workshop on Digital Cities, Kyoto, Japan, October 18-20, 2001. Revised Papers* (*Vol. 2362*). Springer.

Tao, F., Cheng, J., Qi, Q., Zhang, M., Zhang, H., & Sui, F. (2018). Digital twin-driven product design, manufacturing and service with big data. *International Journal of Advanced Manufacturing Technology*, *94*(9-12), 3563–3576. doi:10.100700170-017-0233-1

TeamEbony. (2019 September 12). *Mae Jemison Fights For Diversity In Space And In The Classroom*. Ebony. https://www.ebony.com/mae-jemison-fights-for-diversity-in-space-377/

The LEGO Group. (n.d.). https://www.prnewswire.com/news-releases/lego-group-kicks-off-global-program-to-inspire-the-next-generation-of-space-explorers-as-nasa-celebrates-50-years-of-moon-landing-300885423.html

Theodoridis, E., Mylonas, G., & Chatzigiannakis, I. (2013, July). Developing An IoT Smart City Framework. In *IISA 2013* (pp. 1–6). IEEE. doi:10.1109/IISA.2013.6623710

Tinholt, D., & van der Linden, N. (2013). *Public services online 'Digital by default or by detour?'–Assessing user centric eGovernment performance in Europe–eGovernment benchmark 2012*. Academic Press.

Tomaras, D. (2018, March). Crowdsourcing Techniques For Smart Urban Mobility. In *2018 IEEE International Conference on Pervasive Computing and Communications Workshops (PerCom Workshops)* (pp. 460-461). IEEE. 10.1109/PERCOMW.2018.8480244

Tomičić Pupek, K., Pihir, I., & Tomičić Furjan, M. (2019). Smart City Initiatives In The Context Of Digital Transformation–Scope, Services And Technologies. *Management*, *24*(1), 39–54. doi:10.30924/mjcmi.24.1.3

Torky, M., Darwish, A., & Hassanien, A. E. (2022). Integrated Digital twin technology and Blockchain Framework for Privacy and Security of Applications. In Digital twin technology for Digital Transformation: Innovation in Industry (pp. 99–112). Springer.

UN E-Government Survey 2012: E-Government for the People. (2012). United Nations.

UN, DESA. (2012, June). *Compendium of Innovative E-government Practices*. United Nations.

United Nations. (2001). *Epilogue The City Development Index (CDI)*. Retrieved from https://www.un.org/ga/Istanbul+5/116.pdf

United Nations. (2016). *New Urban Agenda. United Nations Conference on Housing and Sustainable Urban Development (Habitat III)*. https://www.urbanagendaplatform.org/about

Unnikrishnan, S., & Blair, C. (2019). *Want to Boost the Global Economy by $5 Trillion? Support Women as Entrepreneurs*. Boston Consulting Group. https://www.bcg.com/publications/2019/boost-global-economy-5-trillion-dollar-support-women-entrepreneurs

Vaidya, S., Ambad, P., & Bhosle, S. (2018). Industry 4.0–A Glimpse. *Procedia Manufacturing*, *20*, 233–238. doi:10.1016/j.promfg.2018.02.034

van den Besselaar, P., & Koizumi, S. (2005). Digital Cities III. Information Technologies for Social Capital: Cross-cultural Perspectives. In *Third International Digital Cities Workshop, Amsterdam, The Netherlands, September 18-19, 2003, Revised Selected Papers* (*Vol. 3081*). Springer.

Van der Auweraer, H. (2021). *Digital Twins – Simulations by Siemens*. Siemens Publishing. https://new.siemens.com/de/de/unternehmen/stories/forschung-technologien/digitaler-zwilling/digital-twin.html

Venkatesh, V., Morris, M. G., Davis, G. B., & Davis, F. D. (2003). User acceptance of information technology: Toward a unified view. *Management Information Systems Quarterly*, *27*(3), 425–478. doi:10.2307/30036540

Venkatesh, V., Thong, J. Y. L., Chan, F. K. Y., Hu, P. J. H., & Brown, S. A. (2011). Extending the two-stage information systems continuance model: Incorporating UTAUT predictors and the role of context. *Information Systems Journal*, *21*(6), 527–555. doi:10.1111/j.1365-2575.2011.00373.x

Verma, R., Patel, S., Minj, S. V., & Bhat, A. (2022). Healthcare 5.0: A Study on Improving Personalized Care. *2022 6th International Conference on Intelligent Computing and Control Systems (ICICCS)*, 1815–1818.

Volkov, I., Radchenko, G., & Tchernykh, A. (2021). Digital twin technology, Internet of Things and Mobile Medicine: A Review of Current Platforms to Support Smart Healthcare. *Programming and Computer Software*, *47*(8), 578–590. doi:10.1134/S0361768821080284

Walsh, B., Pandey, E., & Muller, J. (2021 June 21). *Axios What's Next? The age of the career-jumpers*. https://www.axios.com/newsletters/axios-whats-next-24766a4c-92cf-43cf-8ae9-540980b5f3f4.html?chunk=1&utm_term=emshare#st
ory1

Wang, G., El Saddik, A., Lai, X., Perez, G. M., & Choo, K. K. R. (Eds.). (2019). Smart City and Informatization. In *Proceedings of the 7th International Conference, iSCI 2019, Guangzhou, China, November 12–15, 2019* (*Vol. 1122*). Springer Nature.

Wang, L. J., Zhang, Y., Li, Q., & Ruan, P. N. (2016). Under The Background Of Smart City Development Of E-Government Performance Evaluation: Study In Every Districts Of Beijing. *Proceedings of the 6th International Conference on Information Technology for Manufacturing Systems.*

WCAG 2 Overview. (2012). W3C. Retrieved from https://www.w3.org/WAI/intro/wcag

Weber-Lewerenz, B. (2021a). CDR and Ethical Aspects in AI in Construction. The Disruptive Role of Data, AI and Ecosystems in Services. *Proceedings of the 31st RESER Conference*. https://publica.fraunhofer.de/dokumente/N-642928.html

Weber-Lewerenz, B. (2021b). Gender Innovation in Construction in Digital Era. The Disruptive Role of Data, AI and Ecosystems in Services. *Proceedings of the 31st RESER Conference*. https://publica.fraunhofer.de/dokumente/N-642928.html

Weber-Lewerenz, B. (2021c). Corporate digital responsibility (CDR) in construction engineering - Ethical guidelines for the application of digital transformation and artificial intelligence (AI) in user practice. *Springer Nature SN Applied Sciences, 3*(10). DOI: doi:10.1007/s42452-021-04776-1

Weber-Lewerenz, B. (2021c). Ethical Aspects in AI in Construction. In *Conference Presentation and Proceedings of Reser Conference*. Fraunhofer Institute Publishing. https://publica.fraunhofer.de/dokumente/N-642928.html

Weber-Lewerenz, B. (2021d). Die unternehmerisch verantwortungsvolle Digitalisierung im Bauwesen. *Journal Der Bauingenieur, 96*(1-2), 19-25. DOI: doi:10.37544/0005-6650-2021-01-02-45

Weber-Lewerenz, B. (2021e). Unternehmerische Compliance im digitalen Transformationsprozess Teil 1 [Corporate Compliance in Digital Transformation Process - Part 1]. *Journal of Risk, Fraud and Compliance*. DOI: doi:10.37307/j.1867-8394.2021.06

Weber-Lewerenz, B. (2021f). *Ethik in der KI im Bauwesen in Deutschland und KI Methoden und Erfahrungswerte* [Study by Fraunhofer Institute Stuttgart IAO on AI in Construction in Germany]. Fraunhofer Institute Publishing. https://publica.fraunhofer.de/eprints/urn_nbn_de_0011-n-6306697.pdf

Weber-Lewerenz, B. (2021g). *KI in Lehre und Ausbildung, BIM und KI in Wissenschaft und Unternehmenspraxis.* buildingSMART Deutschland bSD Publishing

Weber-Lewerenz, B. (2021e). Corporate Digital Responsibility in Construction Engineering. *International Journal of Responsible Leadership and Ethical Decision-Making*, *2*(1). Advance online publication. doi:10.4018/ijrledm.2020010103

Weber-Lewerenz, B. (2022a). *Accents of Added Value in Construction 4.0 – Ethical Observations in Digitization and AI.* Springer Nature Publishing.

Weber-Lewerenz, B. (2022b). Technological Dream or Safety Traumata? Fire Protection in Smart Cities-Digitization and AI ensure burning ideas and a New Culture of Thinking in Construction 4.0. In I. Vasiliu-Feltes (Ed.), *Impact of Digital Twins in Smart Cities Development*. IGI Global Publishing. doi:10.4018/978-1-6684-3833-6

Weber-Lewerenz, B. (2022c). Impact of education and training for diversity, inclusion and sustainable digital innovations. In I. Vasiliu-Feltes (Ed.), *Impact of Women Empowerment on SDGs in the Digital Era.* IGI Global Publishing. doi:10.4018/978-1-6684-3637-0

Webmasters. (n.d.). Google. Retrieved from https://www.google.com/webmasters/

Weinzierl, M. (2018). Space, the Final Economic Frontier. *The Journal of Economic Perspectives*, *32*(2), 173–192. doi:10.1257/jep.32.2.173

West, D. (2007). Global perspectives on e-government. *Governance and information technology: From electronic government to information government*, 17-32.

West, D. M., Deitch, A., Vigón, M. A. P., & Sobisch, F. (2004). *Development of e-government strategies in Chile Canada and Brazil*. Academic Press.

West, D. M. (2004). E-government and the transformation of service delivery and citizen attitudes. *Public Administration Review*, *64*(1), 15–27. doi:10.1111/j.1540-6210.2004.00343.x

What is a smart city? Definition and Examples. (n.d.). TWI-Global. Retrieved from https://www.twi-global.com/technical-knowledge/faqs/what-is-a-smart-city

What is Gaia-X ? (n.d.). Gaia-X. Retrieved from https://www.data-infrastructure.eu/GAIAX/Navigation/EN/Home/home.html

What is Plain Language? (2014). Plain Language Action and Information Network (PLAIN). Retrieved from https://www.plainlanguage.gov/about/definitions/

Wickramasinghe, N., Jayaraman, P. P., Forkan, A. R. M., Ulapane, N., Kaul, R., Vaughan, S., & Zelcer, J. (2021). A Vision for Leveraging the Concept of Digital twin technology to Support the Provision of Personalized Cancer Care. *IEEE Internet Computing*, *26*(5), 17–24. doi:10.1109/MIC.2021.3065381

Wiggers, K. (2022). *PassiveLogic, which creates digital twins of buildings, lands $34M*. https://venturebeat.com/2022/01/11/passivelogic-which-creates-digital-twins-of-buildings-lands-34m/

Wilkinson, M. D., Dumontier, M., Aalbersberg, I. J., Appleton, G., Axton, M., Baak, A., Blomberg, N., Boiten, J.-W., da Silva Santos, L. B., Bourne, P. E., Bouwman, J., Brookes, A. J., Clark, T., Crosas, M., Dillo, I., Dumon, O., Edmunds, S., Evelo, C. T., Finkers, R., ... Mons, B. (2016). The FAIR Guiding Principles for scientific data management and stewardship. *Scientific Data*, *3*(1), 1–9. doi:10.1038data.2016.18 PMID:26978244

Windley, P. J. (2002). *eGovernment maturity*. USA Wind. Technolometria.

Winkowska, J. (2019). Smart city concept in the light of the literature review. *Journal of Engineering Management in Production and Services.*, *11*(2), 70–86.

Wongbumru, T., & Dewancker, B. (2014). *Smart Communities for Future Development: Lessons from Japan*. Academic Press.

World Economic Forum. (2021). *Governing smart cities: policy benchmarks for ethical and responsible smart city development* [Paper presentation]. WEF White Paper. https://www.weforum.org/whitepapers/governing-smart-cities-policy-benchmarks-for-ethical-and-responsible-smart-city-development

World Economic Forum. (2021). *Governing Smart Cities: Policy Benchmarks for Ethical and Responsible Smart City Development*. White Paper. https://www3.weforum.org/docs/WEF_Governing_Smart_Cities_2021.pdf

Wrick, T. (2020). *Wenn die Digitalisierung den Brandschutz fordert und fördert – 5 Beispiele* [When digitization demands and promotes fire protection - 5 examples]. https://industrie-wegweiser.de/wenn-die-digitalisierung-den-brandschutz-fordert-und-foerdert-5-beispiele/

Wu, C., Lorenzo, G., Hormuth, D. A. II, Lima, E. A., Slavkova, K. P., DiCarlo, J. C., Virostko, J., Phillips, C. M., Patt, D., Chung, C., & Yankeelov, T. E. (2022). Integrating mechanism-based modeling with biomedical imaging to build practical digital twin technology for clinical oncology. *Biophysical Reviews*, *3*(2), 021304. doi:10.1063/5.0086789 PMID:35602761

Xing, X., Del Ser, J., Wu, Y., Li, Y., Xia, J., Xu, L., . . . Yang, G. (2022). *HDL: Hybrid Deep Learning for the Synthesis of Myocardial Velocity Maps in Digital twin technology for Cardiac Analysis.* ArXiv Preprint ArXiv:2203.05564.

X-Road. (2014). E-Estonia. Retrieved from https://e-estonia.com/components/x-road

Xu, L., de Vrieze, P., Lu, X., & Wang, W. (2022). Digital twin technology Approach for Sustainable Industry. *International Conference on Advanced Information Systems Engineering*, 126–134.

Yahoo Yslow for Chrome. (2015). Yahoo. Retrieved from https://developer.yahoo.com/yslow/

Yamashita, J. (2022). Smart City Initiatives in Japan: Achievements and Remaining Issues. In R. K. Mishra, C. L. Kumari, S. Chachra, P. S. J. Krishna, A. Dubey, & R. B. Singh (Eds.), *Smart Cities for Sustainable Development. Advances in Geographical and Environmental Sciences*. Springer. doi:10.1007/978-981-16-7410-5_6

Yan, J., Zlatanova, S., & Aleksandrov, M. (2019). Integration of 3D objects and terrain for 3D modelling supporting the digital twin. *14th 3D GeoInfo Conference.*

Yang, D., Karimi, H. R., Kaynak, O., & Yin, S. (2021). Developments of digital twin technologies in industrial, smart city and healthcare sectors: A survey. *Complex Engineering Systems*, *1*(1), 3. doi:10.20517/ces.2021.06

Yaqoob, I., Salah, K., Uddin, M., Jayaraman, R., Omar, M., & Imran, M. (2020). Blockchain for digital twins: Recent advances and future research challenges. *IEEE Network*, *34*(5), 290–298. doi:10.1109/MNET.001.1900661

Ye, Z., Yang, J., Zhong, N., Tu, X., Jia, J., & Wang, J. (2020). Tackling environmental challenges in pollution controls using artificial intelligence: A review. *The Science of the Total Environment*, *699*, 134279. doi:10.1016/j.scitotenv.2019.134279 PMID:33736193

Yigitcanlar, T., Kamruzzaman, M., Buys, L., Ioppolo, G., Marques, J., Costa, E., & Yun, J. (2018). *Understanding 'smart cities': Intertwining development drivers with desired outcomes in a multidimensional framework.* Cities. doi:10.1016/j.cities.2018.04.003

Yu, P., Yu, M., & Sampat, M. (2022n). Smart Management for Digital Transformation in China. In B. Barbosa, S. Filipe, & C. Santos (Ed.), Handbook of Research on Smart Management for Digital Transformation (pp. 411-438). IGI Global. doi:10.4018/978-1-7998-9008-9.ch019

Yuefei, C., Sisi, W., Mingqi, T., & Chuan, C. (2021). Application of Digital twin technology in Medical and Health Fields and Related Research Progress. 计量科学与技术, *65*(10), 6–9.

Yu, P., Chen, D., & Ahuja, A. (2022a). Smart and Sustainable Economy: How COVID-19 Has Acted as a Catalyst for China's Digital Transformation. In S. Kautish & G. Kaur (Eds.), *AI-Enabled Agile Internet of Things for Sustainable FinTech Ecosystems* (pp. 106–146). IGI Global. doi:10.4018/978-1-6684-4176-3.ch006

Yu, P., Chen, J., Sampat, M., & Misuko, N. (2022b). The Digital Transformation of Rural Agricultural Business Management: A Case Study of China. In S. Bilgaiyan, J. Singh, & H. Das (Eds.), *Empirical Research for Futuristic E-Commerce Systems: Foundations and Applications* (pp. 23–52). IGI Global. doi:10.4018/978-1-6684-4969-1.ch002

Yu, P., Gong, R., & Sampat, M. (2022c). Blockchain Technology in China's Digital Economy: Balancing Regulation and Innovation. In P. Tehrani (Ed.), *Regulatory Aspects of Artificial Intelligence on Blockchain* (pp. 132–157). IGI Global. doi:10.4018/978-1-7998-7927-5.ch007

Yu, P., Gu, H., Zhao, Y., & Ahuja, A. (2022d). Digital Transformation Driven by Internet Data Center: Case Studies on China. In D. Piaggesi, H. Landazuri, & B. Jia (Eds.), *Cases on Applying Knowledge Economy Principles for Economic Growth in Developing Nations* (pp. 203–230). IGI Global. doi:10.4018/978-1-7998-8417-0.ch011

Yu, P., He, K., & Sampat, M. (2022e). COVID-19's Impact on the Chinese Tourism and Hospitality Sector. In V. Costa & C. Costa (Eds.), *Sustainability and Competitiveness in the Hospitality Industry* (pp. 23–66). IGI Global. doi:10.4018/978-1-7998-9285-4.ch002

Yu, P., Jiao, A., & Sampat, M. (2022f). The Effect of Chinese Green Transformation on Competitiveness and the Environment. In P. Ordóñez de Pablos, X. Zhang, & M. Almunawar (Eds.), *Handbook of Research on Green, Circular, and Digital Economies as Tools for Recovery and Sustainability* (pp. 257–279). IGI Global. doi:10.4018/978-1-7998-9664-7.ch014

Yu, P., Liu, Z., & Hanes, E. (2022g). Supply Chain Resiliency, Efficiency, and Visibility in the Post-Pandemic Era in China: Case Studies of MeiTuan Waimai, and Ele.me. In Y. Ramakrishna (Ed.), *Handbook of Research on Supply Chain Resiliency, Efficiency, and Visibility in the Post-Pandemic Era* (pp. 195–225). IGI Global. doi:10.4018/978-1-7998-9506-0.ch011

Yu, P., Liu, Z., Hanes, E., & Mumtaz, J. (2022h). Integration of IoT and Blockchain for Smart and Secured Supply Chain Management: Case Studies of China. In S. Goyal, N. Pradeep, P. Shukla, M. Ghonge, & R. Ravi (Eds.), *Utilizing Blockchain Technologies in Manufacturing and Logistics Management* (pp. 179–207). IGI Global. doi:10.4018/978-1-7998-8697-6.ch010

Yu, P., Lu, S., Hanes, E., & Chen, Y. (2022i). The Role of Blockchain Technology in Harnessing the Sustainability of Chinese Digital Finance. In P. Swarnalatha & S. Prabu (Eds.), *Blockchain Technologies for Sustainable Development in Smart Cities* (pp. 155–186). IGI Global. doi:10.4018/978-1-7998-9274-8.ch009

Yu, P., Lu, S., Sampat, M., Li, R., & Ahuja, A. (2022j). How AI-Enabled Agile Internet of Things Can Enhance the Business Efficiency of China's FinTech Ecosystem. In S. Kautish & G. Kaur (Eds.), *AI-Enabled Agile Internet of Things for Sustainable FinTech Ecosystems* (pp. 190–223). IGI Global. doi:10.4018/978-1-6684-4176-3.ch009

Yu, P., Tang, H., Tang, H., & Hanes, E. (2022k). Using "Digitalization + Intellectual Property" Management to Realize Cultural Economy: Case Study of Chinese Museums. In R. Pettinger, B. Gupta, A. Roja, & D. Cozmiuc (Eds.), *Handbook of Research on Digital Transformation Management and Tools* (pp. 236–264). IGI Global. doi:10.4018/978-1-7998-9764-4.ch011

Yu, P., Weng, Y., & Ahuja, A. (2022l). Carbon Financing and the Sustainable Development Mechanism: The Case of China. In A. Rafay (Ed.), *Handbook of Research on Energy and Environmental Finance 4.0* (pp. 301–332). IGI Global. doi:10.4018/978-1-7998-8210-7.ch012

Yu, P., Xue, W., & Mahendran, R. (2022m). The Development and Impact of China's Digital Transformation in the Medical Industry. In M. Rodrigues & J. Proença (Eds.), *Impact of Digital Transformation on the Development of New Business Models and Consumer Experience* (pp. 97–128). IGI Global. doi:10.4018/978-1-7998-9179-6.ch006

Yu, Z., Wang, K., Wan, Z., Xie, S., & Lv, Z. (2022). *FMCPNN in Digital twin technology Smart Healthcare*. IEEE Consumer Electronics Magazine.

Zhang, J., Bates, J., & Abbott, P. (2022). State-Steered Smart Mentality In Chinese Smart Urbanism. *Urban Studies (Edinburgh, Scotland)*, (14), 2933–2950. Advance online publication. doi:10.1177/00420980211062888

Zhang, Y. (2021). Game And Coordination Of Diverse Benefits In Smart City PPP Projects Based On Evolutionary Game. *Annals of Operations Research*, 1–24. doi:10.100710479-021-04290-y PMID:34776572

Zhao, Q., Yu, L., Du, Z., Peng, D., Hao, P., Zhang, Y., & Gong, P. (2022). An Overview of the Applications of Earth Observation Satellite Data: Impacts and Future Trends. *Remote Sensing*, *14*(8), 1863. doi:10.3390/rs14081863

Zhao, Y., Zhang, H., Wu, Y., & Liu, Y. (2015). Public Service Demand Of Residents For Smart City Construction: Shijiazhuang City In Hebei Province As An Example. *Progress in Geography*, *34*(4), 473–481.

Zimmer, J. (2022, May 9). *Data spaces provide a new way to share information*. https://www.engineering.com/story/data-spaces-provide-a-new-way-to-share-information

Zohdi, T. I. (2021). A digital twin framework for machine learning optimization of aerial fire fighting and pilot safety. *Journal of Computer Methods in Applied Mechanics and Engineering.*, *373*, 113446. Advance online publication. doi:10.1016/j.cma.2020.113446

About the Contributors

Ingrid Vasiliu-Feltes (Dr.) is Forbes Business Council Member and expert advisor to the EU Blockchain Observatory Forum. Founder and CEO of The Science, Entrepreneurship and Investments Institute.Country Director for WBAF USA, Senator of WBAF, Faculty Member of the WBAF Business School-Division of Entrepreneurship and teaching the Executive MBA Business Technology Course at the UM Business School. Chief Ethics Officer for GBA Global, Senator and as the Chief Quality and Safety Officer at MEDNAX and provides leadership to the MEDNAX Center for Research, Education, Quality and Safety. During her academic tenure she taught 50 + courses and served Chief of Compliance and Director of Research. Member of numerous prestigious professional organizations. She holds several certifications, such as Bioethics from Harvard, AI and Business Strategy from MIT Sloan, Blockchain Technology and Business Innovation from MIT Sloan, Finance from Harvard Business School, Negotiation from Harvard Law School, Innovation and Entrepreneurship from Stanford Graduate School of Business, Fellow of the American College of Healthcare Executives, Master Black Belt in Lean and Six Sigma Management.

* * *

Alain Abran holds a Ph.D. in Electrical and Computer Engineering (1994) and master degrees in Management Sciences (1974) and Electrical Engineering (1975). He is a professor at the École de Technologie Supérieure (ETS) – Université du Québec (Canada). He has over 20 years of experience in teaching in a university environment as well as more than 20 years of industry experience in information systems development and software engineering. His research interests include software productivity and estimation models, software engineering foundations, software quality and software size measurement. He has published over 400 peer-reviewed publications and books, including: 'Software Project Estimation', 'Software Metrics and Software Metrology', 'Software Maintenance Management' (Wiley & IEEE-CS Press). Dr. Abran is co-editor of the 2004 version of the Guide to the Software Engineering Body of Knowledge – SWEBOK (see ISO 19759 and) and a co-founder and the Secretary of the Common Software Measurement International Consortium (COSMIC) – cosmic-sizing.org

Rafa E. Al-Qutaish is an associate professor of Software Engineering. He has a Ph.D. degree in Software Engineering from the Department of Software Engineering at the École de Technologie Supérieure (ÉTS) - University of Québce in Canada. Also, he has M.Sc. degree in Software Engineering from Putra University in Malaysia, and B.Sc. degree in Computer Science from Yarmouk University in Jordan. In addition, Al-Qutaish is a senior member of the IEEE & IEEE-CS, and an active member of the ACM & ASQ. Specialties: Software Measurement (Metrics), Software Quality, Engineering, Software Product Quality Measurement & Evaluation, Software Engineering ISO Standards, Software Engineering Maturity Models, CARE (Computer Aided Re-Engineering) Tools, Software Comprehension & Maintenance, Software Testing, Reversal Engineering, E-Government, and Smart-Government.

Mahendra Bhandari, CEO-ILTES, is Director-Infinity Online Global Academy (IOGA), Founder CEO- Infinity Law Tech Educational Services LL.P- India (A pivotal Law-Tech Edu-Tech Initiative); Visiting Professor in leading Universities in India, Fmr. Director -Law- NMIMS -Mumbai and Dean and Head at Amity University- Mumbai IMS Unison University- Dehradun and JNVU University- Jodhpur. Life Members Indian Law Institute- New Delhi; Fellow – Salzburg Seminar-Austria.

Shelli Brunswick brings rich perspective and deep vision of the global space ecosystem to her work as chief operating officer of Space Foundation. She advocates for space technology innovation, entrepreneurship, diversity, inclusion, and global partnership while overseeing Space Foundation's three primary divisions: Symposium 365, Center for Innovation and Education, and Global Alliance. In her work, Shelli collaborates with organizations around the world to connect commercial, government and educational sectors. This includes speaking at more than 100 global engagements each year to advance space technology. Formerly a space acquisition and program management leader and congressional liaison for the U.S. Air Force, Shelli is currently one of 55 women selected by the United Nations Office of Outer Space Affairs (UNOOSA) as a mentor for its Space4Women program. She was also named the 2021 Global Technology Leader by WomenTech Network and the 2020 Diversity & Inclusion Officer and Role Model of the Year.

Laila Cheikhi is a professor at Ecole Nationale Superieure d'Informatique et d'Analyse des Systemes (ENSIAS), Mohammed V University in Rabat, Morocco. She received her M.Sc. degree in 2004 in Informatique from University of Montréal, and Ph.D. degree in 2008 in software engineering from Ecole de Technologie Superieure (ETS), University of Québec, Montreal Canada. She has over eight years of experience in computer engineering at the Ministry of Finance of Morocco. Her research interests include software quality models, software metrics, software engineering ISO standards, software product and process quality, software engineering principles and data analysis.

Abdoullah Fath-Allah holds a bachelor degree in computer science (2005) and a master degree in computer networks with a minor in software engineering (2007) from Alakhawayn University in Ifrane, Morocco (AUI) and a PhD in computer science from ENSIAS.

Nor Hazlina Hashim (Lina) commenced her Doctor of Philosophy in year 2010 at University of Wollongong (UOW), Australia. Earlier, Lina graduated Masters of Arts in Communication at Universiti Sains Malaysia (USM) in 2003 and Bachelors of Arts (in Public Relations) at Universiti of Techonologi MARA (UiTM) in 1997. Lina also has 4.5 years of working experience as Supply Management Analysts (SMA) at Intel Technology Sdn. Bhd., Penang (1997-2001) and was a senior lecturer for 8 years at School of Communication, USM, Penang (2010-2018). Lina has taught Message and Design, and Digital Marketing for Master in Integrated Marketing Communication (IMC) at USM@KL, Wisma Sejarah, K.Lumpur from 2013 until 2018. Lina went for her Industrial Training at McCann-Erickson (M) Sdn. Bhd. and exposed to advertising and copywriting. Her research interest is on websites and online studies especially Integrated Marketing Communication (IMC), Digital Marketing, and Product Development and Innovations.

Ali Idri is a Full Professor at Computer Science and Systems Analysis School (ENSIAS, University Mohammed V, Rabat, Morocco). He received his Master and Doctorate of 3rd Cycle in Computer Science from the University of Mohamed V in 1994 and 1997 respectively. He received his Ph.D. in Cognitive and Computer Sciences from the University of Quebec at Montreal in 2003. He is the head of the Software Project Management research team since 2010 and the Chair of the department Web and Mobile Engineering for the period 2014-2020. He has a long experience in dealing with the use of machine learning techniques in software engineering and medical informatics. He was the principal investigator of several leading national and international projects. He was ranked at the 3rd position of the Top-Ten researchers in the field of software effort estimation according to the study “Research Patterns and Trends in Software Effort Estimation” published in the journal Information and Software Technology (Information and Software Technology 91 (2017) 1–21). He was also ranked 2nd of the Top-Ten researchers in doing Systematic Mapping Studies in Software Engineering according to the study “Landscaping systematic mapping studies in software engineering: A tertiary study” published in the journal of Systems and Software 149 (2019) 396–436 . He is the academic responsible of ERASMUS+, KA1 with the University of Murcia (Spain) since 2011. He is the responsible of the Erasmus+, Capacity Building “Including Students with Impairments in Distance Education” (InSIDE), Erasmus+ KA2, Coordinator: University of Macedonia, Greece. He is an Associate Editor of BMC Medical Informatics and Decision Making, JCR 2016, IF = 1.643 (). He is an Editor of the Scientific Africain Journal (). He is an expert evaluator at the CNRST () which consists of evaluating the research projects handled by this organism. He is very active in the fields of software engineering, machine learning and medical informatics and has published more that 200 papers in well recognized journals and conferences such as Information and Software Technology, Journal of Software and Systems, International Journal of Medical Informatics, Computer Methods and Programs in Biomedicine and Journal of Applied Soft Computing.

Saravanan Nathan Lurudusamy is a Solution Consultant for ICT at a leading telecommunication company in Malaysia. He is also an active member of the Institute of Engineers Malaysia (IEM) and the Board of Engineers Malaysia (BEM) and has been awarded Professional Engineer status since 2009 under the Electronics division. He is also a HRDF certified trainer under the Ministry of Human Resources. He has been delivering technical training and academic courses for various institutions such as private companies, colleges and universities, targeted for adult learners. He has also written numerous papers which have been published in journals and has attended conferences at both local and international levels. He has completed his Ph.D. in the field of Technology Management in Universiti Sains Malaysia (USM), Penang, Malaysia and his research area is on information science and technology adoption. He has also involved in a research grant awarded by Malaysian Communications and Multimedia Commission (MCMC) to study the impact of Smart Community initiatives under the patronage of MCMC at various locations in Malaysia.

Normalini Md Kassim is currently senior lecturer at the School of Management, University Science of Malaysia and Visiting Professor Management & Science University (Malaysia). She is Chartered Member of Chartered Institute of Logistics & Transport (CMILT) and Malaysia Chapter of the Association for Information Systems (MyAIS). Her publications have appeared in IGI Global Handbook, Procedia- Social and Behavioral Sciences (Elsevier), International Journal of Productivity and Performance Management (Emerald), Global Business Review (SAGE), Taylor and Francis, Social Indicators Research, International Journal of Communication Systems, International Journal of Enterprise Information Systems, Industrial Engineering & Management Systems, Global Business and Management Research and Springer. She has passed experience with industry like Maybank Berhad as system engineer for 4 years and Hewlett Packard Singapore as Project Manager for Asia Pacific Project for 10 years. She was completed her Master of Business Administration (MBA) in University Science of Malaysia (2005) and completed her PhD in Technology Management from the same university (2012). Now she has embarked to new research area which is smart community, smart cities and business analytics. With experience in banking, manufacturing, communication and financial industry, she would like to collaborate and share her experience in technology management, business analytic and risk management area. Her full profile can be accessed from .

Prafulla Mishra is Dean, School of Law, KIIT University; Former visiting professor to University of Melan, under the Indo-Italian Cultural Exchange Programme by UGC-MHRD. He is experienced in teaching and research since 1978. Besides being member of different bodies and research institutions he was also awarded Scroll of Honour at the International Conference of Jurists on Terrorism by International Council of Jurist, New Delhi.He has also served as a Member of Research Board of Advisors, American Biographical Institute, North Carolina, USA.

Michael Sampat is an Independent Researcher working in Asian International Business and Biblical Studies.

Diya Sarkar is a Research Scholar & Former Assistant Professor (Law), Kalinga Institute of Industrial Technology, India; Certified International Blockchain Advisor; Member, IOGA; Global Goodwill Ambassador, Peace Justice Humanities and Relief Foundation (PJHRF); Foreign Language Student, University of Munich; Germany.

Ramayah Thurasamy is currently a Professor of Technology Management, School of Management, Universiti Sains Malaysia, Visiting Professor Minjiang University (China), Daffodil International University (DIU) Bangladesh, Universiti Malaysia Sarawak (UNIMAS), Universiti Kebangsaan Malaysia (UKM) and Universiti Teknologi MARA (UiTM), Adjunct Professor at Sunway University and Universiti Tenaga Nasional (UNITEN), Malaysia. He was also a Visiting Professor at King Saud University (Kingdom of Saudi Arabia) and Adjunct Professor at Multimedia University previously. He is also currently the Chief Editor of the Asian Academy of Management Journal (AAMJ) and Journal of Applied Structural Equation Modeling (JASEM). He has an h-index of 87 and citation of 31,241 in Google Scholar and i-10 index of 432, his h-index in SCOPUS is 49, with 9041 citations while his h-index in ISI/Clarivate is 36 with 5844 citations. His publications have appeared in Information & Management, Journal of Cleaner Production, International Journal of Operations & Production Management, Tourism Management, Journal of Travel Research, International Journal of Contemporary Hospitality Management, Journal of Environmental Management, Technovation, International Journal of Information Management, Safety Science, Industrial Management and Data Systems, Social Indicators Research, Quantity & Quality, Service Business, Knowledge Management Research & Practice, Journal of Medical System, International Journal of Production Economics, Personnel Review, and Telematics and Informatics among others. His full profile can be accessed from .

Ana Turcan has 10+ years of experience in International Development in the public and private sectors and has driven various major development projects during this time. Currently, Mrs. Turcan serves as Data Spaces Growth, Development & Partnerships Officer at the DIO - Data Intelligence Offensive in Vienna, Austria. At DIO, she is responsible for establishing and maximize the impacts of Data Spaces, ideation of use-cases together with DIO members and development of DIO's partnerships on national and European levels. During her time with the Institute for Advanced Research in Artificial Intelligence (IARAI) in Vienna, she has designed, developed, and managed the international agenda of the Institute. Mrs. Turcan developed institutes strategic partnerships e.g. ESA (European Space Agency), with objectives ranging from data sharing, joint research projects, research exchange, data competitions and workshops, and scientists exchange. Her particular field of expertise lies in the area of researching funding opportunities, funding applications, technical programs/grants, and managing EU calls for tenders. Ana was driving strategic community engagement activities with EU institutions, Governments, Corporate, and Academia via building relationships within the artificial intelligence and machine learning industry. Prior to joining IARAI, Mrs. Turcan was a Policy Advisor within the World Artificial Intelligence Organization in New

York. Previously, she was a Business Developer and Researcher with Metis Vienna and a Project Manager in Steinbeis Innovations Zentrum in Stuttgart. Ana graduated the Diplomatic Academy of Vienna. Mrs. Turcan speaks 7 languages English, German, Russian, Spanish, Romanian, Bulgarian, and Italian.

Bianca Weber-Lewerenz is an Independent Researcher. Founder of the "Excellence Initiative for sustainable, human-led AI in Construction" in 2020 -in cooperation with the Construction Industry, National and International Institutions, in the domain of Research, Development and Application. First Representative of Ethics in AI in Construction Industry Research on "Corporate Digital Responsibility in Construction - Ethical Principles in dealing with Digitalization and AI." Member of the German Association of Engineers VDI and of the VDI Women in Engineering. Member of the Association of the Top Business Women of Baden-Württemberg, Germany. Member of Women in AI & Robotics. Member of Women in AI Ethics.

Jasmine A. L. Yeap holds a PhD in technology management, an MBA and a Bachelor of Management degree specialising in Marketing from Universiti Sains Malaysia (USM). She joined the School of Management, USM as a lecturer in 2015 and teaches marketing-related subjects. Prior to that, she has worked with numerous researchers under diverse research projects ranging from topics in marketing, organisational behaviour, psychometrics and technology management to strategic planning in education.She is actively involved in research having most of her works published in journals of repute as well as presented at both local and international conferences. Her research areas include technology adoption and diffusion, social media and its impact, retailing as well as consumer behaviour.

Poshan (Sam) Yu is a Lecturer in Accounting and Finance in the International Cooperative Education Program of Soochow University (China). He is also an External Professor of FinTech and Finance at SKEMA Business School (China), a Visiting Professor at Krirk University (Thailand) and a Visiting Researcher at the Australian Studies Centre of Shanghai University (China). Sam leads FasterCapital (Dubai, UAE) as a Regional Partner (China) and serves as a Startup Mentor for AIC RAISE (Coimbatore, India). His research interests include financial technology, regulatory technology, public-private partnerships, mergers and acquisitions, private equity, venture capital, start-ups, intellectual property, art finance, and China's "One Belt One Road" policy.

Zixuan Zhao is an independent researcher. Her research interests include FinTech, Digital Twin and business analytics.

Index

W

www.ingramcontent.com/pod-product-compliance
Lightning Source LLC
LaVergne TN
LVHW081339300926
849161LV00008B/2016
* 9 7 8 1 6 6 8 4 3 8 3 3 6 *